Springer Water

Series Editor
Andrey G. Kostianoy, Russian Academy of Sciences, P.P. Shirshov Institute of Oceanology, Moscow, Russia

The book series Springer Water comprises a broad portfolio of multi- and interdisciplinary scientific books, aiming at researchers, students, and everyone interested in water-related science. The series includes peer-reviewed monographs, edited volumes, textbooks, and conference proceedings. Its volumes combine all kinds of water-related research areas, such as: the movement, distribution and quality of freshwater; water resources; the quality and pollution of water and its influence on health; the water industry including drinking water, wastewater, and desalination services and technologies; water history; as well as water management and the governmental, political, developmental, and ethical aspects of water.

More information about this series at http://www.springer.com/series/13419

Roland Werchota

Empty Buckets and Overflowing Pits

Urban Water and Sanitation Reforms in Sub-Saharan Africa – Acknowledging Decline, Preparing for the Unprecedented Wave of Demand

Springer

Roland Werchota
Kottingbrunn, Austria

ISSN 2364-6934 ISSN 2364-8198 (electronic)
Springer Water
ISBN 978-3-030-31385-2 ISBN 978-3-030-31383-8 (eBook)
https://doi.org/10.1007/978-3-030-31383-8

This Springer imprint is published by the registered company Springer Nature Switzerland AG.
The registered company address is: Gewerbestrasse 11, 6330 Cham, Switzerland

Courtesy of Dr Han Seur, publically available at Aquapix, WSTF, Kenya

Figure 2.1, Chap. 2, and Figs. 3.1 and 3.2, Chap. 3: courtesy of GIZ Water Sector Reform Programme, Kenya (Dirk Schaefer and Philipp Feiereisen), publically available at the Ministry of Water and Sanitation, Kenya

[illegible]

Preface

The theme of this book is about the urban water and sanitation development in the low-income world, where the situation is increasingly becoming critical in numerous countries. Water rationing in Sub-Saharan African is, at this point in time, a lasting event in many towns,[1] and cholera outbreaks have started to permanently threaten urban dwellers.[2] The fast-growing urbanisation in the Sub-Saharan region, so far unmatched in the world, is responsible for a breathtaking increase in demand for water and sanitation services driven by industrialisation and a growing middle class. The decline in sanitation affects all income classes, yet the situation for the poor in low-income settlements is much more worrisome. The worldwide situation is equally critical as substantiated indications prove that an increasing number of people have to consume contaminated water.[3] It is in the interest of developing and industrialised countries that the decline in access to safe water and sanitation in the low-income world are reversed urgently because mass migration is fuelled by the unacceptable living condition in the low-income countries.

The negative development is truly surprising because since ancient times, the importance of people having access to clean water and safe sanitation in towns is well known. Throughout history, civilisations have undertaken huge efforts to maintain urban life with the development of water and sanitation. It was equally understood that settlements are the engines of the economy and that they could only function efficiently with such suitable infrastructure. The biggest medical

[1] Refer also to Turton and the dramatic water shortages in Cape Town, South Africa in 2017/2018, https://www.youtube.com/watch?v=FNt9EayG3-g (last visited 01.2018).

[2] Africa has become the 'new homeland of cholera'; Gaffga et al. (2007: 705).

[3] According to solid evidence, even the MDG target for water has not been reached despite UN reports of achievements. The contradiction arises from differences in monitoring the target 7.C 'Halve, by 2015, the proportion of the population without sustainable access to safe drinking water and sanitation'. The indicators used for the monitoring by JMP are different from those that the target would require. This is responsible for the misleading message that the proportion of the population, which is '…using an improved drinking water source has increased from 76% to 91%, surpassing the MDG target, which was met in 2010'. JMP (2015: 58). Further details to this issue can be found later in this book.

achievement from the year 1840 onwards in the industrialised world is considered to be the presence of water taps accessible to everyone at any time of the day and safe toilets linked to a piped sewer system, so as to protect the urban living space.[4]

Such a successful development in one part of the world would make one think that enough experience and knowledge are available today for Sub-Saharan countries to draw from, in order to move to universal access and protect the living space of people. In addition to this wealth of experience, UN institutions, donors, international NGOs and others have been providing substantial support to low-income countries since the 1950s.[5] Numerous efforts have been undertaken by the international cooperation to move the sector forward, such as the declaration of water decades, adoption of human rights to water and sanitation, substantial financing of infrastructure by donors, global standard setting for water quality and considerable capacity building of sector institutions, just to name a few.

Yet many people in the developing world (often as much as half the urban population) do not have unrestricted and sustainable access to clean water and safe sanitation that secures also adequate collection, treatment and disposal of human waste. In the near future, hundreds of million urban dwellers in Africa will be directly concerned by this unacceptable situation. Convincing information throughout this book suggests that about two billion people in the world have to consume contaminated water, do not have access to acceptable toilets and live in an environment polluted by human waste.[6] These situations being the result of inadequate access to W+S infrastructure and services remain one of the biggest killers worldwide.

Many decision makers wrongly believe that using medical advances in health services and distributing new drugs can substitute water and sanitation infrastructure development. However, medical advances cannot make W+S infrastructure redundant, as the development in countries indicates. It is backfiring sooner or later and is not only very costly but also produces negative effects. Frequent infections are the cause of stunting among children despite sufficient food intake, and treatment of repetitive appearing waterborne diseases will build widespread resistance against drugs (e.g. antibiotics). It is indisputable that the decline of the sector combined with the effects of climate change on water and sanitation is making it increasingly difficult for developing countries to catch up with the industrialised world. It keeps a very large part of the urban population in unbearable living conditions and exposure to avoidable risks.

This short outline leads to the first intriguing question. What are the reasons that are causing the persistent paradox, where on one hand a long-standing wealth of

---

[4] Ferriman (2007: 17).

[5] The proportion of bilateral and multilateral development aid is about 70% to 30%. 'About 80–85% of developmental aid comes from government sources as official development assistance (ODA). The remaining 15–20% comes from private organisations such as "non-governmental organisations" (NGOs), foundations and other development charities (e.g., Oxfam). In addition, remittances received from migrants working or living in diaspora form a significant amount of international transfer.' https://en.wikipedia.org/wiki/Development_aid (last visited 04.2018).

[6] According to JMP (2018), 4.5 billion people have no toilet where excreta are safely managed.

experience as well as substantial support by the international cooperation oppose, on the other hand, a constant decline within the sector in so many developing countries? A literature review reveals already possible answers. Many stakeholders are not aware or simply don't know that a decline is taking place. Global monitoring by UN institutions for water and sanitation access is regularly releasing misleading messages of progress in the world and recently also for the Sub-Saharan region.[7] Thus, many decision makers maintain the perspective that there is no need to double or even maintain efforts in the sector.

Over many decades, global water and sanitation goals have fallen short of their mark without provoking a significant outcry by the public or leading to consequences for the people behind such declarations. Hence, there was so far no reason to change, although there was always solid proof that these political goals cannot be achieved in the given time frame and with the usual approaches. Even recently, the unachievable (safely managed access for all in the world in line with the human rights for W+S by 2030) was launched anew with the SDGs.[8] The repetition of goals which have never been achieved so far provides no further motivation for implementers to continue to strive for them. On the other hand, decision makers responsible for such unachievable declarations increase their insistence and argue anew that now is the time for success. However, most of the decision makers in the international cooperation, which often are no water experts anyway, instruct their water specialists to integrate these goals into their work plans. Because these goals are set at the highest level, most specialists do not resist and choose the easier way out of copy and paste of such goals, which is for them more convenient than facing the consequences of questioning UN set goals and targets.

Another possible answer to the first question might be that there is a gap in knowledge of how to secure access to urban water and sanitation in the low-income countries. This seems to be confirmed by a document analysis and expert interviews used by this work.[9] Also the many contributions in literature could not provide convincing answers to the way forward to reverse the negative trend. Most of these publications concentrate on isolated aspects, such as private sector participation, community management, water quality, non-revenue water (NRW) and water tariffs at public outlets, or cross-cutting issues on water and sanitation. Many authors present findings and suggest proposals which are too superficial because, for instance, they do not even differentiate between urban and rural realities.

Few contributions are written about progress in the sector in the developing world with a holistic view. Although some contributions describe sector reforms in low-income countries and indicate that they produce quite different results, they are more narratives than studies with findings based on solid data.[10] However, they

---

[7] Increased access to 'improved sources' according to JMP reporting.

[8] Sustainable Development Goals. Refer also to Werchota et al. 2015, From the Millennium Development to the Sustainable Development Goals for Drinking Water and Sanitation – Experience from Sub-Saharan Africa, gwf, Wasser / Abwasser, international issue 2/2015

[9] Based on my dissertation submitted to the University of Vienna in 2017.

[10] Locussol A. and Fall M, (2009), Schiffler (2015), Bertrand, J. and Geli, H. (2015).

suggest that results of reforms can be very different, ranging from outstanding success to complete failure. Hence, this raises another intriguing question. Are reform outcomes really so different? And if yes, why? The answers to both questions, the paradox and the reform outcomes, are necessary to understand the situation and derive from its analysis an effective development concept for the sector in low-income countries. This book is all about closing this knowledge gap, reminding that there is insufficient attention by the international cooperation and the governments of the receiving countries, and recommending a feasible and effective way forward to prepare for the unpresented wave of demand.

The analysis in this work uses a sound methodology as well as several data sources.[11] The collection of data was carried out in four target countries in the Sub-Saharan region.[12] Extensive literature review and own considerations led to a proposal for a model of development (crucial factors) with an adequate sector orientation for Sub–Saharan countries,[13] both forming a theoretical concept. This was tested with qualitative and quantitative data. The analysis was carried out with a thorough multidimensional and multilevel view and revealed that indeed, some reforms can be considered outstanding. The frameworks have been upgraded, regulatory systems have been established, utility performance and information have substantially improved, and funds for infrastructure development have doubled or tripled – all combined leading to increased access in a number of countries. Nevertheless, reflecting on the urbanisation which will take place in the next 30 years in Sub-Saharan Africa and considering the persisting financing gap in infrastructure development, it becomes apparent that even notable progress might no longer be sufficient to meet future demands.

With the analysis, it also became obvious that today, the developing world has to deal with a much more complex situation in the sector than the industrialised countries had to face in the past. This explains to a large extent why the knowledge from the past is insufficient, although important. There are so many more players in the sector today in the developing world. Most can act entirely uncontrolled and claim to know the answers for a successful development despite ignoring academic knowledge. Closer networking from global to local levels adds to these difficulties. Although in general positive, it has in this case also given room to a range of short-lived, constantly reappearing 'innovative' ideas promoted with stories and buzz words like 'waterpreneurs', from aid to trade, a business case for water or sanitation for the poor, etc. Many of these promises on which decision makers started relying are eventually not fulfilled, after waiting 10 and more years for results.

These confounding factors often create an unjustified euphoria or unnecessary hysteria in the sector which hamper development in the partner countries. Many of the proposals of how to proceed are not feasible but reappearing regularly in the

[11] Statistical data, expert interviews and document analysis.

[12] Burkina Faso, Kenya, Tanzania, Zambia.

[13] Factors described in this work are dimensions through which, in combination or alone, sector development can be significantly influenced in order to ensure that people can satisfy their most basic needs. This indicates that urban water and sanitation development is multidimensional and that factors have a strong link with each other.

literature, such as the demand to regulate the often thousands of small-scale informal providers (SIP) in a country. Most contributions have not helped the many underserved poor even on medium or long term. Unrealistic expectations and concepts carried into the low-income countries on policy level by the champions of the discourse distract the decision makers from reality and from following an appropriate development path.

It seems that stakeholders have forgotten that the development in the sector was never a quick fix and also needed many decades in the (rich) industrialised world. Once a comprehensive development approach is adopted, based on lessons learned with global relevance and matching the context and the complexity in the low-income countries, regular fine-tuning is required instead of hectic and erratic steering provoked by impatient donors and 'innovative' ideas. Hence, the book argues for a sweeping rethinking in sector development for low-income countries. The emphasis of sector development must be on bringing controlled services to everyone with at least basic service levels. The poor can no longer be left to the mercy of neighbours, criminal cartels or informal and incompetent small-scale providers which sideline the state structures and explore the deplorable situation of dwellers in the LIAs. Equally, the poor cannot be left with traditional water sources, such as shallow or deep wells in densely populated areas where ground water pollution is widespread. Such 'improved sources',[14] even when situated on the plot, should have no place in the provision of drinking water in the urban setting. A two-tier society in towns should not be acceptable when it comes to the most basic life-supporting services. With the human rights declaration at hand, it is time to overcome the *urban water and sanitation divide* in low-income countries.

Such a rethinking has to take place among decision makers in the receiving countries, which need to drive reforms further, as well as within the international cooperation. Funds for infrastructure development for water and sanitation need to multiply and be regarded as more important than the many projects in other sectors which are nice to have. Donor contributions need to reward low-income countries which try very hard to improve their housekeeping. Analysing some reforms, it became apparent what has to be done in the prevailing contexts. The reasons for the success and failures of sector reforms in developing countries can now be pinpointed with the present work.

Two stories about real people from the LIAs in Nairobi, included in this work, should help to illustrate how they have to 'chase' for (often contaminated) water every day, cannot defecate with ease and dignity and have to live in an environment polluted with human waste. They also show why certain concepts are not helpful and why the recommendations in this book can be regarded as an appropriate way forward. In order to help the unfortunate people without adequate access to move to a better future, it is necessary to expose the bottlenecks and weaknesses which allow this to happen without giving the impression that certain institutions or groups are being pilloried.

[14] Refer to WHO and UNICEF global monitoring (Joint Monitoring Program).

Next to the issues mentioned so far, this book also covers issues which are hardly the subject of discussions but significantly influencing urban water and sanitation development, such as the (negative) effects of decentralisation in the water sector and the substantial differences in financing, implementing and operating last and first mile investments. Equally, the deliberation on the different monitoring systems (global and national) and the adjustments which have to be made when comparing access figures between countries can be considered as new contributions.

This book draws from my dissertation submitted to the University of Vienna in 2017 for which I used the experience acquired during almost 30 years in African countries working as an integrated advisor in national institutions on different levels. It offered me an exceptional insider view of national utilities, regulators, financing baskets and ministries in low-income countries but also on how the international cooperation is impacting. Thus, this book can be considered an important reading where new areas are explored and original ideas are expressed. It should help to make the shortcomings of the water and sanitation discourse more apparent and institutions to move beyond their present perceptions and agendas. Recent events, such as the excruciating water shortages in Cape Town,[15] the widespread cholera epidemic in Haiti or the mass migration from low-income countries, show that this book deals with hot topics. It can be expected that the unprecedented growth of urbanisation in Sub-Saharan Africa and the effects of climate change will increase the frequency and magnitudes of emergencies, which will further heighten attention of the public and literature for Sub-Saharan Africa and water- and sanitation-related issues.

The book is organised in three parts. Part I offers an introduction and a critical review of the literature which are supplemented with my own contributions. Part II provides a model of crucial factors for development in the sector which is tested with the results of sector reforms in four Sub-Saharan countries. It also shows which orientation the sector should take in the region because of the prevailing contexts. This is followed by a brief overview why reforms produced such different results in four countries. Part III provides an indication on how to reach the urban poor in low-income areas with formalised service provision and winds down with conclusions and recommendations on sector reforms.

Kottingbrunn, Austria Roland Werchota

[15] The crisis in Cape Town also indicates how rapidly water shortages lead to a thriving Mafiosi-type market. See 'Wasser ist das neue Öl' Tagesschau.de, 2018 http://www.tagesschau.de/ausland/kapstadt-wassernutzung-105.html (last visited 05.2018).

References

Gaffga NH, Tauxe RV, Mintz ED (2007) Cholera: a new homeland in Africa? The American Society of Tropical Medicine and Hygiene, American Journal of Tropical Medicine and Hygiene 77(4):705–713

Ferriman A (2007) BMJ readers choose the "sanitary revolution" as greatest medical advance since 1840. BMJ Publishing Group Ltd Available: http://www.ncbi.nlm.nih.gov/pmc/articles/PMC1779856/ (last visited 06.2015)

Locussol A, Fall M (2009) Guiding principles for successful reforms of urban water supply and sanitation sectors. World Bank, Note No. 19

Acknowledgement

This work would not have been possible without the availability for dialogue of the many stakeholders consulted and the open exchange with the experts interviewed in the four target countries which provided me with a wealth of information. My special gratitude goes to my counterparts and particularly to Yamba Harouna Ouibiga (Burkina Faso), Osward Chanda (Zambia) and Robert Gakubia (Kenya). As this book is based on a dissertation submitted to the University of Vienna in 2017, the guidance of my supervisor Professor, DDr. Kunibert Raffer, was crucial to overcome the challenges of someone who has mainly worked in the field when undertaking such work. In addition, this book would not have been completed as it is without the valuable comments on my dissertation from Professor Dr. Mark Oelmann, from Hochschule Ruhr West (University of Applied Sciences in Germany).

I also thank all colleagues from the German International Cooperation (GIZ) and their consultants, especially Katrin Bruebach and Dirk Schaefer, as well as from other development agencies, such as the financial cooperation and civil society, who added value in the course of professional dialogues. I am especially grateful to my wife and children to whom I dedicate this work and who helped with the transcription of expert interviews and the design of graphs and never stopped encouraging me to carry on and finalise this undertaking.

Contents

About the Author

Roland Werchota born in 1948 in Vienna, Austria, spent 30 years in the International Cooperation managing water programs in developing countries and working as a long-term advisor in partner institutions (ministries, regulators, financing baskets and utilities) in Kenya, Zambia and Burkina Faso. Short-time interventions took him to several continents, such as Africa, South America and Asia. His work focused mainly on water resource management as well as urban and rural water supply and sanitation. Some of the supported partner institutions have in the meantime reached levels of excellence not only outstanding in comparison to other institutions in the developing world but also comparable to the development of European institutions/companies for instance. One of these examples is the national utility in Burkina Faso, the Office National de l'Eau et de l'Assainissement. Mr. Werchota is a civil engineer specialised in public works and obtained a Masters of Economic and Social Science as well as a PhD in the field of International Development at the University of Vienna. He has also worked for many years in the construction industry being involved in large-scale projects, e.g. in the Middle East (Iraq and Saudi Arabia) and North, South and Central Africa (Tunisia, South Africa, Congo, Mozambique, etc.). He is the author of *The Growing Urban Crises in Africa*, (2013) and a co-author of numerous contributions to water and sanitation, such as *MDG Monitoring for Urban Water Supply and Sanitation: Catching Up with Reality in Sub-Saharan Africa* (2007) and *From the Millennium Development to the Sustainable Development Goals for Drinking Water and Sanitation: Experiences from Sub-Saharan Africa* (2015).

Abbreviations

AHC	Asset Holding Company, Copperbelt, Zambia
AMCOW	African Ministers' Council on Water
ATM	Automated Teller Machine
AU	African Union
BoDs	Board of Directors
CBOs	Community Based Organisations
CEO	Chief Executive Officer
CLTS	Community Led Total Sanitation
CU	Commercial Utilities (Zambia)
DANIDA	Denmark's development cooperation
DTF	Devolution Trust Fund (Zambia)
EWURA	Energy and Water Utilities Regulatory Authority
FC	Financial Cooperation
GDP	Gross Domestic Product
GFA	GFA Consulting Group (Germany)
GIZ	Gesellschaft für Internationale Zusammenarbeit
GSB	Ghana Standard Board
GTZ	Gesellschaft für Technische Zusammenarbeit
HDI	Human Development Index
hh	household connection
HPI	Human Poverty Index
IFC	International Finance Cooperation (WB)
IMF	International Monetary Fund
JMP	Joint Monitoring Programme (UN)
KES	Kenyan Shillings
KEWI	Kenya Water Institute
KfW	Kreditanstalt für Wiederaufbau
LIAs	Low Income Areas
MD	Managing Director
MDGs	Millennium Development Goals (UN)
MLGH	Ministry of Local Government and Housing (Zambia)

MoH	Ministry of Health
MPI	Multidimensional Poverty Index
MWI	Ministry of Water and Irrigation (Kenya, Tanzania)
NGO	Non-Government Organisation
NIS	NWASCO Information System (Zambia)
NRW	Non Revenue Water
NWASCO	National Water Supply and Sanitation Council (Zambia)
NWWSC	North Western Water and Sewerage Company (Zambia)
OECD	Organisation for Economic Co-operation and Development
O+M	Operation and Maintenance
ONEA	Office National de l'Eau et de l'Assainissement (Burkina Faso)
PD	Proposed Definition
PPP	Public Private Partnership
PSP	Private Sector Participation
PPWSA	Phnom Penh Water Supply Authority
PSP	Private Sector Participation
SIP	Sector Investment Plan (Kenya)
SODECI	Société de distribution d'eau de Côte d'Ivoire
SWAP	Sector Wide Approach to Planning
SDGs	Sustainable Development Goals (UN)
TA	Technical Assistance
UfW	Unaccounted for Water
UN	United Nations
UNDESA	United Nations Department of Economic and Social Affairs
UNDP	United Nations Development Programme
UNHABITAT	United Nations Human Settlements Programme
UNICEF	United Nations International Children's Emergency Fund
USD	US Dollars
WATSAN	Water and Sanitation
WAB	Water Appeals Board (Kenya)
WARIS	Water Regulation Information System (Kenya)
WASREB	Water Services Regulatory Board
WB	World Bank
WHO	World Health Organization (UN)
WRM	Water Resource Management
W+S	Water and Sanitation
WSBs	Water Services Boards (Kenya)
WSDP	Water Sector Development Program (Tanzania)
WSPs	Water Service Providers
WSS	Water Services and Sanitation
WSSBs	Water Supply and Sanitation Board (Tanzania)
WSTF	Water Services Trust Fund (Kenya)
WWTP	Waste Water Treatment Plant

Part I
Urban Water and Sanitation (W+S) – Separating Fiction from Reality

Part I intents to raise the key issues relevant to the urban water and sanitation sector and make the problems better known and understood. For this a literature review is complemented with an input of lessons learned from extensive field work and expert interviews. Hence, this part starts with an introduction to urban water and sanitation focused mainly on the developing world which is followed by a critical literature review and deliberations on basic topics for urban water and sanitation. Part I closes with deliberations concerning issues on urban water and sanitation which are beyond the usual debate and a summary.

Chapter 1
Introduction to Urban W+S in the Developing World

Abstract Continuous access to water and sanitation (W+S) in towns is crucial for the survival of the individual but also for society. When infrastructure provides easy access, other basic needs will gain importance for individuals but for society the sustainability of W+S services must remain top priority because of health and other important benefits. These indisputable facts did not convince decision makers to stop the deterioration of access in Africa and also in the world. Since four decades' global goals for access to W + S have been continuously missed, and following the irrefutable data, the same is trough for the MDGs water target. Despite this, the unrealistic goals have now been carried forward with the SDGs. The reasons for this are a knowledge gap and a one-sided and biased discourse in urban W+S development for low-income countries which leads to inadequate development concepts and attention of decision makers. There is an urgency to act and to make more prominently known to the world that an increasing number of people are suffering and unnecessarily dying because they have to consume contaminated water, chase daily to fill their water canisters, look for a place to defecate and constantly fight infections in an environment polluted with human waste.

This book concentrates on Sub-Saharan Africa concerning sustainable development of access to clean drinking water and safe sanitation in the urban setting which has its specific challenges but also success stories. The introduction should help to obtain a brief overview of the situation and brush over a few basic topics before a more elaborate discussion on other issues follows.

R. Werchota, *Empty Buckets and Overflowing Pits*, Springer Water,
https://doi.org/10.1007/978-3-030-31383-8_1

1.1 Importance of W+S

Literature highlights the importance of water with comments such as: 'Water is the primary life-giving resource,'[1] 'Water is one of the most important substances on Earth, being essential to all living cell',[2] or according to Cullet,[3] 'It is so important that it gives the planet its nickname'. Aderinwale and Ajayi emphasise[4]

> …water is the only utility [services] that directly affects [people] and has to be provided for by families and individuals almost on a daily basis. Its acute shortage or inadequacy has a direct impact on the overall well-being of a society with limited alternatives and coping strategies [available to individuals],

The fact that humans need safe water for consumption on a daily basis, several times a day, in order to stay alive, should lead to a common understanding that access to safe water services means satisfying the most important of all other life-securing basic needs, such as food, health, energy, housing, transport and security. This applies, almost to the same extent, to sanitation, which refers to daily defecation, the need for hygiene and safe collection and disposal of human waste and grey water to ensure bearable living condition in the urban setting. People, forced to live in congested conditions, deprived of access to safe water and are obliged to defecate in unsanitary and undignified conditions, will instantly realise the importance of sanitation for their physical survival and for their development as individuals. There seems to be no exaggeration in the statement of Vuorinen et al.[5] that civilisation is built on safe drinking water supply and sound sanitation, and that sustainable access is the real achievement of development. It seems justified to say, that today the fruits of a modest prosperity can hardly be realized in a city without a functioning water and sanitation system. This is also shown by the example of Agnes which is described in the Sect. 2.5.

The crucial importance water plays is not only emphasised by literature, but also by policy and strategy papers of the sector in the developing world. For instance, the (draft) sessional paper of Kenya[6] for the water sector stipulates:

> Water plays [a] significant role in the national development of a country with respect to social, economic and environmental spheres. It is a social and economic good which is critical in sustainable development of the country. As a social good it supports domestic

---

[1] UNESCO (2006: 6) Water-A Shared Responsibility, Paris.

[2] https://www.google.com/search?source=hp&ei=JPv9Wb3XIMLFwQLNm5WACw&q=water+is+the+most+common+substance+on+earth&oq=water+is+most+common+on+earth&gs_l=psy-ab.1.2.0i22i30k1l4j0i22i10i30k1j0i22i30k1.2426.2426.0.16601.1.1.0.0.0.0.283.283.2-1.1.0....0...1.2.64.psy-ab..0.1.282....0.gmqX-MM0pPY (last visited 09.2017).

[3] Cullet (2009: 8).

[4] Aderinwale and Ajayi in Hemson et al. (2008: 67).

[5] Vuorinen et al. (2007: 49).

[6] Second Consolidated Draft of Sessional Paper on National Water Policy, (2017: 8). However, generally national policy papers covering all sectors do not provide such an importance to water and sanitation development (see Sect. 3.3)

needs, life and health, and as economic good water supports agriculture and industry. It is a major input in many productive sectors like agriculture, energy, processing and manufacturing, hospitality, mining, construction and transport.

Also donors underline in their strategies for international cooperation the crucial importance of access to safe water and sound sanitation.[7] Nevertheless, many people, including some of the dwellers in low income area (LIAs), who have access to a toilet to defecate, tend to consider access to safe water more important than sanitation. The reason for this preference for water might be based on the fact that in many cases people can help themselves with building a toilet but they have no alternatives than to obtain safe drinking water from utilities in the urban setting. However, does this limited preference for sanitation mean that safe sanitation can be regarded as less important by society?[8] Certainly not! When people have access to an adequate toilet, the challenge of sanitation in town is by far not overcome. There is a need for a controlled collection and disposal of human waste in order to avoid that the living space of people becomes infested and eventually is threatening urban life.

The users of water and sanitation infrastructure in towns often do not pay too much attention to the services needed after using the toilet. These are generally taken for granted. In the industrialised world it is the piped sewer system ensuring such a safe evacuation and treatment for almost all people. In the developing countries where the majority of people depend on onsite sanitation (non-piped sewer),[9] the human waste is stored in pits or septic tanks. When the septic tank or the pit of latrines are full, in towns the sludge has to be emptied, transported and treated because of limited space which deprives people to dig a new hole and move the toilet. This is generally not necessary in the rural setting where more space for the individual households is available. Thus, in the urban setting, safe sanitation goes beyond an adequate facility at home with a pit where human waste can be stored and left safely in the ground.

It is understandable that once access to a safe sanitation facility in the household and to safe water has been acquired and services are reliable, the individual will shift its attention to other basic needs such as transport, health care, energy, education, etc. However, even if water and sanitation has lost attention of the individuals because they can easily access infrastructure and services, society still depend on clean drinking water and safe toilets throughout each day. Hence, sustainability of services is crucial for society to maintain urban life despite water and sanitation has lost top places in the preference lists of households.

[7] E.g. the US government global water strategy (2017:7) states: 'Finally, improving access to basic services such as water and sanitation can be an important aspect of efforts to strengthen government stability and accountability'.

[8] Expert interviews included in the dissertation submitted to the University of Vienna, International development, Werchota (2017).

[9] Sanitation facilities not linked to a sewer system.

1.2 Benefits Which Justify Highest Priority for Water and Sanitation Development

There is a significant amount of literature documenting the relationship between inadequate access to water and sanitation or inadequate hygiene practices and mortality rates among children under the age of five. According to relevant studies, water-related (intestinal) diseases, such as diarrhoea, worms and schistosomiasis,[10] are one of the main causes of death in low-income countries. These studies in the developing world only confirm what was already well understood in the distant past concerning the link between water and sanitation with health. Esrey et al.[11] compared 144 studies in countries and concluded that child mortality fell by 55% due to improved access to water and sanitation. He also found that 'Better water quality reduced the incidence of dracunculiasis, but its role in diarrhoeal disease control was less important than of sanitation and hygiene'. Furthermore, it is known that the under-five mortality dropped sharply after centralised urban water and sanitation systems were established[12] and water filtration became mandatory in the industrialised world.[13]

In addition, findings from recent studies in India suggest that providing sufficient food will not alone guarantee a healthy development of children. According to Spears et al.[14] safe sanitation seems to be as much important for avoiding childhood stunting and its negative effects on progress far into adulthood as the availability of food. It seems to follow that programs feeding children under the age of five aiming to avoid childhood stunting will not be successful without securing simultaneously access to safe sanitation. Furthermore, the evidence collected to document the increasing resistance to antibiotics in the industrialised world provides a rough indication what might be looming in the low-income countries if active health care is carried on to be considered more important than the passive health care through the development of water and sanitation infrastructure and universal access to its services.

To these outstanding health benefits, non-health benefits can be added when access to water is improved and services comply with minimum standards such as 30 minutes cycle, controlled tariffs, unrestricted access, etc. Shifting to utility services from informal providers' leads to reduced household spending for water[15] and

---

[10] Also known as Bilharziose.

[11] Esrey et al. (1991: 609, 611).

[12] Exner (2015).

[13] According to a presentation at the MATA/GIZ (Mitarbeitertagung), Germany in 2015, Martin Exner, University of Bonn, underlined that before 1892, the 'Under Five' mortality rate in Germany was 4% with a sharp increase in 1892 (rising over 6%), the year of the cholera epidemic in Hamburg. After Koch's proposal to install water filtration, there was a sharp and immediate decrease of 'Under Five' mortality rate to 2% in 1893.

[14] Spears et al. (2013: 3).

[15] The poor often spend in absolute terms more for water than the middle and high-income classes,

a reduction of time wasted from which especially women and children can benefit when fetching water is facilitated and less people fall sick in the family. Furthermore, Devoto et al.[16] describe insufficient access as a source of conflict[17] in the household:

> For example, in Morocco, 66 percent of households without a water connection report that water is a major source of concern; 16 percent have had a water-related conflict within the family; and 12 percent have had water-related conflict with their neighbors.

Therefore, with healthier and less water stressed people, productivity in the country is increasing. It is estimated that insufficient sanitation alone costs countries between 1% and 4% of their GDP grow potential every year.[18] Hence, it is obvious that improved water and sanitation has a positive effect on the economy of countries. Last not least, Reade et al. underline the positive contributions of access to standardised clean water and sound sanitation to poverty alleviation especially helpful to women and girls in the LIAs.[19] Considering the importance of water and sanitation as well as their benefits and the alterative costs to contain waterborne deceases in towns, it should not be difficult to accept that governments need to give special attentions to water and sanitation in their development plans. Unfortunately, this is not the case yet in many countries.

1.3 A Deploring/Declining Situation in Sub-Saharan Africa

According to the Joint Monitoring Programme (JMP),[20] Sub-Saharan Africa is the only region in the world where piped water on premise in the urban setting has declined between 1990 and 2012, from 42% to 34%. In contrast, the same indicator for least developed countries on a global level has increased from 29% to 33% over

---

which hampers the escape out of poverty by using such savings for education, productivity improvements, and etc. According to GIZ's baseline survey 2009 on low-income underserved urban centres in Tanzania, Pauschert et al. (2012: 20) 'On average households in LIAs, which receive water from an ISP pay 13-times the price than they would if they received water from a house connection [of the utility]; and still pay 3-times the price than they would, if they received their water from a [utility] kiosk'.

[16] Devoto et al. (2012, pp. 68–69).

[17] E.g. also the US government global water strategy (2017:7) states: 'Moreover, access to sanitation for women and girls is particularly crucial to…reducing gender-based violence.'

[18] According to the Water and Sanitation Program (WSP) of the World Bank, DeFrancis (2011: 55) the loss due to insufficient sanitation is 3.4% of GDP in India, discounted for 2006. Other documents of WSP (Coombes et al. 2012: 1–6) find 1.1% of GDP loss for Uganda and 1.3% GDP loss for Zambia.

[19] Reade and Ndirangu (2009, survey at WSTF 2010).

[20] The WHO/Unicef Joint Monitoring Programme (JMP) for Water and Sanitation was put in place to monitor the progress towards the relevant MDGs on global level (global monitoring), http://www.wssinfo.org/about-the-jmp/mission-objectives/ (last visited 06.2016). As well as http://www.unwater.org/publication_categories/whounicef-joint-monitoring-programme-for-water-supply-sanitation-hygiene-jmp/ (last visited 10.2017)

the same period.[21] The simultaneous increase in access to other 'improved sources' during the same period, according to JMP, has to be viewed critically as water from many sources considered 'improved' cannot be taken for safe.[22] Furthermore Sub-Saharan Africa topped the list in absolute numbers of people not having access to even 'improved water sources' with 325 million people in 2012 followed by Southern Asia with 149 million, i.e. 'Two out of five people [worldwide] without access to an improved drinking water source live in Africa'.[23] Nevertheless, JMP qualifies the increase in access to 'improved sources' in Sub-Saharan Africa as a positive development despite that it is accompanied with an increase in people who have to consume contaminated water.

This negative development is supported by the findings of Booysen et al.[24] using an asset index for poverty (private and public), which includes water and sanitation. The analysis of the trend over 10–15 years for seven Sub-Saharan countries (including Kenya, Tanzania and Zambia) between 1984 and the late 1990s shows that poverty has declined in five out of seven countries, only because access to private assets compensated the deterioration in access to piped water and to a toilet (flushed and pit latrine) in the home. Thus, it can be said that poverty, as it is measured, might decline in the region, but water and sanitation poverty is increasing.

Gaffga et al.[25] provide another indication on how far behind Sub-Saharan Africa fell in the development of access to water and sanitation compared to the rest of the world. The incredible upsurge in waterborne diseases such as cholera is a direct result of insufficient access to safe water and sanitation services and infrastructure:

> In 2005, 31 [78%] of the 40 countries that reported indigenous cases of cholera to WHO were in Sub-Saharan Africa. The reported incidence of indigenous cholera in sub-Saharan Africa in 2005 (166 cases/million population) was 95 times higher than the reported incidence in Asia (1.74 cases/million population) and 16,600 times higher than the reported incidence in Latin America (0.01 cases/million population).' Furthermore, 'However, the application of well-established public health principles - ensuring universal access to potable water and the separation of human fecal wastes from food and water sources - is sufficient to prevent widespread cholera transmission. Through these measures, epidemic cholera was eliminated from Europe and the United States over a century ago.

Even in towns where infrastructure for water supply is in place and people are considered to have access, often they are deprived of water because of frequent and long lasting rationing by the utilities. However, water rationing, as UNDP notes, is generally not the result of insufficiently available water resources[26]:

---

[21] Progress on Drinking Water and Sanitation, 2014 update, WHO and Unicef, JMP website: http://www.wssinfo.org/ (last visited 07.2016).

[22] This holds true especially in highly populated areas where a controlled system of evacuation of effluent and human waste is missing. See also Sect. 2.5, e.g. the Uganda example.

[23] JMP (2014: 6).

[24] Booysen et al. (2007: 1125–1127).

[25] Gaffga et al. (2007: 705).

[26] http://www.undp.org/content/undp/en/home/ourwork/environmentandenergy/focus_areas/water_and_ocean_governance/water-supply-and-sanitation.html, (last visited 01.2016).

> But the global water and sanitation crisis is mainly rooted in poverty, power and inequality, not in physical availability. It is, first and foremost, a crisis of governance and thus governance reform must be a key pillar of any strategic approach to addressing the crisis.

But this is just half of the story. It is a fact that insufficient access of the poor is a result of sector institutions neglecting LIAs and that the poor are the first to be cut off from services when rationing programs are put in place.[27] Nevertheless, many water utilities are forced to put a rationing program for water distribution in place because governments neglected the need for investments in infrastructure development for raw water storage and abstraction over many years.[28] Numerous fast-growing capital cities in the developing world are placed far from major water sources, which make the development of infrastructure for water storage, abstraction and transport of raw water by pipelines very expensive. Inflicting these costs of developing infrastructure for raw water on the utility will overstretch customers' ability to pay.[29]

Concerning improved sanitation facilities, access in the urban setting was stagnant between 1990 and 2012 at 41% in Sub-Saharan Africa while in the least developing countries it had increased from 21% to 29% in the same period, according to global monitoring.[30] Therefore, Sub-Saharan Africa is not advancing in sanitation and finds itself at the bottom of the ranking in the world, as it does for water. It is interesting to observe that the decline in the urban sector in Sub-Saharan Africa appears to be more accentuated in access to water than in access to sanitation facilities at households.

1.4 Missing Targets and Goals for W+S Since Decades

The UN conference at Mar del Plata in 1977 called for access to water and basic sanitation for all to be achieved by 1990. According to Langford and Winkler,[31] 'This target was reaffirmed in the proclamation of the International Drinking Water Supply and Sanitation Decade for 1981-1990'.[32] Unfortunately, the target was not

[27] Stadt ohne Wasser – Südafrika, Spiegel Nr.5 (2018: 82 and 83)

[28] This is relevant Africa wide and presently especially apparent in Cape Town, South Africa as Anthony Turton from the university of Bloemfontein explains, https://www.youtube.com/watch?v=FNt9EayG3-g, (last visited 01.2018)

[29] As an example: Ouagadougou, Burkina Faso, which experienced a water shortage in 2016 despite the significant investments carried out since the end of the 90s with the construction of the Ziga Dam and the extensive transmission pipelines. Equally, for Lusaka, Zambia where the long transmission pipes from the Kafue River to the town incurs substantial O + M costs.

[30] Progress on Drinking water and sanitation, 2014 update, WHO and Unicef, JMP website: http://www.wssinfo.org/ (last visited 07.2016).

[31] Langford and Winkler (2013: 6).

[32] 1981–1990, http://www.un.org/en/sections/observances/international-decades/ (last visited 06.2016).

met. Without rethinking feasibility, it was carried forward in the Global Consultation of Safe Water and Sanitation in New Delhi in 1990 to be achieved in the year 2000. Again, the target was missed. It was now downscaled with the Millennium Declaration in 2000[33] for water and in 2002 for sanitation with the aim to half the proportion of people who do not have access to safe drinking water and basic sanitation by 2015. There was also a second water decade from 2005 to 2015 (water related issues).

Despite some emerging doubts, which were later confirmed (see Sect. 3.5), the UN released the following statement:

> United Nations Secretary-General Ban Ki-moon said, 'Today we recognize a great achievement for the people of the world. This is one of the first MDG targets to be met. The successful efforts to provide greater access to drinking water are a testament to all who see the MDGs not as a dream, but as a vital tool for improving the lives of millions of the poorest people', declaration by the UN in 2012.[34]

At the same time as the UN proclaimed success, global monitoring (JMP) acknowledged in their reports that 1.8 billion people in 2012 and 1.9 billion in 2015 had to consume contaminated water. Consequently, it can be concluded that the MDG target for water was not achieved[35] and the situation is not improving because an increasing number of people in the world do not have acceptable access to water. Surveys on which JMP based its message measured 'improved' water sources[36] instead of sustainable access to safe water, as the target of the goal was formulated.

---

[33] With a reduction in the ambitions, moving away from the goal of access for all.

[34] Joint news release Unicef/WHO on the 06.03.2012, http://www.who.int/mediacentre/news/releases/2012/drinking_water_20120306/en/, (last visited 01.2016).

[35] According to WHO '748 million people lack access to improved drinking-water and it is estimated that 1.8 billion people use a source of drinking-water that is faecally contaminated' [and therefore, not safe – own remark], JMP report (2014: 42), http://www.who.int/water_sanitation_health/hygiene/en/ (last visited 01.2016). This represented 25% of the world population in 2015. Furthermore, according to the JMP report (2017: 3, 110), 'Three out of four people (5.4 billion) used improved sources free from contamination' in 2015 which leaves 1.9 billion people to consume contaminated water (7.3 billion world population minus 5.4 billion). Therefore the water MDG was missed, with access of 74% instead of 88% (JMP 2015: 4) https://washdata.org/, http://data.un.org/Data.aspx?q=world+population&d=PopDiv&f=variableID:12;crID:900 (last visited 06.2017). In addition, it is to note that the number of people without access to safe water is increasing. In other words: 43% of the world population in 1990 and 57% in 2015 had access to piped water. This progress was mainly achieved in medium and low-income countries because the industrialised world had already achieved universal access before 1990. If the proportion without sustainable access to safe drinking water should have been halved, then the percentage of the underserved people should have been at 28.5% and not at 43% in 1995. Thus, the MDG water target was missed by 14.5%, which represents over one billion people worldwide. It is not certain that progress to other improved sources than piped water in the rural areas has offset the one billion people who have not achieved access to piped water as intended.

[36] 'An improved drinking-water source is defined as one that, by nature of its construction or through active intervention, is protected from outside contamination, in particular from contamination with faecal matter.' http://www.wssinfo.org/definitions-methods/ (last visited 04.2015). Technically 'improved' construction is supposed to prevent access of surface water, but not ground water which is often contaminated in the urban setting.

This explains the contradiction of declaring an achievement while the situation is getting worst (see Sect. 3.5).

The even more ambitious goal of access to drinking water for all, repeated many times over before the MDGs, but never achieved so far, has now found its way back to the SDGs. Taking into consideration the results of the four country comparison, presented later in this book (Part II), it is safe to predict that the water and sanitation SDGs will be missed by large again in the Sub-Saharan region and most likely also worldwide. The challenges in the urban water sector compared to the present situation and the progress in the last 10 years cannot possibly be overcome in the remaining 12 years.[37]

All of these declarations, water decades and the recognition as early as 1990 that sector development was far away from the goals set on the international level, did not lead to actions, which could have stopped the deterioration of access to water and sanitation in most of the African countries.[38] Today many countries are further away from the objective of universal access than they were at the times of these declarations.

Concerning sanitation, the message from the global level is that the MDG target has not been achieved and that access to sanitation is considered to lag far behind that of water development. The progress indicated in the global monitoring reports (until 2015) for access to sanitation can be disputed when the sanitation chain to access for onsite sanitation[39] is included into the equation (storage, transport and treatment of effluent). However, the JMP report of 2017 (based on the SDGs) includes the notion of safely managed sanitation but also shows that data for this indicator is only available for less than half of the world's population.

1.5 Insufficient Understanding About the Challenges

The challenge of moving the sector to a level of access for all has been largely underestimated since the 1970s by national experts but also in international discussions. There is a long history of lack of realism documented by the repetitive failure to meet goals. The repeated commitments to goals, which have been out of reach for decades in most African countries, might indicate a limited understanding about the

[37] SDG 6, https://sustainabledevelopment.un.org/post2015/transformingourworld (last visited 04.2016). Expressing a critical view on achievements is Chakava et al. 2014 as well as Langford and Winkler 2013.

[38] The Joint Monitoring Programme (JMP) of the UN indicates that access to piped water has decreased from 1990 to 2012 in Sub-Saharan Africa in Progress on Drinking water and sanitation, 2014 update, WHO and Unicef, JMP website: www.wssinfo.org (last visited 04.2016).

[39] Onsite sanitation means a toilet / shower etc. which are not connected to a (centralised) sewer system and therefore, need in the urban setting a decentralised chain for sanitation in order to avoid pollution.

sector among experts involved in the water discourse.[40] This and the many unsuccessful approaches raise the question, if the decision makers understand what sustainable access to safe water and sanitation in the urban setting means and what the crucial factors for development are.

In addition, the literature review suggests that in Sub-Saharan Africa social concerns in water and sanitation might superimpose the importance of infrastructure development and economic consideration for sustainability. Nevertheless, the 1980 declaration of the first water decade acknowledged the need for an appropriate and comprehensive policy and institutional framework and the importance of the international development cooperation for financing the sector infrastructure in the developing world.

Many countries in Sub-Saharan Africa have realised that they need to tackle the sector challenges in different ways than in the past and thus, have started in the 1990s reforming urban water. Several approaches were tried and among them specifically the promotion of private sector participation for urban water and sanitation by the Bretton Woods institutions and some bilateral donors.[41] Today these approaches focusing mainly on utility services can hardly be considered a success in Sub-Saharan Africa. It shows that progress in the sector through reforms[42] do not depend on one specific factor and on one particular mode of delivery such as private sector participation (PSP).[43] The results of the first reforms underlined that the dominant actors promoting PSPs with their significant power over decision making did not sufficiently understand the complexity and the (local) context in the sector.

Furthermore, most of the literature on water and sanitation refer to rural water but unfortunately provide the impression that urban water is included in their generalisations. This often leads to ill-conceived conclusions and proposals regarding urban water and sanitation. There are limited contributions on urban water and sanitation sector development in the low-income countries and very few on sector reforms based on solid and quantitative evidence.[44] In the 1990s, the discussions concentrated on the mode of delivery and in the 2000s included as new issue sector regulation in Africa.[45] It is now time to start looking closer into urban water and

[40] See also Sect. 4.2.

[41] E.g. the French cooperation (own experience in Burkina Faso in the 1990s).

[42] A reform can be described as a planned reorganization with the aim to improve existing systems according to the Duden, http://www.duden.de/suchen/dudenonline/reform (last visited 01.2016). However, reforms also mean that sector development takes a new direction, which is given by the introduction of a new set of principles, Cullet (2009: 3), see also to Sect. 5.1. For the present work, reforms mean that significant changes or improvements of a given system are undertaken in order for it to be able to address challenges adequately (i.e., moving towards policy goals). This includes a new dynamic where stakeholders are not necessarily side-lined but receive different roles than in the past, where knowledge is enlarged (e.g. international and national knowledge is considered side-by-side), the framework is adjusted, new mechanisms are introduced, etc.

[43] Schiffler (2015: 1) states that worldwide only an estimated 6% of people served '…are billed by a major privately-owned utility that provides all water and sewer services…'

[44] E.g. Bertrand and Geli (1995).

[45] E.g. Pollem 2008: The regulators in Ghana, the Public Utilities Regulatory Commission (PURC),

sanitation development and reforms undertaken in the developing world in order to close the existing knowledge gap and reverse the negative trend of acceptable access.

Despite the seemingly limitation in knowledge about the sector at the international dialogue, UN institutions have helped to make water and sanitation issues gradually better known in the international discussions, among academics, NGOs and the donor communities. The water discourse initiated a number of literatures on access and service provision for water and sanitation and even more on related (crosscutting) issues such as: water and poverty, water and gender, water and health, water and human rights, water and good governance, etc.

1.6 Pressure to Act and Overcome Limits

Presently, urban areas in Sub-Saharan Africa register the fastest population growth in the world. Worldwide, 54% of the population live in urban areas. From the 2.5 billion people who will be added to the world's urban population of today by 2050, 90% will stem from Asia and Africa. By then, the urban population will have reached 66% of the total world population.[46] In Sub-Saharan Africa, the percentage of the urban population compared to the total population is still significantly lower and presently just over one third of the total population, or as some sources indicate, rather around 40%.[47] Nevertheless, as Kessides[48] lined out, Africa is catching up very fast with the rest of the world.

> …the take off in Africa's urban population growth is yet to come.' Furthermore 'What is historically unprecedented is the absolute rate of urban growth in Africa averaging almost 5 percent per year implying close to a doubling of the urban population in 15 years.

Many towns in Sub-Saharan Africa are growing faster than that such as Ouagadougou, the capital of Burkina Faso, which is situated in a very arid area. It grew from around 60.000 inhabitants in the 1960s to around 2.5 million today.[49] If population density would be considered as key for decision making on urban or

---

created in 1997 as a multi- sectoral regulator, and in Mozambique, Coselho de Regulacao de Aquas (CRA), created as single sector regulator in 1998, can be regarded as the commencement of regulation for water and sanitation service provision in Africa. Thereafter, a number of countries followed suit such as Zambia in 2000, Tanzania in 2002, Kenya in 2004, just to name a few.

[46] UN, World Urbanization Prospect, 2014 revision.

[47] It could even be higher than 40% when considering that the regulator in Kenya obliges utilities to include in their service areas all settlements which have urban characteristics (a certain population density). It follows that 45% of the population in Kenya lives in the service areas of the utilities (2017).

[48] Kessides (2005: x).

[49] According to the 'Schéma Directeur d'Aménagement et d'Urbanisme (SDAU) de la Ville de Bobo-Dioulasso, Horizon 2030, Ministère de l'Habitat et de l'Urbanisme' (2012: 29). The population of Ouagadougou was 59,126 in 1960 and was 2.053 million in 2011 according to http://www.indexmundi.com/burkina_faso/demographics_profile.html (last visited 01.2016).

rural solutions for water supply, most likely more than half of the population in Sub-Saharan Africa would need piped systems.[50] There is obviously more urgency to act in urban than in rural water and sanitation development.

Next to this unprecedented urban growth, the sector in Sub-Saharan Africa has to face the challenges of high poverty and inequality and of overcrowded unplanned settlements for the urban poor.[51] Inequality is very pronounced in the Sub-Saharan region. The richest quintile is sharing around 50% of the income, while the poorest quintile has to live on 3–4%.[52] According to the World Bank, over 40% of the population in the region have to live on less than 1.9 USD per day. The poor in the slums have to live in areas with a high crime rate and provide a large part of their income for low-value accommodation and daily transport. There is no space for the households to produce basic food. The living conditions for the poor in the slums seem to be even more unreasonable than those that the poor face in any other areas in the region (including rural) according to Werchota.[53] Despite these facts, many donors concentrate exclusively on rural water and sanitation development with the argument that they need to help the poor.

This seems to leads to a bias in water supply and sanitation towards the rural areas which is documented by Payen[54] although the future challenges are much more pronounced in the urban setting. He underlines that the dependency on unimproved drinking water sources by the urban population has been growing by 28%, while it has been decreasing for the rural population by 54% in the last 25 years. The dependency on un-piped sources has grown in the urban setting by 76% while it has fallen by 9% in the rural setting also during the last 25 years. The evolution of the situation seems even to be worse than these figures indicate when considering that the risks of contamination linked to un-piped sources in the urban setting is tremendously higher than in the rural areas. The same can be said for sanitation where the use of unimproved sanitation facilities in the urban areas has grown 165%, but only 11% in the rural areas over the last 25 years. It is also to note, that the need to evacuate human waste and effluent with a sanitation chain is in the rural setting limited to market centres while it is a necessity in all areas of the towns.

Adding to the challenges outlined above, national governments struggle to balance their budgets and therefore, can barely contribute to water and sanitation infrastructure development. They mainly have to rely on international cooperation for its financing, which brings its own unique challenges.[55] In addition, local politicians

[50] Refer to Kenya where the regulator defines the services areas of the utilities according to the population density and does not follow administrative boarders or the census figures. The result is that more than halve of the population in the country live in the utility areas (Impact Report 2016 from WASREB) although the census is reporting that around 2/3 of the population lives in rural areas.

[51] UNDP 2016.

[52] http://povertydata.worldbank.org/poverty/region/SSA (last visited 01.2016).

[53] Werchota (2013: 7).

[54] Payen (2015: 1, 2).

[55] See Sect. 5.8.

deprive the sector of a higher and achievable self-financing level by consumers because of political interests and some stakeholders use the underserved areas as a market to introduce solutions for water and sanitation services which hamper long-term sustainable development. Therefore, it is not surprising that the sector in the Sub-Saharan countries can hardly cope with these unprecedented increases in demand under such difficult conditions. However, there are examples in Sub-Saharan Africa which document that the negative trend in access can be reversed and that the increase in the numbers of underserved people can be stopped through comprehensive reforms, even under the above mentioned challenging circumstances.

1.7 The Suffering from Insufficient W+S Is Yet Too Silent

Access to water and sanitation is securing life for individuals and society. It supports all other basic needs for human and the wealth building of nations. Consequently, water and sanitation development must be given at least the same priority than access to food, health services, shelters, security, and recognised as more important than energy, education, transport and communication. Inadequate access to water and sanitation is one of the main causes that people remain in misery and countries limit their development. And yet, powerful images of badly emaciated children's faces and bodies are as good as always associated with hunger or the pictures of dying patients in the third world are linked to insufficient medical care or missing drugs. This are the general images presented to the public by the global discourse and the media in the industrial world. Very seldom such shocking images are explained with dehydration due to the consumption of contaminated water or the repeated infection due to poor sanitation.[56] Government slogans on highest levels have the same bias. Their contribution to make the world a better place for many is expressed in 'One World – No Hunger'[57] instead of 'One World – Everyone Safe W+S'. Such a message released on the same policy level would make a big difference for the sector and help to shift funds from projects which are 'nice to have' to effective measures improving living condition for the poor.

The revolting pictures mobilize spontaneously billions of euros for food security or medical research and drug supply when a large famine is proclaimed or an epidemic is looming which could potentially threaten the people in the industrialized world (Ebola, HIV/AIDS, etc.). The waterborne diseases in the developing world do not pose a similar threat to the well-off people in advanced countries and therefore is less of a concern for donors. In addition, politicians can polish their image when pronouncing support for the distribution of food and drugs. Water and sanitation

---

[56] With the exception of some NGOs involved in water and sanitation development.

[57] BMZ (Federal Ministry of Economic Cooperation and Development) global initiative 2017, www.bmz.de

issues do not mobilize the same empathy or provide the boost for an image although there might be even more people exposed to their deficiencies than to insufficient food and missing drugs. The water and sanitation related suffering and deaths seem to be still too silent.

It is time that the global community recognises the outstanding importance of water and sanitation, its decline in too many of the low income countries and its effects on the build-up of social pressure, especially in the Sub-Saharan region. It is overdue to end misleading messages about water and sanitation development in the world, set more realistic goals and deal adequately with the complexity and local context in the developing world. The coming wave of demand in Sub-Saharan Africa will put enormous stress on the water sector to do more and accelerate development. It is in the interest of the decision makers everywhere in this world to avoid a further decline in the low-income countries because inadequate access to urban water and sanitation means for many people unbearable living condition which fuels social unrest and adds further pressure on global migration.

References

Aderinwale A, Ajayi O (2008) The link between poverty and water supply: the Nigerian example. In: Hemson et al (eds) Poverty and water – exploration of the reciprocal relationship CROP (Comparative Research Programme on Poverty). Zen Books, London

Bertrand J, Geli H (1995) From insolvency to excellence – a tale of transformation, office National de l'Eau et Assainissement, 1995, National Office for Water and Sanitation, Burkina Faso, ID Bleue, Mars 2015, AgroParisTech-ONEA

Booysen F, van der Berg S, Burger R, von Maltitz M, Du Rand G (2007) Using an asset index to assess trends in povery in seven sub-Saharan African còuntries. World Dev 36(6):1113–1130. Elsevier 2008

Chakava Y, Franceys R, Parker A (2014) Private boreholes for Nairobi's urban poor: the stop-gap or the solution'. Habitat Int 43:108–116

Coombes, Y., Hutton, G., Hickling, S., Sara, L. and Kiama, A. (2012) Economic impacts of poor sanitation in Africa (Uganda and Zambia), WSP – Water and Sanitation Program, World Bank Group

Cullet P (2009) Water law, poverty, and development – water sector reforms in India. Oxford University Press, Oxford

DeFrancis, M., (ed) (2011) Economic impacts of inadequate sanitation in India – Flagship Report, WSP – Water and Sanitation Program, World Bank Group

Devoto F, Duflo E, Dupas P, Pariente W, Pons V (2012) Happiness on tap: piped water adoption in urban Morocco. Am Econ J Econ Pol 4(4):68–99. 2012, MIT Open Access Articles, American Economic Association

Esrey SA, Potash JB, Roberts L, Shiff C (1991) Effects of improved water supply and sanitation on the ascariasis, diarrhoea, dracunculiasis, hookworms infection, schistosomiasis, and trachoma. Bull World Health Org 69(5):609–621

Exner M (2015) Regelungen zur Sicherstellung der Trinkwasserqualität in Deutschland. Lauterbach Germany, presentation at MATA/GIZ (Mitarbeiter Tagung der GIZ), July 2015

Gaffga NH, Tauxe RV, Mintz ED (2007) Cholera: a new homeland in Africa? AM J Med Hyg 77(4):705–713

JMP-Joint Monitoring Programme on MDG (2014) Progress on drinking water and sanitation, WHO and UNICEF, Available: http://www.wssinfo.org/fileadmin/user_upload/resources/JMP_report_2014_webEng.pdf (last visited 04.2014)

JMP Annual Report WHO and UNICEF (2015) https://d26p6gt0m19hor.cloudfront.net/whywater/JMP-2015-Annual-Report.pdf

JMP (2017) Progress on drinking water, sanitation and hygiene. https://www.who.int/mediacentre/news/releases/2017/launch-version-report-jmp-water-sanitation-hygiene.pdf

Kessides CH (2005) The Urban Transition in Sub-Saharan Africa: Implications for Economic Growth and Poverty Reduction, African Region, Working Paper Series No. 97, Urban Development Unit, Transport and Urban Development Department, The World Bank

Langford M, Winkler IT (2013) Quantifying water and sanitation in developing cooperation: Power or perspective? Working Paper Series, Harvard School of Public Health, Available: http://fxb.harvard.edu/wp-content/uploads/sites/5/2013/09/Langford-and-Winkler_Final-Working-Paper-92413.pdf

Pauschert D, Gronemeier K, Jebens D (2012) Informal service providers in Tanzania, Eschborn, GIZ (Deutsche Gesellschaft für Internationale Zusammenarbeit GmbH), 2012, Available: http://www.giz.de/fachexpertise/downloads/giz2012-en-informal-service-provider-tanzania.pdf (last visited 06.2015)

Payen G (2015) Unsatisfactory access to drinking water and sanitation trends derived from JMP estimates for 2015 Source: "Progress on Sanitation and Drinking Water: 2015 Update and MDG Assessment" UNICEF-WHO JMP report launched on 30 June 2015. http://www.wssinfo.org/filcadmin/user_upload/resources/JMPUpdate-report-2015_English.pdf

Pollem O (2008) Regulierungsbehörden für den Wassersektor in Low-Income Countries – Eine vergleichendc Untersuchung der Regulierungsbehörden in Ghana, Sambia, Mosambik und Mali', Dissertation, Carl von Ossietzky Universität Oldenburg, Fakultät II, 2008

Reade N, Ndirangu N (2009) Interim evaluation 2008, Water Sector Reform Program, Centre for Evaluation (CEval) for Eschborn, Deutsche Gesellschaft für Technische Zusammenarbeit (GTZ) GmbH (Publisher)

Schiffler M (2015) Water, politics and money a reality check on privatisation. Springer International Publishing Switzerland, Springer Cham Heidelberg/New York/Dordrecht/London

Spears D, Ghosh A, Cumming O (2013) Open defecation and childhood stunting in India: an ecological analysis of new data from 112 districts. PLoS One 8 (9), September, e73784 (pp 1–9)

UNDP (2016) Human development report 2016, Human development for everyone. http://hdr.undp.org/sites/default/files/2016_human_development_report.pdf

UNESCO (2006: 6) Water-A shared responsibility, Paris

US Government Global Water Strategy (2017) https://www.usaid.gov/sites/default/files/documents/1865/Global_Water_Strategy_2017_final_508v2.pdf

Vuorinen HS, Juuti PS, Katko TS (2007) History of water and health from ancient civilisations to modern time. Water Sci Tech Water Supp 7(1):49–57

Werchota R (2013) The growing urban crises in Africa, Eschborn, Deutsche Gesellschaft für Internationale Zusammenarbeit (GIZ), Germany

Werchota R (2017) 'The crucial factors for a development of the urban water and sanitation sector towards sustainable access for all in Sub-Saharan Africa' Band 1 und 2, University of Vienna, Austria

WSTF – Water Services Trust Fund (2010) Survey on the impact of formalised water kiosks on living conditions in Athi River and Ongata Rongai, Water Services Trust Fund, Kenya

Chapter 2
Considerations for General W+S Issues

Abstract There are lessons learned from history which gained global relevance. Growing settlements need centralized W+S systems with standardized service provision organized by professionals. However, because of specific challenges in low-income countries, these lessons learned and academic knowledge have to be blended in with local knowledge. Though modern medicine can help people to fight diseases, the health sector cannot substitute W+S infrastructure nor can it make rural W+S solutions appropriate for towns. The global and national monitoring systems produce very different signals. The desperate situation in many countries contradict the positive messages from JMP which make some donors pull out of the sector despite its deterioration. This aggravates the chronic investment gap. W+S is also mis-used by politicians and the service vacuum left by utilities in urban low-income areas attracts many narrow minded and unqualified actors without them being called to order. Centralized systems help to substantially reduce gender inequalities in urban W+S. However, the discussions show how little is known about mitigating them. Equally, the existing multi-dimensional poverty indexes can hardly capture the brutality which comes with (water) poverty in towns. The contribution of W+S to lift and keep people out of poverty is not recognised sufficiently compared to health and education.

In this chapter the pertinent topics for urban water and sanitation and their relevance to the low-income countries in Sub-Saharan Africa are discussed. This reading offers a critical review of perceptions, approaches and definitions used in the discussions.

2.1 Lessons from the Past with Global Relevance

Several thousand years ago, people started to settle turning from hunters and gatherers into agriculturalists. The lifestyle of people changed gradually into urban life. It generated first wealth for elites through the collection of taxes which had to be paid

R. Werchota, *Empty Buckets and Overflowing Pits*, Springer Water,
https://doi.org/10.1007/978-3-030-31383-8_2

by the often enslaved common people. They had to work in the surrounding fields of settlements to produce food for the town population and serve as labourers or craftsmen for constructions being paid with the produced food. To live in growing settlements had advantages and disadvantages for its dwellers. Obviously, the advantages outstrip the disadvantages which made the towns dominate the rural areas and set the pace for the development of the society. With urbanisation state building began.

However, this dominance of settlements could only last as long as the risk of life threatening disadvantages in the form of epidemics was contained. Population density and people living in a close environment with domesticated animals needed sufficient sanitation practices. Most of the epidemics could be controlled by the availability of adequate water for the households. This was in the interest of all dwellers, equally of the elite, which had to maintain the productivity of its people in order to accumulate its own wealth and power. There are several cases in history where civilisation seemed to have disappeared because of repeated outbreaks of epidemics which most likely were linked to water borne diseases or shortages of water for domestic use. Hence, safeguarding water quality and practicing sound sanitation in densely populated environments was crucial and had to have the highest priority for the state. Infrastructure for water and sanitation unearthed from the earliest days of settlements bear witness to this concern and indicate the important efforts undertaken to provide potable water for the urban dwellers and handle safely their effluent and human waste.

Later, with the formation of citizen states, living conditions for many dwellers progressed, which led to an increase in per capita demand for domestic water use and installations for sanitation. This produced more effluent which despite existing sanitation facilities started to pollute not only surface but also ground water. It had a negative effect on the water quality of single water points in towns. The individual urban households could no longer rely on traditional water points in their vicinity as a source for drinking water without its member falling sick and causing waterborne epidemics.

Growing Settlements Need Centralised[1] W+S Systems

Both, the polluted water sources in towns and the increased per capita demand for water, forced communities to start searching for other safe water sources outside towns and to stretch effluent evacuation installations beyond town boundaries. From this point of development, even more intervention of the state was needed because of the high costs inflicted by the construction of infrastructure and the complex organisation which was required to establish and maintain a bigger water supply and sanitation system. Water and sanitation provision became the responsibility of governments and could no longer be left to the individual households or a small local community.

[1] Centralized water and sanitation systems means systems which are in technical and managerial terms centralized with raw water from intakes or boreholes, treatment plants, storage facilities and distribution networks managed by a utility with well-trained professionals.

In Ancient Rome[2] for instance, authorities recognised that a thriving urban lifestyle could only be sustained through substantial investments in the aqueducts (each around 10–20 km long), which became part of a large distribution system for water. In addition, it needed a collection system for the effluent, called sewer system. For ancient Rome, the Cloaca Maxima is a symbol of developing a sanitation chain, which was part of an extended drainage system for human waste, storm water and water from swamps to be dried up. Individuals in the urban setting could shift away from the use of open wells as traditional water points, which however, remained appropriate for the rural setting where space for protection of sources was available.

Water in towns was supplied by an urban centralised water infrastructure system. According to Kramer,[3] in Rome, 13 aqueducts provided 750,000 m^3 water daily to a population of around 1 million people[4] in the fourth century AD. This would have been a staggering water availability of an average of 750 l per capita per day compared to an average consumption in Europe of around 120 l per capita per day presently. The management of this infrastructure was delegated to agents appointed by the public administration.

Nevertheless, very few households had running water and toilets in their homes in Roman times. Most of the urban dwellers collected water from public fountains and used public toilets. Despite this sharing of facilities, the risk of epidemics in towns was sufficiently managed to maintain urban life. Notwithstanding the improvement of providing water of good quality, waste management remained poor, thus resulting in high levels of infections and child mortality according to Vuorinen et al.[5]

Centralised Systems Brought Standardised Services for All

The next noteworthy step in urban water and sanitation development was achieved in the nineteenth century during the time of growing industrialisation. Multiple widespread cholera epidemics were the trigger for the construction of large-scale centralised water and sewer systems built for all urban dwellers. This strong focus on infrastructure development for the benefit of all, in the quest to curb waterborne epidemics and improve living conditions in the urban setting, was initiated by Chadwick in 1842[6] in Britain and was swiftly copied by the rest of Europe and North America.

In the early seventeenth century, long before Chadwick, piped water supply systems were built for commercial purposes or for status and convenience of some people and groups. In certain cases, the trigger was the need to provide water for

---

[2] The development of large water supply systems in Roman times was not limited to the city of Rome – Exner 2015, stating the case of the Colonia Claudia Ara Agrippinensium (Eifelwasserleitung to Köln).

[3] Kramer (1997: 23–24).

[4] http://davidgalbraith.org/trivia/graph-of-the-population-of-rome-through-history/2189/ (last visited 06.2017).

[5] Vuorinen et al. (2007: 51).

[6] http://www.historyhome.co.uk/peel/p-health/sanrep.htm, (last visited 08.2015).

firefighting, like Bohman indicated.[7] However, most of the systems built in Europe before the nineteenth century served the upper class, some government institutions and industries, but not the general middle and lower income classes.

Finding water, a place to defecate/urinate and someone to evacuate human waste and effluent does not automatically mean that people have access to safe water and sanitation since consuming polluted water or spilling untreated human waste into the environment will make urban dwellers fall sick or even lose their lives. Open defecation and human waste released uncontrolled by many households in towns which can use (safe) sanitation facilities contaminates surface and ground water and thereby, when consumed by the poor, causes outbreaks of waterborne diseases on a large scale. Thus, access to safe water and sanitation must be linked to minimum standards.

Not long after Chadwick, this was already recognised among many specialists and regulations for water service provision were introduced[8] in order to secure the safety of drinking water and effluent evacuation and treatment. It made providers fulfil minimum requirements. Gradually, further improvements in water and sanitation service provision took place as better treatment techniques, such as the chlorination of raw water and ultraviolet water purification, were applied. Within a few decades, the entire urban population was connected to the networks of controlled drinking water and sewer systems although shared water and sanitation facilities within blocks of flats were widespread long after. This significantly improved public health and contributed to an increasing productivity of the population.[9] That a system of control is needed for water and sanitation service provision is widely acknowledged today among experts, but unfortunately still not always taken on board in the water and sanitation discourse and when activities are carried out.[10]

The widespread waterborne epidemics in the nineteenth century in Europe forced authorities to scale-up water and sanitation infrastructure because there was no effective medical treatment to cure most of these diseases. This scaling up of infrastructure in the industrialised world also helped to replace the small and decentralised (often-private) water and sanitation systems with modern large-scale

---

[7] Bohman (2010: 4–10).

[8] Filtration for drinking water production was made mandatory by the Prussian government in 1859, which prevented the massive cholera outbreaks in Altona (Hamburg) which emerged on the other side of the administrative boarder in (Hanseatic) Hamburg. http://mulewf.rlp.de/fileadmin/mufv/img/inhalte/wasser/ressortforum/01a_Prof__Dr__Martin_Exner__Teil_1_.pdf, (last visited 08.2015).

[9] The proposal by Robert Koch in 1893 to install water filtration led to a drop in the children mortality rate per 100 live births in Hamburg from 1881 to 1927 to around 2% in 1893 (before 1892 it stood at 4%, during the cholera epidemic in Hamburg). Martin Exner, University of Bonn, explains: 'The general lesson which still holds today is that passive health protection is the best way to improve population health' (presentation at the GIZ MATA 2015).

[10] E.g. springs and open and drilled wells in the urban setting, referred to in this work as single (traditional) water points or single water sources, are often contaminated in densely populated areas, but counted as 'improved' access by the global monitoring. In addition, global monitoring does not make a difference between formalised or uncontrolled (informal) service provision.

centralised[11] infrastructure. It did away with a two-class system where the upper class was connected to the water supply and sewage systems and the underserved people, the lower end of the society, depended on high-risk traditional water sources or on uncontrolled service providers. With up-scaled systems generating economies of scale, water and sanitation services became high quality and low-cost commodities for all.

Another leap forward in water and sanitation development took place during the twentieth century as urban networks in towns stretched gradually into the rural areas either by extending existing urban systems or by establishing piped systems with an association of several municipalities using preferentially low cost (gravity) solutions.[12] Such extensions of service provision was especially urgent where the population was fluctuating substantially, e.g. in areas with seasonal (mass) tourism and possible because of increasing economic wealth in the area. Untreated effluent caused environmental damage hindering further development in the areas and damaging commercial interests.[13] From the moment on where the state had the means to offer the highest service level for everyone and the vast majority of the people could afford to pay for it the distinction between urban and rural water and sanitation was no longer relevant in the industrialised world. Nonetheless, it took Germany around 100 years to reach this point of covering the rural areas with the modern systems initially established for towns.

This development seems to be one of the fundamental differences in the rural sector between industrialised and low-income countries. The promotion of piped systems in the rural areas in the low-income countries was more driven by humanitarian than sustainability concerns mainly pushed by civil society and philanthropic organisations. They offered to the rural communities an urban water supply system with a service level people could not or would not maintain because they still remained with the alternative to go back to traditional water sources, which the urban dwellers no longer had. Today many systems in the rural areas suffer from insufficient sustainability because NGO employees, supported by donors, transported their understanding of development to the partner countries from their own comfortable upbringing in the industrialised world. They forgot to ask their parents or grandparents how water and sanitation were organised in their households. Many were also self-proclaimed experts in water and sanitation development. Today many low-income countries struggle with these negative effects of well-intended but

---

[11] Centralized water and sanitation systems means systems which are in technical and managerial terms centralized with intakes, treatment plants, storage facilities and distribution networks managed by a utility with well-trained professionals.

[12] E.g. Germany, with water connection at 100% in urban and 99% in rural areas and improved sanitation at 100% in urban and rural areas (JMP report 2014). Bain et al. (2014a, b: 922, Table 2) indicated that access to piped water in premises in urban Europe (High Income – HI) is 99.5% and in rural Europe (HI) 98.8%. For the Americas the respective figures are 97.3% and 98.8%.

[13] Schulz and Schulz 1977: 1577–178; E.g. around the lakes in Carinthia, Austria in the 1960s where tourism in the summer was threatened because algae developed in the lakes due to the discharge of untreated effluents. http://www.zobodat.at/pdf/CAR_167_87_0157-0178.pdf (last visited 05.2016).

incompetent support and oblige the utilities, responsible for urban areas, to take over rural piped systems (e.g. Uganda). Thereby, they compromise urban water and sanitation development in their countries as they would have in the industrialised world with the same short cuts and neglect of sustainability, appropriate service levels, etc.

However, for urban water it can be concluded that the use of traditional water points on large scale and many small scale systems operating in one town were no solutions that prevailed in thriving cities. The knowledge obtained in ancient and modern time was that growing towns which want to secure their survival and want to offer decent living conditions to their population need to establish large scale centralised systems for water and sanitation which use safe water sources outside the towns. Later, the awareness grew that such systems need to be overseen by the state or their agents to guarantee quality of drinking water and services.

Developments in W+S Are Closely Interlinked

In the nineteenth and twentieth century, sector development in the industrialised world ensured in many cases simultaneous progress in access to urban water and sanitation but not necessarily with a well-functioning sanitation chain. According to Kuks and Kissling,[14] the town of Brussels, disposing of a sewer system, diverted its untreated sewage[15] into the sea until 1998 more than 100 years after it was established. Equally in the low-income countries today, there is common understanding that especially in densely populated LIAs, water and sanitation has to be developed simultaneously in order to reduce health risks. However, piped water infrastructure usually precedes sewer systems in low-income countries, as it was the case in many countries in the industrialised world according to Kramer.[16] The establishment of piped sewage is more costly than systems for drinking water. Urban dwellers are usually interested in having a safe toilet to use and their concern about a sanitation chain ends there when connected to a sewer system or with the emptying of the pit when it is full. Therefore, someone has to take care of the remaining sanitation chain.

Without state interventions in sanitation an increasing water supply will lead to a growing sanitation challenge. Improvements in water supply attract new residents, and therefore lead, in many cases, to an increase in population density. This in turn produces more waste, which in the absence of an organised effluent evacuation

---

[14] Kuks and Kissling (2004: 30).

[15] In this book the following expressions for sanitation are used: A sewer system is a centralized piped system where customers connect their sanitation facilities. Sewerage means a system consisting of a full centralized sanitation chain which includes, in the case of off-side sanitation, inlets for customers, pipes to transport the waste (including elements such as pumping stations, man holes, etc.) to the treatment facilities and a controlled disposal of the remaining sludge. In the case of household sanitation facilities which are not connected to a piped system (called onsite sanitation) sewerage includes a decentralized sanitation chain which ensures collection and transport of the human waste by other means than pipes (trucks, hand pulled carts, an intermediate storage if needed, the deposit at treatment facilities) and a controlled disposal of the remaining sludge. Sewage is the human waste often mixed with effluent.

[16] Kramer (1997: 37).

system, is diverted onto the shrinking public space. Hence, increases in water supply without sanitation development are forcibly counterproductive in the urban setting. Furthermore, increasing population density in LIAs reduces the distance between traditional water points and toilets but also the space where toilets can be placed. Contaminated groundwater spreads to traditional water points and the absence of toilets to use, forces people to defecate in the open on public ground or use solutions such as 'flying toilets' (see also the case of Betty in Kawangware slum, Sect. 2.5).[17] Open defecation and human waste emptied from toilet pits and discharged uncontrolled as well as overflowing toilets worsen the pollution in and around the settlements. It now becomes urgent for the community/state to establish a controlled evacuation of human waste in the LIAs where seldom sewer network exist.

It can be said that without substantial progress in sanitation, the benefits of extended water development can either not fully materialise or even worsen living conditions in settlements. Because sanitation infrastructure development is generally lagging behind water and thereby, increasing the risk of ground water contamination, the use of traditional water points (even those which are considered 'improved') by households should never be an option in urban development.

Modern Medicine Offers the Opportunity to Fight Symptoms

The option to treat waterborne deceases with medication seems to have changed the way the water sector is functioning compared to the nineteenth century in the industrialised world. The decision makers can call upon the health sector to counteract waterborne epidemics. This is not necessarily an advantage because it is becoming apparent that neglecting water and sanitation infrastructure development people are prone to be constantly exposed to waterborne diseases and therefore need continuous active health protection. For instance, the 'bye bye cholera' campaigns carried out by the health sector in the 2000s in Zambia, which celebrated the end of cholera in the country, has not prevented repeated outbreaks of waterborne diseases thereafter. Contrary, they seem to accelerate and become a permanent event since a number of years. This is an indication that with the modern medical system there is the option to treat the symptoms but it cannot solve the problem at the root.

For many people in the slums like Betty or even Agnes from the lower middle class (see Sect. 2.5) being forced to obtain water from boreholes or neighbourhood sales, spending for water will either eat unnecessarily into their already insufficient or limited household budget or increase their expenditure for health care because of the missing sanitation chain. Hence, regular treatment of waterborne diseases inflicts high costs on the individuals or the society (health system). Over a longer period, it is likely that the costs of active health care, necessary to contain waterborne diseases, will outstrip the spending for water and sanitation infrastructure development as passive health protection in the low-income countries.[18] In addition,

[17]A designation in Kenya for defecation in a plastic bag, which is then thrown onto roofs or into yards of households or public ground.

[18]WSP Programme (World Bank Group) on the Economic impact of poor sanitation in Africa 2012. Infrastructure in this paper means 'hardware' such as treatment plants, pipe network, etc.

regular treatment by modern medicine is likely to damage people's health over time, according to Spears.[19] Furthermore, relying on active healthcare to the detriment of water and sanitation infrastructure hinders the development of the countries because they cannot make use of the important non-health benefits of water and sanitation service provision for all.

It seems that politicians in low-income countries face a dilemma in the decision of where to place public money. Active healthcare can solve a problem linked to insufficient access to water and sanitation as a quick fix, which might help in the upcoming elections (e.g. 'bye bye cholera'). Contrary to this, if politicians invest in costly long-term infrastructure development for water and sanitation they might not receive sufficient credit from their electorate while being in office.

The five lessons from the past with global relevance

- Urban water and sanitation systems need to grow in size with increasing demand and be operated sustainably to maintain urban life.
- With increasing population density traditional water points in towns are no longer safe because the risk of ground water pollution is growing. Even with protection from surface water (considered 'improved') they cannot be part of an urban supply system.
- Safety standards are necessary to be enforces in order to guarantee quality of drinking water and adequate effluent treatment.
- In towns, the development of water and sanitation systems needs to go hand in hand.
- The medical treatment of waterborne diseases cannot be a long-term substitute for water and sanitation infrastructure development.

These five lessons are very obvious for urban water and sanitation experts. However, as explained in the following chapters, a number of stakeholders active in urban water and sanitation development in low-income countries still ignore these lessons from the past which gained global relevance.

2.2 Academic Text-Book Knowledge

Water and sanitation service provision has an economic and a social dimension. A utility which cannot cover costs is not in a position to secure sustainability of services provision and respond adequately to increasing demand. If this economical condition is not ensured, the service providers cannot fulfil social goals. The fact that social goals are often considered very important in the political discussions and when addressing an electorate does not mean that economic goals can be neglected.

[19] Spears et al. (2013: 29).

W+S: A Natural Monopoly[20]

Centralised systems for water and sanitation service provision have several asset components. The biggest cost element is usually the pipes used for the networks, which are buried in the ground, and more for sewage than for water. This need for huge amounts of investments to enter the market keeps possible competitors away. In addition, a provider engaged in the market cannot easily pull out and take away pipes from the network and connections sunk into the ground for the use in other areas.

Furthermore, the large providers can produce with falling average costs, when networks increase in size and pipes in diameter, because of economy of scale.[21] It follows that for water and sanitation service provision one provider can produce services at lower costs than several would be able to do, according to Nickson and Franceys.[22] There are other sectors, which need large networks to provide services such as electricity, telecommunication, gas, railways, etc. The difference to the water and sanitation sector is that they can promote competition in the use of the networks and therefore, allow for more providers when separating production and distribution (vertical unbundling). In the case of water and sanitation systems, this is not possible without substantial risks for water quality. Hence, the network, asset ownership and the difficulties to vertically unbundle functions are bottlenecks limiting the options for service provision. Consequently, the urban water and sanitation service provision is regarded as 'the natural monopoly'.

In economics, it is recognised that services produced by monopolies come with disadvantages for consumers compared to services produced under perfect market competition. Monopolies lead to higher prices and an output which does not satisfy demand in terms of quantity. These will exclude a number of people even for a minimum consumption[23] which is unacceptable in the case of basic goods necessary for survival. In addition, monopolies lead to quality inefficiencies which need to be corrected with the enforcement of minimum quality standards for safe drinking water, complaint handling, water pressure, and etc. Monopolies also lead to technical inefficiencies retarding the application of innovative technologies and manage-

---

[20] 'A *natural monopoly* is a distinct type of monopoly that may arise when there are extremely high fixed costs of distribution, such as exist when large-scale infrastructure is required to ensure supply. Examples of infrastructure include cables and grids for electricity supply, pipelines for gas and water supply, and networks for rail and underground. These costs are also sunk costs and they deter entry and exit. It may be more efficient to allow only one firm to supply to the market because allowing competition would mean a *wasteful duplication of resources*.' http://www.economicsonline.co.uk/Business_economics/Natural_monopolies.html (last visited 05.2017).

[21] According to Europe-economics (2003: 33) 'If an increase in output leads to a reduction in input per unit of outputs, productivity is increased through economies of scale' and 'Economies of scale are found in many sectors of the economy, but particularly so in network infrastructure companies such as the regulated utilities.'

[22] Nickson and Franceys (2003: 4).

[23] The WHO indicates that for drinking and cooking 7–9lt/c/d of water is acceptable. When adding the water needed for hygiene, a water quantity of 20lt/c/d minimum is accepted as appropriate according to Howard and Bartram (2003: 22).

ment technics. Next to reasons of economy of scale and inefficiencies, Nickson and Franceys[24] suggest the asymmetry of information, and water and sanitation as 'merit goods', as two other factors in market failure.

In addition, in the industrialised world, the funding of infrastructure of larger network systems for basic goods were generally provided through the public budget, at least in the initial phase of large scale systems. In the developing world, the states are not in a position to do so. Hence, in Sub-Saharan Africa, the lion's share of funding for infrastructure is since decades provided by donors (see Part II). Reimbursement of loans should be stretched over generations as several of them will benefit from a part of the water and sanitation infrastructure. Since affordability is a major concern in the low-income countries the state has a role to play in the funding of assets even if it cannot provide sufficient funds for the sector through general and other taxes. This external dependency of low-income countries combined with the chronic financing gap in the sector, makes that the national decision makers need to pay increased attention to asset development and funding. Consequently, state interventions not only need to cover the regulation of operation, but equally financing of infrastructure.

Pollem,[25] while discussing the different theoretical models of regulation, indicates that regulation was not limited alone to sectors where market failure takes place. Hence, there are reasons other than economical, which make regulation necessary such as protecting interest groups, redistribution of wealth (cross-subsidisation), raising welfare, and etc. Also Baldwin et al[26] support this view by emphasising the fact that there can be a combination of rationales for justifying regulation beyond economic reasons (see more details in Sect. 4.3). To these various reasons why the state must be involved in the development and provision of water and sanitation services another has been added recently. Since 2015, the UN member states have recognised water and sanitation as two separate human rights.[27] Therefore, states are not only advised but now legally obliged to move towards universal access by respecting minimum requirements according to rights.

W+S: A Public or Private Good?

Water and sanitation assets are generally publicly owned with few exceptions, such as in England and Wales, as Van den Berg explains.[28] Equally in Chile, some cities in the US, and small-scale providers in low-income countries assets for water are privately owned. This ownership of assets must be distinguished from the public ownership of a corporatized service provider. Such companies can be held 100% by municipalities, counties, central state structures, etc. on behalf of the public and be responsible for service provision with or without the responsibility for the development of infrastructure. Hence, asset ownership can be separated from asset holding/

[24] Nickson and Franceys (2003: 5).

[25] Pollem (2008: 46–56).

[26] Baldwin et al. (2012: 23).

[27] United Nations, adoption of General Assembly, A/C.3/70/L.55/Rev.1, 18.11.2015.

[28] Van den Berg (1997: 1).

development. Asset holding and development can easily be delegated by the (public/civil service) owners to asset holding and development agencies which is one way of professionalising planning, fund mobilisation and asset development as it is the case in several countries in Sub-Saharan Africa such as in Mali, Senegal, Kenya,[29] and etc.

The challenge with separating operation and asset development lies in the coordination between asset developer and service providers in order to avoid work being carried out which does not fulfil the need and priorities of the operator. Furthermore, to find a clear separation line between maintenance work to be carried out by the service provider and investments by the asset holder and developer is generally a challenge especially in emergency cases where a time consuming clarification of which institution is responsible for specific works is not helpful. Another challenge may arise when such a professionalization of asset development is regionalised and not overseen by professionals on national level like in Kenya where eight Water Services Boards have been established and coordinated by civil servants. The result is that national planning, information systems on investments and standards setting for asset development are insufficient or missing and there is no link between funding and tariff adjustments. This limits sector performance, fund mobilisation and aid effectiveness.[30] Another risk is emerging when asset developer become 'little republics' and start holding back lucrative assets and starting their own operation instead of handing them over to the utilities.[31]

Because assets are owned by the public and services are of social concern, some sector stakeholders argue that water and sanitation services should be supplied free of charge, at least for the poor. Such stakeholders often consider water and sanitation services as public goods. Nickson and Franceys[32] on the other hand argue that water policy and organisation of water and sanitation services in practice does not follow the characteristics of a public good such as non-excludability and non-rivalry. They explain that in most of the low-income countries, sector policies follow the principle that polluters and users of services pay. It is common practice, with very few exceptions in the world,[33] that service provision is invoiced to consumers and disconnection from service provision takes place in cases of non-payment. These practices, based on international principles and national policies, do not fulfil the two criteria for public goods and, therefore, urban water and sanitation service provision has to be considered as private and not as public good, Nickson and Franceys argue.[34]

---

[29] In Kenya until recently. Devolution of water and sanitation service provision to the counties made the asset holders and developers obsolete (Water Services Boards).

[30] Stern et al. (2008).

[31] As is sometimes the case with the WSBs in Kenya (own observations).

[32] Nickson and Franceys (2003: 4).

[33] For instance, water is free of charge for domestic water use in Turkmenistan and was in Ireland until 2013. Ireland has in the meantime moved to metered water consumption and to billing (Environment, Community and Local Government Paper of Ireland, 2012).

[34] Nickson and Franceys (2003: 3–5).

Another opinion is expressed by Global Water Forum,[35] explaining that these services are not exclusively marketable goods because people have a right to consume them as a basic service. They argue that up to this basic consumption level, which guarantees the survival of people and a minimum hygiene in households, water must be considered a public good. After this consumption level,[36] water services can be regarded a private good. Thus, water can be viewed a social and private good where consumer behaviour differs both below and above a minimum consumption level and where the state has an obligation to guarantee access for all at a minimum quantity and standards.

However, economic theories do not uniquely distinguish between public or private goods but allow for mixed cases such as the near-public good where the two characteristics of public goods apply but in addition, non-payers can be excluded. This exclusion is considered important because the non-containment of the free-rider phenomena of public goods will lead to insufficient service provision. Hence, it can be concluded that water and sanitation are near public goods. Therefore, referring to Nickson and Franceys again, water policy and organisation of services are not relevant in the distinction of the different types of goods. Next to the risks of public goods to be over-consumed (wastage), it is recognised that underfunding is another challenge to be faced.

W+S as Social or Economic Goods

The Dublin declaration [37] considered water as having an economic value. This was the reference used for many stakeholders to demand cost-recovery in the use of water and for its service provision. Also the operationalisation of human rights suggests that the right to water does not mean that water and sanitation services have to be free of charge. Others argue that water is a social good and therefore, should be supplied at least with a subsidised (social) tariff for the poor. This polarising and often heated debate of either subsidised tariffs for water or cost coverage has little meaning where an urban system includes several income groups within one or a number of towns. In such a situation, both objectives can be accommodated at the same time by a tariff policy. An average tariff can be set to cover all costs (e.g. 150% of O+M) and the tariff structure can distribute the load to the different consumer groups in such a way that it is affordable for everyone. Such cross-subsidisation is not possible when there are two separate systems in a town, the utility as a formalised provider serving mainly the upper and middle classes and the small scale providers (e.g. community or private small-scale systems) serving only the poor in the LIAs.

[35] White C. in Global Water Forum, page 1–7, http://www.globalwaterforum.org/2015/04/27/understanding-water-markets-public-vs-private-goods/?pdf=12237 (last visited 04.2017).

[36] Around 20 l per capita per day.

[37] 1992: Principle 4 of the Statement on Water and the Environment (ICWE) in Dublin, Ireland, January 1992, proposed to the assemble at the United Nations Conference on Environment and Development (UNCED) in Rio de Janeiro in June 1992, https://www.wmo.int/pages/prog/hwrp/documents/english/icwedece.html – (last visited 04.2015).

Private Sector Participation: Pro and Contra

Already in the 1970s, it became obvious to many experts that the sector in the developing world could not respond adequately to the growing demand for water and sanitation. The UN declaration of the first water decade in 1980 specifically mentioned the inadequate policy and institutional framework and the insufficient funding of infrastructure as causes.[38] By the mid-1990s, donors injected a substantial amount of funds into the sector during the first water decade for rehabilitation and extensions of infrastructure. With this considerable flow of funds provided by donors the sector was not able to reverse the negative trend in access. It became obvious that other factors are important for development such as the performance of service providers, which never reached the necessary level in Sub-Saharan countries at this time. The disappointing results of the first water decade and a continuous wastage of funds for premature rehabilitation due to insufficient maintenance of assets led to an understandable deception among donors. As Schiffler[39] explained, it triggered a worldwide wave of promoting privatisation/PSP[40] in water and sanitation service provision by the IMF, World Bank and the IFC, and especially in the highly underperforming Sub Saharan region.[41]

There is extensive literature on the privatisation and liberalisation of utility services. In general, the argument for privatisation/PSP is that private enterprises will obtain a higher level of performance and at the same time, are more likely to bring down costs than public utilities are able to do. In addition, it is said that privatisation reduces the risks of a 'fiscal burden' for governments. However, Newbery is pointing out that[42] privatisation needs to be combined with liberalisation (vertical unbundling) in order to pass on the benefits of lower costs of service provision to the consumers in the form of lower prices. However, liberalisation in the case of water and sewerage is unlikely, he underlines, because the costs of moving water any distance is high compared to its value and because the owner of the network, who does not produce water would have to guarantee the quality of water delivered to the consumer. In addition, Oelmann[43] indicates that the options for a possible competition in the sector also depends on the situation in the country (e.g. the numbers of utilities and their sizes) and a vertical unbundling would lead to an increase in the

[38] 55th UN plenary meeting 10 November 1980, 35/18. Proclamation of the International Drinking Water Supply and Sanitation Decade, http://www.un.org/en/ga/search/view_doc.asp?symbol=A/RES/35/18 (last visited 012016).

[39] Schiffler (2015).

[40] Schiffler (2015: 4–6) briefly explains the many types of private sector participation. The present work refers to private sector participation and not privatisation in Sub-Saharan Africa with the sale of public assets. The term private sector participation is used in the sense that the utility management is taken over partially or with full control in various forms of contracts (from management to concession) by foreign companies (generally multinationals).

[41] But also other donors such as the French Cooperation (own observation in Burkina Faso during the 1990s). It should also be mentioned that many French multi nationals received contracts during these times combined with financing of the WB for asset development.

[42] Newbery (2004: 2, 4, 6, 26).

[43] Oelmann (2005: 40–48).

number of companies and employees in the sector, which in many Sub-Saharan countries would further complicate the institutional framework with a negative impact on costs and organisation. Hence, the argument for privatisation in the sector to bring down the water tariffs was unrealistic from the start and therefore, can be considered as an instrument of propaganda.

Furthermore, water privatisation in the sense of transferring ownership of assets from the state (central or local government) to private owners has little relevance in Sub-Saharan Africa. Governments are hardly in a position to sell assets either through the public capital market or to a few investors. Megginson's et al[44] emphasise that selling assets through the public equity market by shares in a country with a high level of inequality (income/wealth) would mean that the government would have to accept a substantially lower price for the shares. Thus, there is little incentive to sell a public company with its assets when the contribution to the budget is not significant. Selling the assets of a water and sanitation utility to a few investors is also not an option as long as there is a risk of later expropriation (re-nationalisation), which is easier to carry out when the number of owners is small. In any case, today, most of the utilities in Sub-Saharan Africa and also globally, are publicly owned and operated, according to Schiffler.[45] Hence, water is different from other utility services because it has a low value, is costly to move and it is difficult to ensure quality at the point of consumption when different water is mixed in one network. According to Newbery,[46] 'The case for water privatisation is therefore weaker than for most other utilities [sectors].'

There is a need to distinguish privatisation in the sense of selling assets to the private from Public-Private Partnership (PPP) or Private Sector Participation (PSP) intended to improve operation of assets and generate funds for investments. There are many forms of PSP and some of them found in Sub-Saharan Africa. However, it can be said that there are some successful PSP arrangements, but on large scale, PSP did not solve the problem of underperformance of the sector in the developing world. As is seems, this approach concentrated only on one or two crucial factors for development (utility performance and investments) and focused almost exclusively on big cities. Side-lining medium and small towns, which often make the bigger part of the market, PSP cannot be considered a general solution for the urban water and sanitation sector in low-income countries. Despite these shortcomings, privatisation/PSP was often a condition linked to receiving loans from the international cooperation as Schiffler explains:[47]

> Their role in the global wave of privatisation had an ideological element. At least at one time, no matter what the problem and the local conditions were, the solution was always more private sector participation – the question was only of what kind it would be. Over

---

[44] Megginson et al. (2001: 1, 2, 20).

[45] Schiffler (2015: 1) 'About 90% of water and sanitation utilities in the world are publicly owned and managed.'

[46] Newbery (2004: 26).

[47] Schiffler (2015: 4).

> time, its enthusiasm for privatisation waned, mainly during the 2000s. While the idea never completely disappeared…

While underperformance of utilities and the wastage of (insufficient) funds due to premature rehabilitation was the centre of discussions, the orientation in the sector and the way funds were used were not or rarely questioned at this moment. These issues are still relevant in many countries today (e.g. Tanzania).[48] Donors like the WB sometimes not only pressured national governments to accept PSP but also used their concessional funds to attract a private operator. In the 1990s, the WB even agreed to subsidise the water tariffs with payments to a French private operator for each m3 sold in Guiney Conakry in order to make PSP possible.[49] Some experts see this 'blending of funds' very critically. According to Jomo et al:[50]

> PPPs have often tended to be more expensive than the alternative of policy procurement while in a number of instances they have failed to deliver… PPPs are better suited for economic infrastructure such as transport and electricity... within Europe, PPPs represent little more than 5 per cent of all infrastructure investment.' Furthermore: '…donor support for public sector capacity building in developing countries may be better spent than the current trend of blended financing, which frequently channels aid money directly to the private sector, including for PPPs.

Finally, it became clear that the results of such a push for PSP in Sub-Saharan Africa are mixed and do not allow the conclusion that private sector participation produces a higher performance level than public utilities which have undergone a socially responsible commercialisation process.[51] The case of the Asset Holding Company (AHC/MMS Ldt.) for the former mine towns in the Copperbelt in Zambia is another good example. After 5 years of a management contract with a French multinational corporation, combined with a significant investment program funded by the WB, the regulator concluded that AHC's overall performance is not better than the other eight (commercialised) utilities operating in the country. It has to be noted that some of these utilities had no or very minor support from the international cooperation. In fact, AHC has never made it to the top position in the regulator's benchmarking during this time. This led to the decision of the minister (Ministry of Local Government and Housing – MLGH) in 2004 not to extent the management contract although the WB offered an additional package of investments with it.[52]

The reasons why PSP often does not offer a substantial advantage in performance in developing countries might be that it cannot have more influence on tariff

---

[48] In 7 years in Tanzania, a financing basket channelled around 650 million USD into the urban water and sanitation sector without reversing the negative trend in access.

[49] A decreasing subsidy fee was paid by the WB to the multi-national private operator for each cubic meter sold (Own observation, Guiney Conakry 1998).

[50] Jomo (2016: 15, 16, 22).

[51] The expression socially responsible commercialization means that a utility is thriving for cost recovery and at the same time is under regulations in order to protect the consumers and the needs of the poor.

[52] NWASCO annual report (2005: 11) and own experience, see also Sect. 4.3.

adjustments than the public utilities falling under the same (regulatory) regime and that foreign management might have difficulties making full use of local knowledge. In addition, PSP is not popular in public opinions. Nevertheless, it seems that private sector participation can help in two ways. Once in place, it can shield the utility better than public utilities from undue political interference in their daily operation and, being under consideration among decision makers, it can provide a threat to public utilities to become a reality, which helps to provide incentives for performance increases.

According to Europe-Economics,[53] this threat is limited in water and sanitation service provision because of missing liberalisation (rivalry) and the nature of natural monopoly. However, Megginson and Netter emphasise the positive effect of putting state owned enterprises under pressure of a possible privatisation in order to make them 'subject … to market discipline'. This was certainly the case in Burkina Faso in the 1990s when the big investments for the Ziga dam to provide raw water to the system for Ouagadougou was under discussions. That was a moment when the reforms gained additional speed and substance. Equally, the short engagement of a multi-national corporation in Dar es Salam, Tanzania at the end of the 1990s exerted enough pressure to implement a reform in the sector, although with limited results as it seems that the pressure for 'discipline' gradually subsided when the PSP arrangement ended abruptly and pre-maturely.

Parallel to the increased promotion of PSP, the support for community systems in the LIAs flourished.[54] This was often seen as complementary to PSP as the private operators were not eager to move utility services to the LIAs, arguing that the NGOs know better how to serve the poor. Hence, PSP focused foremost on financial improvements in operation and as often observed by literature in the enhancement of customer care in order to overcome scepticism of the public, but concentrated less on difficult LIAs. Especially connected consumers as well as new customers with household connections seemed to benefit from the improvements. However, a decade later, it became apparent that the level of utility performance does not depend on one specific mode of delivery, e.g. either private sector participation or public management[55] and that communities or civil servants even on local level are not the best suited to develop and manage service provision in the urban setting. It is obvious that the combination of PSP for high and middle income areas and community systems for the poor in one town enlarges the *'urban water and sanitation divide'* (see Sect. 3.3).

The prominence of the mode of delivery in the sector dialogue was overcome in many countries when decision makers looked beyond the utility in the discussion on how to reform the sector. This was a quality jump because it led to a more compre-

---

[53] Europe-Economics (2003: 27, 28).

[54] E.g. the establishments of water trusts in Lusaka promoted by Jica and an international NGO which reached an impressive size serving up to 50.000 people.

[55] Schiffler (2015: 179).

hensive thinking. Megginson and Netter[56] found that (*sector*) reforms can be generally considered an alternative to privatisation, particularly in sectors with a monopolistic market, where the product is a public or near public good and where there is broad opposition to privatisation. In the majority of Sub-Saharan countries, commercialisation and professionalization (corporatisation) of service provision (from municipal water departments to water companies) has been the answer to underperformance in many countries. However, the lessons learned are that reengineering utilities by private or public management is not the complete answer to overcome inadequate sector development.

Acceptance of State Interventions and Delegation of Functions
The deliberation in this section underlines that the state is responsible to ensure sustainable access to safe water and sanitation for all and that a regulatory regime is needed in order to counteract the negative effects of a (natural) monopoly and non-rivalry. But regulators counteracting inefficiencies and other negative effects should also be able to make use of the effective organisation of the private sector. Hence, because access to water and sanitation for all is generating huge positive external effects, the state should professionalise service provision with commercialisation or PSP in combination with (professional) regulation. Such a set-up is in this book referred to as *'socially responsible commercialisation'*. However, next to professionalising the operation of infrastructure and regulation, there is also a need to professionalise fund raising and infrastructure development. A decision for the latter seems to be more difficult to obtain from politicians because of the high amounts of money involved.

2.3 The Low Income Country Context

The term 'urban' is associated with cities where dwellers lead an urban lifestyle. To secure sustainability of this lifestyle the towns need infrastructure for basic services. Furthermore, urban areas are also the backbone of the country's economy. This engine for the development can only be effective with well-functioning infrastructure and services and in particular for water and sanitation because of its importance for the individual and society. Nevertheless, the placement of infrastructure need sufficient public space, its functioning need protection of the assets and its design and management has to be tailor made to the specific context and demands.

Informal and Formal, Planned and Unplanned Urban Settlements
In low-income countries, it is common to find both, planned and unplanned settlements within or outside the town administrative borders. Unplanned settlements generally start as illegal settlement, which attract newcomers searching for cheap accommodation. The poor flocking into the towns swiftly overcrowd these dwell-

[56] Megginson and Netter (2001: 324, 329, 337).

ings. Many countries have started to recognise illegal settlements realising that its population is a pool of voters who need to be registered in order to participate in elections. This lifted previous restrictions or arguments of utilities not to serve such illegal settlements. In addition, former planned areas offering rooms for rent at low costs sometimes take on an unplanned character over time because of illegal constructions undertaken by the owners of the plots.

In this work, both types of settlements for the poor and some of the lower middle income class are called LIAs. In the developing world, the urban LIAs are the fastest growing and often the most densely populated areas within or at the edge of towns. According to Ravallion et al[57] 'Africa is the most rapidly urbanizing region, with poverty in urban areas increasing. Using the $1/day line, he suggests that about 40 percent of urban residents are poor and the $2/day line, close to 70 percent is poor.' It is to note that LIAs can be very different in their layout, the social fabric, the prevalence in infrastructure, etc. which has an influence of how the dwellers can be reached with water and sanitation services. A utility which ignores the security situation, the restriction in available public ground to place its infrastructure, etc. will risk to lose control over its assets and the water delivered in critical areas. Hence, a strategy on how to serve areas with different contexts is needed instead of one approach for all or excluding unplanned LIAs from formalised service provision all together.

Referring to the two cases (Betty and Agnes) described in Sect. 2.5, it is obvious that a utility will have less difficulties to control their infrastructure in the planned area with multi-story building where Agnes and many other low-income households are living than in the slums where Betty is residing and were narrow unpaved, winding food paths and high security risks with the presence of organised gangs are prevalent.

Unclear Separation Line Between Urban and Rural

Despite the high population density, sometimes LIAs are classified as rural areas because either they do not have a certain minimum of public and private infrastructure (sometimes included in the definition of urban) or are considered as part of a larger rural administration area situated outside the towns. In order to give more precision in the way to count the urban and rural population, some countries have introduced administrative units smaller than towns, such as sub-locations in Kenya,[58] which are now classified as either rural, urban or mixed. A verification of sub-locations in Kenya, according to the 2009 census, documents that the classification of urban and rural areas does not always follow population density. There are very densely populated sub-locations, which are classified as rural areas although their density is equal to that of medium-income residential or even LIAs in towns.[59]

[57] Ravallion et al. in Backer (2008: 8, 21).

[58] Over 7,000 sub-locations in Kenya (KNBS – Kenyan National Bureau of Statistics).

[59] The sub-location Kizigitini, labelled rural by the 2009 census, has a density of 5,095 people per km^2 compared to urban-labelled sub-locations in the heart of Nairobi town such as Kileleshwa (3,210), Embakasi (1,444) and Mwiki (2,084).

There are also very crowded settlements with a homogeneous population density, which stretch across administrative borders, where one part is classified as urban and the other as rural. This indicates that the separation line between the classification of urban and rural population is not always clear or its definition not always harmonised among institutions.

It is important to note that national statistics often define areas as urban when the number of people reach a certain threshold within an administrative border. A population of over 10,000 in a settlement for instance, would fall under the term 'urban'.[60] This does not refer to population density, which is expressed as a number of people per square kilometre. However, for water and sanitation service solutions population density is more important than the number of people living within an administrative boundary. Also sanitary conditions and with it the risks to contaminate water sources are foremost linked to population density.

As settlements grow beyond towns, utilities are sometimes obliged by national regulation to extend urban water and sanitation services outside administrative borders because of the population density and the need for urban systems.[61] Thus, the number of urban dwellers to be served by utilities can go substantially beyond the number of urban population given by the census. In Kenya for instance, the urban population in 2012[62] was estimated at 15.3 million, while the population living within the service areas, as reported by the utilities, was estimated at 20.6 million.[63] This indicates that even figures provided by the census on the urban population need to be taken with care when considering water and sanitation development. Therefore, statistics on coverage of urban water and sanitation can be distorted when census data is used which are not linked to population density.

Also the expression peri-urban is used by literature which can have two different meanings. It can either be a settlement, which is in a transition from rural to urban

[60] There are also other definitions of urban areas influenced by politicians such as the one in Zambia in the 1990s where urban was defined by the office for statistics as areas where basic services are supplied such as water, electricity, etc. This led to the phenomena that the population in the rural areas were growing faster (on paper) than the urban as many LIAs in towns missing basic services were consequently classified as rural. Hence, the more rapid development (on paper) of rural than urban population in Zambia according to statistics was contrary to the development in all other countries in Sub-Saharan Africa, where urbanization progressed.

[61] In 2014, WASREB, the regulator in Kenya started to define the service areas of the utilities regardless of administrative borders but according to population density. This exercise is based on the Service Provision Agreements signed by the WSBs and the utilities in their area of responsibility at the start of the implementation of the water sector reform in 2005, which also often went beyond town boundaries. Therefore, the population in the service areas of the utilities reported by WASREB is significantly higher than the urban population reported by the census (own experience 2014). The census 2009, Kenya National Bureau of Statistics, 2009 Kenya Population and Housing Census, Volume 1A (2010: 194–197) indicates a core urban population of 7,579,820 and peri-urban population of 6,108,607, which is a total of 13,688,427 people compared to the indication of the WASREB Impact report 2008/9 with a population in the service areas of utilities of 17,320,031).

[62] National census data for 2009 (urban 7,6 million and peri-urban 6.1 million) also estimated 3.9% growth rate in urban and peri-urban.

[63] Wasreb Report No. 6, http://www.wasreb.go.ke/impact-reports (last visited 03.2015).

indicating a medium population density or it is, in most cases, referred to a densely populated urban LIA for the poor, which has no or insufficient infrastructure. Often, such LIAs where newcomers from the rural areas are rushing to, are established in high-risk zones prone to regular inundation or situated on or near hazardous waste dumps or industries as space is becoming scarce in the urban setting.

Population Density Determines W+S Solutions

Penrose[64] shows, that there is a significant correlation between cholera incidences and poverty, absence of sanitation systems, number of informal residents and population density. Most of the urban LIAs are characterised by a combination of these four variables, which amplifies the risks for its dwellers to a much higher level than the rural population is exposed to. 'Flying toilets' and overflowing pits from traditional toilets worsen the pollution in and around the settlements and thereby, degrade the living environment and the raw water sources.[65] According to Esrey,[66] in the urban setting sanitation plays a more important role in disease control than water. In contrast, according to Clasen et al,[67] an increased coverage of toilets alone in the rural setting has no impact on child mortality or the prevalence of 7-day diarrhoea in children younger than 5 years. The superstructure of the toilets can be shifted when the pit is full. This is frequently not an option in the crowded LIAs where the (private) living space for the households is often reduced to a few square metres per person. The remaining options for the urban poor are either open defecation/flying toilets or the use of a toilet in the neighbourhood outside the plot which exposes women and children to further risks (especially at night), inconveniences and higher spending – see the case of Betty in Sect. 2.5 where very few households in the slums have a toilet and most of the people depend on private toilets in the neighbourhood to be paid for per call because public toilets provided by municipalities are scarce and in unacceptable conditions.

Hence, an onsite toilet in the densely populated areas needs a provision to be emptied and the sludge to be safely disposed. While traditional toilets without such provisions can be regarded as safely managed in the rural areas they are not safely manageable in the urban setting without a provision for emptying and a link to a controlled sanitation chain. In both cases the superstructure might be regarded safe for use but in the case of the urban environment the other people living in the areas are no longer safe when the pit is full and overflowing. Therefore, there is little need to offer sanitation services through professionals in the rural areas (except for market centres) but they are very crucial in towns. This supports the argument that densely populated areas in towns create a very different reality compared to the rural settings. Despite these facts there are still donors financing inappropriate toilets in LIAs today.[68]

[64] Penrose et al. (2010: 6).

[65] Bain et al. (2014a, b); Uhuo et al. (2014); Sabo et al. (2013).

[66] Esrey et al. (1991: 617).

[67] Clasen et al. (2014: 649, 650).

[68] E.g. northern Uganda (own experience in 2018).

However, other factors than population density are also responsible for the different realities. The unplanned character of settlements in a dense environment brings with it many constrains which have an impact on the living conditions of dwellers. The development of infrastructure is limited, individual households have no place to grow crops, people are more exposed to insecurity and gang violence, generally dwellers have to pay high rent for low quality accommodation often provided without water and a toilet, there are no alternatives to certain services (e.g. water) because traditional water points can no longer be used, the risk of pollution and spread of epidemics is very high, and etc. This urban context, very different from the rural situation, influences the way access to water and sanitation has to be organised. It also indicates that living conditions for the urban poor can be especially repulsive in the urban LIA settlements as Werchota[69] explains. Many of these factors are ignored when poverty is only measured in terms of income and asset ownership by the households such as owning a radio, bicycle, etc. (see also Sect. 2.7).

Furthermore, complex centralised systems need sophisticated technology and management by professionals. Such systems also require huge amounts of funds for investments at certain moments compared to the smaller (rural) systems, where the users can finance small-scale investments and operatc a simpler technology. Thus, for rural water and sanitation, often individual or community management is regarded as appropriate and local administration can help with the dcvelopment of water and sanitation infrastructure. This is diffcrent for urban water and sanitation where utilitics with well-trained professionals are needed and the local administration rapidly faces limitations whcn being involved in the planning and implementation of infrastructure.

Neglecting these different realities in the rural and urban settings in a situation where insufficient funds do not allow for everyone to be reached by centralised systems, leads to confusion in the water sector and an unacceptable generalisation. In cases where rural technology and management solutions are transferred to towns, e.g. community systems, the discrimination of the urban poor depending on informal service provision is further prolonged (see also Sect. 3.3).

Insufficient Priority and Misleading Categorisation

The contrast in daily lifestyles between low-income and industrialised countries can hardly be explained better than with statistics and images related to access to water and sanitation. Access to a water tap in the house, where residents can drink water with guaranteed quality at any time of the day has become a basic standard for all in the industrialised world for over 100 years. This normality of lifestyle includes access to a toilet connected to a system, which ensures evacuation and safe disposal of human waste and used water. Unfortunately, the normality of access to water and sanitation in the industrialised world is in stark contrast to living conditions in low-income countries where sustainable access to water and sanitation is not available at all or a costly luxury for a significant proportion of urban dwellers. For the poor in

[69] Werchota (2013: 7).

LIAs in Sub-Saharan Africa, 'Chasing for Water', as described by Peloso[70] is normality in daily life due to the absence of physical access to utility systems and/or toilets linked to a sanitation chain.

In the industrialised world the outstanding benefits of access to water and sanitation for the development of a country was immediately recognised when the first systems where established. Consequently, with increasing urbanisation water and sanitation development received an overarching priority in the development of the countries. This seems not to be the case in the low-income countries where water and sanitation development is second in line compared to other sectors (health, education, agriculture, etc., refer also to Kolker et al).[71] The allocation of funds in the national budget reflects this insufficient priority of water and sanitation.[72]

Furthermore, many prominent national policy documents do not suitably mention water and sanitation development, not even in the chapters of water. Drinking water and sanitation, if considered at all, is in many cases treated as an appendix of water resource management.[73] The recommendation of the African Ministers' Council on Water (AMCOW) in 2008, as an institution of the African Union (AU), to spend 0.5% of GDP for water and sanitation development is obviously not sufficiently heard. Only eight countries in Africa have met this goal. The four target countries of the comparative analysis used by this book are not among them.[74]

International cooperation does also not necessarily help to lift water and sanitation development higher on the national agenda. Sanitation, although recognised as more important in the urban setting than water by many health experts, was not even mentioned in the first declaration of the Johannesburg summit on sustainable development. Water and Sanitation was not one of the MDGs but one of the targets under

[70] Peloso (2014: 121–139), in this case, urban LIAs in Ghana.

[71] Kolker et al. (2016:3). 'While water ODA grew by 90% during this period [from 1995 to 2014] overall ODA increased by more than 230%.' Furthermore, 'The water sector has historically attracted smaller amounts of ODA than other [social] sectors, including education health, population planning, and governance and civil society.'

[72] In Kenya for instance, the government investments in urban water and sanitation was in 2012/13 around 2% of the national budget (against the recommended 5% by AMCOW) whereby the contribution by the state from this 2% of the national budget represented only 4% of the total investments in the sector (around 95% were provided by the donor community). The average over 8 years was even lower than in 2012/13 (Annual Water Sector Review 2013/14, MEWNR March 2015, and 2013/2014 Estimates of Development Expenditure of the Government of Kenya, June 2013). Also, refer to Sect. 4.6.

[73] E.g. the Vison 2040 from Uganda, signed by the President of the state, does not provide a single paragraph for water and sanitation development although recognizing that over 70% of diseases are water and sanitation related, that 86% of household use a simple pit latrine and that 82% of toilets have no hand washing facilities. In addition, the need to build piped water and sanitation systems are briefly mentioned in one of the 15 paragraphs under the chapter water resources. Under the chapter improvement of the quality of the population, water and sanitation is mentioned as the last point after health, nutrition, literacy, numeracy and housing. (https://www.jlos.go.ug:442/index.php/document-centre/government-of-uganda-planning-strategies/274-uganda-vision-2040/file (last visited 06.2016).

[74] The 2014 Africa Water and Sanitation Sector Report, African Union, Annual Report, (2015: 12).

the goal of environment. Health, with three of eight MDG goals, food and education were given much more priority. Another example is the declaration on human rights where a specific mentioning of water and sanitation as right was absent until 2015. Before this, it had to be derived from other rights.[75]

There are even signs that the attention for water and sanitation is declining. For instance some donors have started to shift from water development programs (combined Technical Assistance (TA) and Financial Cooperation (FC)) to water components placed under TA programs of other sectors such as governance, agriculture, etc.[76] This is worrying for two reasons: firstly, a water component under a different sector cannot help to initiate or improve a comprehensive reform in the water sector with for instance the improvement of the framework. Secondly, it cannot replace joint programs of TA and FC which makes advisory services lose the link to infrastructure development. The result is that donor countries can still claim that they are engaged in the water sector but on the other hand, it does not help much the urban water and sanitation sector when there is a need for a profound reform. Such a reorientation of cooperation in the sector will not lead to an increase of additional access on scale for the poor. TA provided by donor countries for urban water and sanitation spearheaded by advisors from other sector will have similar effects in low-income countries as the global discourse on water and sanitation has presently because it is dominated by public health and rural water specialists.

The strong message in the 'Chadwick's Report on Sanitary Conditions', London 1842, to concentrate on infrastructure development for water and sanitation seems to have less relevance today among decision makers in the international development cooperation when allocating funds for aid to the different sectors.

> As to the means by which the present sanitary condition of the labouring classes may be improved: …The primary and most important measures, and at the same time the most practicable [sic], and within the recognized province of public administration, are drainage, the removal of all refuse of habitations, streets, and roads, and the improvement of the supplies of water.

It looks like that decision makers in the international cooperation are no longer aware of how important the development of water and sanitation services in their countries has been. What could be the reasons that water and sanitation do not receive the required priorities in low-income countries although it is recognised as so important for sustaining life and ensuring sustainable development of a country?

One answer was provided already. Access to water and sanitation is no longer an issue in the industrialised world where the donors are from. Another is that the discourse on water and sanitation is spearheaded by experts from other than the water sector. The water and sanitation discussions on the international level is primarily housed at the WHO/UN and their decisions based on their primary (active health)

---

[75] General Comment of UN Committee on Economics, Social and Cultural Right regarding the right to water in January 2003.

[76] E.g. German Cooperation closing down water programs (South Sudan, Kenya, Uganda, Palestine, etc.) and establishing components of water development under governance, agriculture, etc. (Jordan, South Africa, Somalia) – own observation 2017 and 2018.

interests trickle down to partner governments through the ministries of health rather than to the line ministries for water and sanitation development. Hence, often the ministries of health are considered to be 'the mother of sanitation' (expert interviews Tanzania, 2016) and not the ministries responsible for infrastructure development and services delivery. Therefore, more emphasis is given to behavioural changes in water and sanitation than to professional service provision and infrastructure development.

When health experts are dominant in the discussion on water and sanitation development, a conflict of interests between active health care and water and sanitation infrastructure development as passive health protection is likely to happen. It seems to be an inherited challenge that WHO and the national ministries responsible for health are facing. Active health care receives more priority than infrastructure development and service provision for water and sanitation. Only a better understanding of the differences between public health and water and sanitation infrastructure development and its operation will help to overcome such conflict of interests in water and sanitation which tremendously disturbs sector development. Relying more on active healthcare instead on infrastructure development for water and sanitation can be regarded as a costly shortcut in a country's development.

Another reason seems to be that global goal setting for water and sanitation as well as monitoring of these goals are influenced by WHO and Unicef. Therefore, the unfavourable priority setting in the developing countries has most likely its root in the global discourse,[77] which influences the decision taken by national governments, civil society and donors. The positive but misleading messages provided by the global monitoring signal that development in the sector is on track to reach global targets (see also Sect. 3.5). This provides donors with the arguments to either pull out of the sector or reduce their contributions for infrastructure development in water and sanitation. Furthermore, the insufficient priority might also be the result of the low value of water. It is a fact, that generally the water bill for a household is much lower than for instance the electricity or telephone bill. There are cases where in the developing world the costs of a monthly water bill from the utility is less the value of two bottles of beer or a packet of cigarettes purchased in restaurants.

In the industrialised world, the focus of urban water and sanitation was on the development of infrastructure not only because of the health issues but also its social and economic benefits. In Sub-Saharan countries, it seems that the results of the water discourse have diverted this strong focus on infrastructure development. Water and sanitation development is much more seen as a social/cultural and health issues and more as a problem of behaviour of individuals. Consequently, the importance of infrastructure development for water and sanitation was pushed into the background. Water and sanitation in Sub-Saharan countries are usually classified under the 'social pillar',[78] often grouped together in the national budget with health,

[77] Particularly WHO and Unicef as host of the MDG/SDG for water and sanitation monitoring with the JMP. Both work mainly with the health structure in the developing world and not with the line ministries responsible for infrastructure development of water and sanitation.

[78] Vison 2030 of the Government of Kenya (Chap. 5: 86) for instance. Water and sanitation is not part of chapter 3, 'foundation for national transformation', which concentrates on infrastructure development such as roads, railways, maritime/water ways, air, energy, etc.

education, etc. Consequently, 'software dominates hardware' in the water and sanitation discussions.

However, there is some light in the tunnel which brings hope to the sector and indicates that there might be change in progress. Regarding the priority setting on global level by the UN, water and sanitation has been given the same level of priority as food and health with the SDGs. Now it is time to copy and paste this priority onto the national frameworks and budgets, donor contributions and decision making by the international cooperation in order to provide more emphasis to infrastructure development for urban water and sanitation.

Politicians (mis-) Use W+S Services Despite State Accountability

With the ratification of basic (human) rights,[79] the state became formally accountable for ensuring sustainable access to safe water and sanitation for all. The provision of basic goods, such as water and sanitation, became part of the state building process.[80] It is the recognition that society has an interest that everyone has access because an outbreak of water borne diseases will ultimately affect all urban dwellers regardless of their social and economic status.[81] Accountability includes the development of infrastructure for physical access, the sustainability of services and standard setting.

To set and enforce minimum requirements for water and sanitation services might be fairly recent in developing countries but is not very new as an idea. Koch[82] in 1893 already proposed that it should be the state to ensure the quality of drinking water and suggested the supervision of treatment plants by the state.[83] Thus, the

[79] 'The right to water is clearly part of international human rights law. In 1977, the international community recognized the right to have access to drinking water in the Mar del Plata Declaration. Other international treaties and declarations have recognized the right to water. In 2002, the United Nations Committee on Economic, Social and Cultural Rights released a General Comment on the Right to Water that clearly stated the basis for the right to water in international human rights law. The Committee has also consistently treated access to water as part of other human rights, such as the rights to health and housing.' The United Nations Committee on Economic, Social and Cultural Rights released a General Comment on the Right to Water in January 2003,file:///F:/1%20Diss%205%202016/1%20DISS%20writing/1%20Writing/10%20 Literatur%20allgemein/Human%20rights/right%20to%20water.pdf (last visited 05.2016).

[80] OECD (2008: #). Refer also to Graevingholt (2012: 18).

[81] E.g., in Kenya, two ministers where admitted to hospital being affected by a reported cholera outbreak after attending a meeting at the Kenyatta International Conventional Centre in 2017, the Star and the Nation newspaper reported on July 14, 2017.

[82] Koch (1893: 201).

[83] 'Aber wer soll die Überwachung übernehmen? Nur der Staat kann es tun. Er kann es nicht nur, sondern er muss es übernehmen; es ist seine Pflicht. Was wird nicht schon alles überwacht und revidiert? Apotheken, Krankenanstalten, Dampfkessel, Fabriken mit ihren Arbeitsschutzvorkehrungen usw. stehen unter staatlicher Aufsicht, um zu verhüten, dass einzelne Menschen durch Ungeschicklichkeit und Fahrlässigkeit zu Schaden kommen. Bei einem Wasserwerk handelt es sich aber, wenn ein Unglück passiert, nicht um einzelne Menschen, sondern um die Gesundheit und Leben von Tausenden. Es ist höchste Zeit, dass man die zuwartende Haltung aufgibt und sich zu energischem Eingreifen entschließt.' (1893: 201).

states in the nineteenth century had already concerns about standards, equal access and sustainability of services. Governments realised very early that service providers hold a natural monopoly for goods which are essential for life. Therefore, regulation in the sector was already a concern more than 120 years ago.

However, the necessary involvement of the government in water and sanitation, especially in the developing world, was often used as a pretext by the state administration to provide such services and decide on investments for infrastructure development through its civil service system. Behind this claim for a direct involvement in the day-to-day operation and in decision making for investments was most likely the interest of politicians from central and local governments to maintain influence over service provision which has such a tremendous impact on people for political aims e.g. water for free, water and sanitation for a specific constituency and political elite, water to attract votes in general elections, etc. It generally still serves as a vote catcher in political programs.[84] Promising water for free for the poor was very popular in the developing world although the state did not provide the necessary means for a long time if at all and did not allow adequate tariff increases either. [85] The resulting degradation of services penalised instead of helping the poor (expert interviews, Kenya).[86]

Furthermore, when politicians want to progress in their careers they need to forge and nurse alliances with other politicians, business people, etc. and serve their institution (i.e. party). Consequently, the temptation to make use of public institutions such as water service providers for such (personal) undertakings is not so far-fetched. There is a widespread risk of patronage, nepotism, imposition of individual interests at board level of utilities, etc. This limits sector performance and thus the mobilisation and use of funds for investments, according to expert interviews and Nickson and Franceys.[87] Unfortunately, it is common practice in low-income areas that members of the Board of Directors (BoD) of utilities, asset developers and regulators are political appointees who often do not shy from bold moves in securing their personal interests. Considering the limited importance politicians generally give to the sector and often less to sustainability concerns, there is hardly a different explanation why politicians keep insisting that the state has the right to intervene directly in water and sanitation service provision than personal or political party interests.

One expert from an NGO in Kenya expressed the multiple challenges the sector is facing due to government interference (local and national), NGOs not taking up

[84] Jubilee manifest of the Kenyan coalition in 'The shared Manifesto of the Coalition between the National Alliance (TNA), the United Republican Party (URP), the National Rainbow Coalition (NARC) and the Republican Congress Party (RC)' (2012: 31), 'Guarantee free water supplies to all those living in informal settlements pending slum upgrading.'

[85] Refer also to Rouse (2013: 26) 'water services cannot be free'.

[86] Dissertation from Werchota (2017).

[87] Nickson and Franceys (2003: 48).

their role as a watch dog and international cooperation not exercise its obligation for shared responsibilities:

> I think it [donor support] is very helpful…I think in terms of financing, not enough is being done…ensuring these institutions are becoming sustainable. The same thing at the NGO level…that huge Mombasa project…how many billions in the system…next thing, they [TA from NGO] see challenges, because they refused to acknowledge there was a governance issue within MWASCO [utility]…there's not enough acknowledgement of realities. There must...take in the governance challenge, head on,…it needs to be addressed. But organization [international cooperation] like xxx refused to…get too involved in some of these issues…they just did not really want to acknowledge that there was some serious governance issues which would have serious repercussions…then it becomes political. Yes, they [Technical Cooperation] are not integrated enough…you know, but then after they will sit at dinner and complain about how corrupt everybody is.[88]

Therefore, the separation of politics but also of policy making in the sector from service provision is crucial, especially in a situation where newcomers in politics have the tendency to use water as an entrance point for their carrier by promising the impossible and ignore urgencies raised by experts.[89] Unfortunately, such behaviour usually does not provoke a critical response from the civil society or the press because other daily issues seem to catch more attention by the public. Countries, which do not introduce and enforce this principle of separation, will face limited sector developments and success in reforms. The sector does not need 'handouts' by politicians but also no parallel actions by development partners to government institutions. It is of little help in trying to hold governments accountable while at the same time side-lining their institutions and sector policies. The professionals need to be given sufficient autonomy so that the sector can become resilient against external influences which threaten the achievements of economic and social goals. Nevertheless, only governments can establish an enabling sector framework and ensure sustainability of good governance in the sector.

Higher Complexity of the Sector

Both, the possibility to contain waterborne deceases with active health care and the significant influence of the international development cooperation[90] since the 1950s, has made the sector development in low-income countries much more complex than it was in the industrialised world in the nineteenth century. There is an army of NGOs, philanthropic organisations, donors, academic institutions etc. which directly or indirectly intervene in the sector today in low-income countries and which have not been needed in the industrialised world in the past.

These many national and international stakeholders also have conflicting interests and therefore, are difficult to streamline towards one single national sector

[88] Expert interview 2016 in Nairobi.

[89] E.g. Turton who raised the alarm many years before the dramatic water shortages in Cape Town, South Africa emerged in 2017/18

[90] The international development cooperation is considered to have been triggered with Harry S. Truman's 1949 inaugural address, https://www.trumanlibrary.org/whistlestop/50yr_archive/inagural20jan1949.htm (last visited 06.2016).

policy. Many of these stakeholders often overstep their limits of competence and roles which public opinion or authorities have assigned to them. Also other factors increase the complexity such as the unprecedented urbanisation, the many settlements with unplanned character, the high level of poverty, etc. Thus, a sector development concept today needs to respond to this increased complexity. A simple copy past approach from the past will not be effective in the low-income countries today.

2.4 Replacing Unprofessionalism and Informality

As outlined, the urban water and sanitation sector in the developing world is still in a transitional phase where many stakeholders step in with their own ideas and approaches of how to provide access for the different population groups in towns. Some don't trust state institutions and establish parallel systems for the poor; others concentrate on first mile investments and use the provision of funds to promote their own ideas. This leads to different proposals for water and sanitation service provision in low-income countries. It is not difficult to see what confusion it brings to the sector and how cumbersome it is for the states to carry out reforms aiming at the long-term goal of providing access with regulated services for everyone.

Different Proposals for Urban W+S Development

Although history and academic knowledge indicate that the most appropriate way to achieve access for all in towns is with one formalised system, there are still many different approaches promoted by the various stakeholders. It also makes the sector a laboratory for 'new innovative ideas' or products sold as 'the solution' to solve sector challenges. The problem for the line ministries with such players is that their interventions often do not match sector policies and strategies or comply with legislation (e.g. formalised service provision). The following two tables provide a brief overview of the most common approaches (Table 2.1).

It is obvious that some of these approaches are not feasible for the moment because of the huge and growing investment gap. Others ignore the lessons learned from history or cannot be considered as substitute for safe urban water supply and therefore are prone to fail. Also for sanitation there are different proposals to reach access for all (Table 2.2).

Also with sanitation it becomes obvious that some of the approaches are either ignoring the reality in respect to the available means or are missing a holistic view. Many stakeholders are not aware about the pressing need of a functioning sanitation chain for onsite sanitation in the urban setting in the developing world. For them, a safe toilet is a place where human can defecate and avoid contact with their waste and once the waste is dropped in a pit, sanitation is safely managed. This is sufficient for the rural but certainly not for the urban areas.

Table 2.1 Different proposals to provide access to water for all

Proposal	Comments	Urban/Rural
Access on the plot should have absolute priority	The JMP, other UN institutions, a number of experts as well as many NGOs give access to water on the plot (improved sources) priority to water sources outside the plot.[a] However, some water sources on the plot are known to be contaminated in towns.[b] History suggests that drinking water quality has to have a higher priority than the place where people fetch water	The use of water point sources is largely appropriate for rural but not for urban areas
Household connections for all (paradigm)	This approach is often shared by politicians, some NGOs and many utilities. It is supply driven and therefore, prone to fail.[c] It delays the development of access for all in a situation where a financing gap prevents the necessary infrastructure development and in situations where people cannot afford to maintain an individual connection	Transitional solutions are relevant for the urban and rural setting. For towns there must comply with regulations
Small scale informal services owned and operated by private or by community systems	This approach is often promoted by NGOs, in the case of community systems, or some donors in the case of private sector participation. However, small scale providers operate generally informally. With the importance of safe water and sanitation, this is unacceptable. Private operators are not necessarily more efficient than other modes of delivery or their consumers benefit from efficiency improvements. As alternative, the advantages in extending an already existing utility system in town to underserved areas seem compelling	Not appropriate for urban
Utility water for free for the poor	Water for free for the poor is usually proposed by politicians or some representatives of the civil society. It leads most likely to insufficient Operation and Maintenance (O+M) costs coverage. This compromises services and expansion of infrastructure. It also leads to high water losses and opens the door for corruption. Therefore, it will not work in the interest of the poor as well as the other income groups	Not appropriate for urban and rural
Bottled water counted as access to water supply	There are a number of actors (NGOs and philanthropic) involved in producing and distributing treated water to the poor in bottles or branded containers. Although they solve the challenge of non-treated water being distributed by other informal providers, the amount of water distributed per consumers (often around 1–3 l per capita per day) indicate that it cannot be considered as urban water supply for a town	Constitute no alternative for urban and rural water supply – acceptable as supplement to water service provision only
Utility services for all	This is the solution which has prevailed in the industrialised world and is aimed at by sector reforms in the developing world when governments want to improve the sector. However, in general utilities have not managed to extent services according to the increase of demand	Relevant for urban setting

(continued)

Table 2.1 (continued)

[a]Kayaga and Kadimba (2006) for instance propose that drawing water from shallow wells is an acceptable option to access water from formalised service providers in urban areas: 'It therefore makes social and economic sense that there is a smaller proportion of households in the project areas drawing water from shallow wells, compared to the area served by LWSC' [utility]

[b]This is the case with the proxy of protected water sources for safe water used by the MDGs monitoring. The SDGs also gives priority to the water sources on the plot but introduced safe management, which implies that surveys have to evaluate safe management in order to judge the water quality consumed by the household. This does not rule out the use of water points on the plot in the urban areas, which are often contaminated. https://unstats.un.org/sdgs/files/metadata-compilation/Metadata-Goal-6.pdf (last visited 03.2017)

[c]Nilsson (2006: 36, 2013)

Furthermore, institutions claim to be responsible for sanitation services which are not the best placed to do so.[91] Presently, in the developing world the responsibilities for sanitation are not as clearly distributed among the many institutions as it is for water although change is on the way. In several countries the line ministries for water have received the addition of sanitation in their names (e.g. Burkina Faso, Kenya, Zambia, etc.).

The Discourse on W+S Contributes to Confusion

A number of stakeholders on global, national and local levels still ignore the different realities in urban and rural areas stemming from population density[92] and the lessons learned from history.[93] For instance, Sasaki et al[94] propose the construction of deep-well facilities for local (community) systems of water supply[95] in the unplanned settlements in Lusaka instead of the extension of the nearby utility network. JMP proposes the same 'improved sources' for rural and urban areas which are protected from surface but not contaminated ground water. Mutono et al recommend even the use of hand pumps in the urban LIAs:[96]

> Assist local Water Authorities, WSSBs, and local private operators and individuals to integrate the management of handpumped supplies and piped schemes. Handpumped supplies are the main technology serving the urban poor and economically disadvantaged.

[91] E.g. municipality departments for the establishment and maintenance of the chain for onsite sanitation and the provision of public toilets.

[92] In Kenya, when defining service areas for water service providers in 2014 the water sector regulator included settlements with 400 households per km^2 in the service areas of licensed utilities. This means that with 30% of land reserved for public ground, individual plots for families of four are 1750 m^2 which corresponds to the upper middle class residential areas in towns (own experience).

[93] E.g., the classification of improved water sources by JMP is the same for the urban and rural setting.

[94] Sasaki et al. (2008: 420).

[95] Where often sustainable water treatment is not secured with community or small scale private management.

[96] Mutono et al. (2015: 63).

Table 2.2 Different proposals to provide access to sanitation

Proposal	Comments	Urban/Rural
Sanitation is uniquely a household affair	This is the position shared by the MoHs and by Unicef, for instance. It is based on the assumption that sensitisation will make the households build their own toilet according to minimum standards. According to their understanding, subsidisation is only needed for the poorest of the poor. Before the SDGs, the sanitation chain was not included in this approach.[a] However, today it is recognised that sanitation in town also needs state interventions	Appropriate approach for rural areas but not for towns
Sanitation service does not need to be linked to state structures	These are interventions like Peepoo, Sanergy, Sanivation[b] for instance, trying to improve the situation of people in the LIAs. However, such undertakings do not make use of existing sanitation infrastructure or cooperate with state institutions of the sector. They are usually not aligned with sector policies and legislation and come to an end when external support is phased out	More appropriate for emergency situations.
Shared toilets are unacceptable	The global monitoring of MDGs/SDGs excludes shared toilets in counting access. However, facilities shared by a limited number of households were practiced long into the twentieth century in the industrialised world. Experts interviewed have not rejected shared facilities as long as the number of people using them is limited	Can be considered as acceptable approach in urban and rural areas under certain conditions
Sewer connections for all	More than for water, this proposal is presently out of reach in many low-income countries and even in some emerging states. Since decades, access to sewer systems is declining in the developing world	Out of reach in urban and rural areas
Utility services for all (onsite and offsite sanitation)	It is gradually becoming apparent that the principles of professionalization and formalisation of service provision is as valid for sanitation as it is for water. Because a growing majority depend on onsite sanitation there is an urgent need to develop infrastructure and controlled services for this type of sanitation in the urban setting	Relevant for the urban setting.

[a]Refer to the national sanitation policies elaborated by the Ministries of Health in the different countries.

[b]Sanergy, Nairobi and Sanivation, Naivasha, both in Kenya are NGO driven projects which depend on services where human waste has to be collected from households at least weekly. For Peepoo refer to http://www.peepoople.com/peepoo/start-thinking-peepoo/ (last visited 05.2017)

This missing consideration of different realities in rural and urban areas has its relevance also for onsite sanitation (refer to Sect. 2.3) where in towns traditional latrines often with no provision for emptying the pits are promoted.

It is obvious that the message from Snow in 1849 (see Sect. 3.2) is still not sufficiently remembered among contemporary stakeholders. Many of them on global and national levels still transfer rural solutions into the crowded LIAs for the poor discarding the risks involved and the limits community members to manage infrastructure.[97] The consequences of ignoring different realities under the condition of insufficient funds are especially dire for the urban poor in the LIAs. It puts them under high risks and delays sector development with ill-conceived concepts.[98] Evidence in Sub-Saharan Africa indicates that a very large part[99] of the urban population is thereby negatively affected.

Governments Facing Stakeholders with Contradicting Interests
The high number of stakeholders pulling in several directions by following their own interests makes it very difficult for governments in the developing world to fulfil their obligation for human rights to water and sanitation. It seems sheer impossible to align all actors to national policies. This challenge was less obvious before sector reforms because line ministries and utilities accepted a laissez-fair attitude. There was no clear vision and policies about sustainable urban water and sanitation development to the benefit of all. Modern reforms had to change this with the adoption of a new framework because of the recognition of rights by the states. However, sector institutions have not yet managed to change the ideas and behaviour of the many non-state stakeholders and donors.

In addition, the sector has to face the interest of other sectors in their country which have a link to water and sanitation because some responsibilities cut across. Some of these sectors have very different ideas about access to water and sanitation which are often influenced by the discussions on global level. For instance, in low-income countries, generally the MoH is responsible for controlling the water quality. However, in reality there is little indication that they are able to carrying out regular quality controls of water from formal and informal service providers or traditional water points. Usually, there is not even an inventory available of the many Small-Scale Informal Providers (SIPs) and also not of the water points in use.

---

[97] The Water and Environment Sector Performance Report from Uganda (2015: 104, 128) states: 'Also, in response to a Typhoid fever outbreak in Kampala, over 700 different drinking water sources from schools, organised communities, informal settlements, suburbs and municipalities were sampled and mainly tested for bacteria of faecal origin (E.coli). Different drinking water source types, namely bottled water, locally packed water sources, boreholes, shallow wells and protected springs were sampled and assessed for compliance to the National Drinking Water Quality Standards… There were notable differences in the level of contamination of water from different water abstraction technology types. Results showed that protected springs are the most prone to contamination followed by shallow wells. Deep boreholes showed less contamination possibly due to the deep nature of the ground water aquifer as compared to the shallow aquifers for other technologies…. Open wells were replaced with stand pipe water supply and public building owners (arcades) were instructed to connect water supply to their buildings using the NWSC water supply network.' According to Annex 17, the result of the assessment on contamination of the 631 improved single water sources are: Protected springs are contaminated to 80% (out of 474), protected shallow wells to 87% (out of 69), protected boreholes to 24% (out of 88).

[98] E.g. small-scale community management in parts of towns which resists later the integration of their assets into a bigger system, like observed with the Water Trusts in Lusaka.

[99] According to JMP's report (2014), the coverage in the urban setting in Tanzania was in 2013 around 8% with public taps and 28% with pipes on premise leading to a total piped water of around 36%. The same source indicates that 78% are covered with improved sources, which means that around 42% depend on high risks single water sources.

It would be the function of the ministries responsible for water resource management or for water and sanitation service provision to establish such inventories.

Very often, the MoHs are also responsible for elaborating the national sanitation policies[100] although their involvement in sanitation development and service provision is limited (for further deliberations see Sect. 3.1). Thus, the line ministries for water and sanitation development have to overcome interests of the stakeholders from the global level, civil society, philanthropic institutions, donors, politicians in the country and the government institutions from other sectors which are frequently counteracting sector policies. To harmonise approaches and activities from the many actors is obviously not a quick-fix and demands a clear long term vision, strong leadership, good coordination and passions.

Towards a Stepwise Process Within a Regulated System

In the developing world not only urban and rural water demand different solutions for access, but also within the urban setting different service levels (or service ladder) are needed in order to reach a high level of access within an acceptable timeframe. However, there are still many experts who insist that even in low-income countries, everyone should have a water tap and a sewage connection in the house or in the yard and thereby, rejecting some shared facilities for water and sanitation.[101] But, insufficient funds for investments in infrastructure, a high poverty level and an unplanned character of settlements makes it impossible to reach everyone in the urban setting in the near future by the means of household connections only. Holding on to this paradigm prolongs the exclusion of the poor and a growing number of other dwellers to controlled services and thereby, ignores their strong desire to escape informal service provision, even when it means that they have to enter utility services at the lowest level (water kiosks, shared toilets).[102] In a nutshell, the absence of shared facilities under regulation as lowest acceptable service level is holding back progress in access for all.

In some low-income countries (e.g. Kenya), a decline or stagnation in access can be linked to the (premature) closure of shared facilities such as water kiosks. With the promotion of social connection programs utilities cannot cover all households in certain LIAs for reasons explained in this book.[103] Within a short time many poor are disconnected and consequently drop out of the utility services because the acceptable alternatives (water kiosks) are no longer there. Figure 2.1 indicates that the closure of water kiosks is generally not fully complemented by the expansion of

[100] E.g. Kenya, Tanzania, Zambia, Uganda, and etc.

[101] E.g. the utility in Dar es Salaam (own experience in 2015 during an interview with top management).

[102] All interviewed dwellers in the LIAs have accepted shared water and sanitation facilities (expert interviews in 4 target countries) and indicated the preference for utility kiosks compared to informal service provision. Refer also to Nilsson (2010 and 2013).

[103] A brief survey in Kenya by GIZ (2018) in six towns showed that when utilities closed down water kiosks in LIAs without ensuring full coverage by yard taps or household connections 5–90% of dwellers are pushed back to traditional sources, neighbourhood sales, mobile vendors and use of surface water.

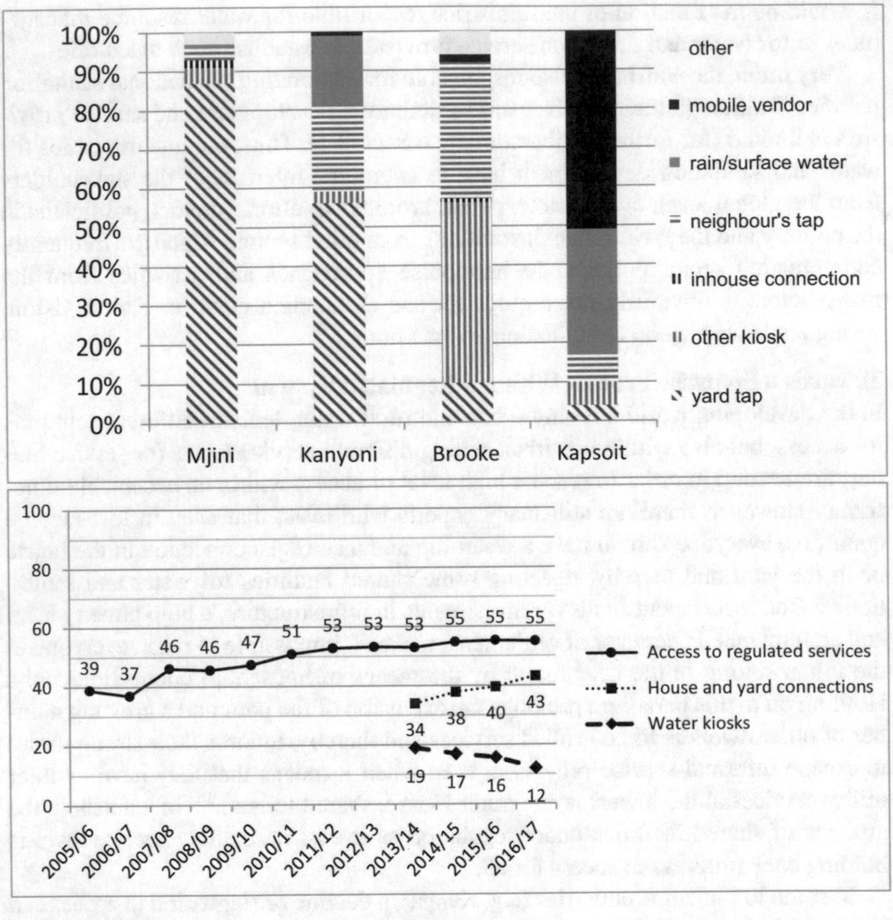

Fig. 2.1 Effect on access when utilities close water kiosk in LIAs, Kenya. (The bottom chart indicates that with the reduction of water kiosks the utilities in Kenya (2018) could not succeed to reduce the number of underserved (refer to Sect. 5.4.) and barely cope with the population growth by the means of increasing the numbers of yard taps and household connections. This only kept access expressed in percentage of population served on a constant level.)

yard taps and individual household connections in LIAs. Progress in access is thereby compromised because many poor are often pushed back to the use of rain and surface water, contaminated ground water sources, mobile vendors and unacceptable neighbourhood sales.

It should not be forgotten that even in the industrialised world, the development of water and sanitation was a stepwise process, which moved along different service levels. Water taps in corridors in multi-storey buildings were shared by several households (called 'Basena' in Vienna for instance) and shared toilets were no exception in the first half of the twentieth century in Europe. These service levels seemed to have fulfilled the aim of preventing waterborne epidemics and ensuring acceptable living standards.

The responsibility of the state in the provision of basic services/access[104] is not to eliminate all risks, which might extend right into the households (by fetching water at shared facilities), and secure the highest service level for everyone. The state has to offer a delivery system, which can, combined with proper hygiene practices by the consumers, reduce the risks to an acceptable level and at the same time ensure affordability and equity for all. Once increasing financial resources are available and the unplanned settlements have been reorganised, the priority for public health, equality and economic condition will permit that each urban household has its individual connection and toilet. Thereafter, as we have seen from history, the urban systems can even stretch into less crowded areas. However, considering the more recent development in the low-income countries in Africa, the distinction between urban and rural water and sanitation and between different service levels in towns will have its relevance in the decades to come. This requires technologies and service levels which are contextual.

Delegation from the State to Professionals[105]

In Sub-Saharan Africa, since some time, there is a general trend that the ministries, municipalities, counties, etc. responsible for water and sanitation, delegate service provision to (commercial/corporatized) utilities, which operate as public entities. In the cases of Ghana, Kenya, Mozambique, Tanzania, Zambia, etc. the line ministries also delegated regulation of service provision to national institutions.[106] The delegation of service provision to utilities helps to separate service provision from politics[107] and thereby, opens the door for an enhanced autonomy for utility management. However, corporatisation alone does not secure sufficient momentum for utilities to 'break through the performance ceiling' as long as politicians can still interfere in tariffs setting as Nickson and Franceys explain.[108]

Delegating urban water and sanitation service provision to professional agents not only intends to improve the performance of service provision and protect consumer rights but also to enhance the mobilisation of funds and their efficient use for infrastructure development. It should also increase sustainability of service provi-

[104] The urban population depend on controlled urban water and sanitation services because water point sources (e.g. wells) cannot be considered safe. Hence, in towns basic access means basic utility services. This is different in the rural setting where the population can in general use water points such as protected wells without great risks. Therefore, basic access means basic technology for the rural population such as hand pumps for which a maintenance service is required but not a water service from a supplier.

[105] Professionalization of development of infrastructure and services implies that there is a separation line between responsibilities of civil service structures and the 'industry' where professionals for water and sanitation service provision receive delegated responsibilities regarding access to water and sanitation services.

[106] Other countries, especially when one national utility is in place, opted for the introduction of mechanism and procedures which have similar effects as tools for regulation.

[107] Especially on local level because of the proximity of local politicians to the utility management

[108] Nickson and Franceys (2003: 49).

sion through progress in cost recovery by water bills. Often, decision makers are forced to delegate functions because service provision is at the brink of collapse and central or local governments can no longer provide subsidies to keep operation functioning. In such a desperate situation politicians are more inclined to ease their grip on the public providers but it is extremely difficult to produce a turnaround because the confidence of the consumers has to be regained in a situation where tariff increases are difficult to sell (e.g. Cape Town 2017).

However, delegation does not mean that authorities should no longer be accountable for ensuring progress towards the fulfilment of rights. It follows that service provision cannot be delegated to (informal) providers over which the state cannot exercise its authority to reach policy objectives. In addition, an effective control is more unlikely where the number of service providers is very high (e.g. over 30 service providers in a country) and the enforcement by the state structure is weak. In the industrialised countries, national governments introduced urban centralised water and sanitation systems and delegated operation to fairly autonomous utilities, established in a number that its (state) structure is able to control. Thereby, the state formalised service provision and ensured that development moved towards access to standardised water and sanitation for all. No stakeholder in the sector contested or undermined this strategy of governments to formalise and professionalise service provision based on a centralised water and sewer systems. This is not the case yet in the developing world where a number of national and international stakeholders pledge understanding and support for informal and small-scale (e.g. community) systems or praise an undifferentiated PSP approach for the urban setting.

2.5 The People Versus Provider Perspective

What people want and what providers offer is often very different. The unacceptable challenges people face when they depend on informal providers have been briefly indicated and will be further explained in Sect. 3.3. However, also with utility services people in the LIAs have often substantial issues. The differences arise due to insufficient knowledge of the situation people have to face in the LIAs and the lack of information concerning the outcome of utility activities in serving the poor. Next to the utilities, regulators and consumer representations, especially when situated on national level, often do not understand what the poor in the LIAs need and desire.

W+S in the LIAs: Two Cases

This book uses two examples from LIAs to illustrate what people experience in daily life and need and want in respect to access to water and sanitation in low-income countries. These representative cases have been selected among the people living in LIAs in Nairobi, Kenya (2017) because they are the underserved. The middle and upper class income groups living in residential areas are already as good as covered by utilities with household connections.

One of the cases exemplifies what people need to go through when forced to 'chase' for water daily, look for a toilet outside the private living space and fight their way through an unsanitary and insecure living environment dotted with human excreta. In this case the woman interviewed does not have access to water and sanitation according to human rights requirement. In the other case, acceptable physical access to safe water and sanitation has been acquired but clean drinking water is only accessible with additional efforts. Both cases are genuine stories for which the names (two ladies) have been changed. From time to time, this book refers to these two cases in order to demonstrate what findings and proposals mean in real life for the majority of urban dwellers in the low-income countries. Both examples belong to population groups (poor and lower middle class) which are likely to represent around 30–50% of the urban population in the Sub-Saharan countries.[109]

The first case, Betty Wambui, 24 years old, lives in the Kawangware slum (around 130,000 habitants, 2009) situated about 20 min' drive from the city centre (out of traffic peak hours). Houses in this unplanned settlement are mainly constructed of wood and metal sheets having makeshift character. The second story is Agnes Omolo, 32 years, who lives in Zimmerman (next to Kasarani), situated about 50 min' drive from the city centre (out of traffic peak hours). Zimmerman is mostly composed of multi-storey concrete and brick buildings situated along the Thika/Embu – Nairobi highway. Betty and Agnes underlined that their living conditions and backgrounds are by no means exceptional among the people in their living areas. Betty is poor while Agnes can be considered as lower middle income class. Both use matatus[110] for transport in town, have electricity at home, use internet regularly and have a mobile phone.

Betty is renting her small room for 35 USD per month where she lives since 4 years with her 6 year old son. She does not know her father and her mother passed away when she was of age 11. This left her dependent on one of her aunts until she dropped out of school with 17. Her aunt has two children and a husband who is in prison since 5 years. The father of Betty's son left her before she gave birth and was recently sent to prison for 3 years. He beat her with a bottle on the head and almost killed her because she refused to join him again. Betty has no living brothers and sisters, is unemployed and earns her living with informal activities, helping out occasionally at her aunts' vegetable stand at the (informal) market and with sporadic prostitution. Her monthly income is usually between 60 and 80 USD.

Betty and her son have to use a makeshift toilet in the vicinity which is used by around 20 other households. For using this private toilet, she pays 0.10 USD per call. She explained that at night, many of her neighbours use some corners of the winding non-paved pathways for defecation or urination where the droppings are visible next morning when she leaves the slum. As far as she knows, there are three

[109] Including a part of the rural population (as counted by the census) which lives in already densely populated areas.

[110] 'In Kenya and neighbouring nations, matatu (or matatus) are privately owned minibuses serving as share taxis, As of 2014, there are more than 20,000 individual *matatu* in Kenya', https://en.wikipedia.org/wiki/Matatu (last visited 05.2018).

public toilets from the municipality in Kawangware which are much more uncomfortable to use than the private makeshift structures. Betty is fetching water from a neighbour who has bought a household connection from the utility and became a reseller in the neighbourhood.

She estimates that around 50 households fetch water from the same reseller who demands usually 0.05 USD for a 25 lt canister compared to 0.02 USD regulated price for the same utility water. However, when there is no water for two days, resellers in the area charge 0.10 USD and in cases of prolonged cuts of supply the price can rise up to 0.50 USD per canister immediately after water is available again.[111] She indicated that the resellers in the area always harmonise their prices and therefore, act as cartel. There are boreholes in Kawangware which however, cannot be used by the households because the water is salty. Betty has no medical cover. There is also a water kiosk from the utility in Kawangware demanding 0.03 USD for a container which however, is too far away from Bettys' room to go and collect water.

The second example is Agnes Omolo, 32 years of age. She is not married, has no children and lived when the interview took place on the fifth floor in a two room flat with a tiny toilet and integrated shower. She is renting the flat since 6 months for 120 USD per month. Agnes is a qualified nurse (diploma) employed in a hospital where she often works shifts. She takes 1–2 h to reach her work place. Her father has passed away when she was 29 and this was the time when her mother returned to the village because she could no longer financially support urban life. She has four brothers and two sisters, the youngest, pregnant and unemployed, is presently living with her because her boyfriend, already married with three children, said that he cannot look after her and his child. Three of her brothers and the older sister are living in the three-bedroom house the father left behind, two of them with their wives and four children among them.

She receives, as she says, water for free, which seems to be included in the house rent. The caretaker of the housing complex explained to her that there are two water systems in the house. Utility water in the kitchen (according to the building code) and water from a borehole in the yard is serving the toilets and shower in the flats. In addition, there is also a tap in the yard connected to the utility water which she started to use to draw water for drinking, cooking and cloth washing. There are two reasons for this, she explained. First, she discovered recently that the borehole water has spoiled a number of her cloths (turning brown). Second, she noticed that when she opens the tap in the kitchen that there is no water coming out of the shower and at the toilet. Hence, she suspects that outlets in her flat must be communicating and therefore, there is only borehole water distributed within the house. Consequently, she concluded that drinking water is only available at the tap in the yard which has already been established for the tenants as a place for washing cloth. Now she is carrying water with a ten litre container from the yard tap to the fifth floor whenever

[111] The regulator for water and sanitation service provision in Kenya, WASREB, limits the price at the water kiosks from utilities for a 20 l canister to one or two KES in the country.

she comes home and the yard tap is open and stores this drinking water in a 100 l plastic drum in the flat.

As the yard tap is only opened by the caretaker at certain hours, it is difficult for her to find the time for cloth washing when she is working shifts in the hospital. Agnes earns between 350 and 480 USD per month depending on the shifts she is working. Concerning sanitation, she could not say if the toilets were connected to a sewer network from the utility or the house has a septic tank. She has never seen an emptying truck in the house or the streets around it. However, she observed that the streets are inundated with sewer when it rains which might be an indication that there is a sewer network from the utility covering the area and overflowing regularly in the rainy seasons.

Next to her work in the hospital, Agnes was attending a course on intensive care at the hospital in order to increase her professional qualification and her income options. To go into town with a matatu Agnes has to pay 1 USD but when it rains 1.50 USD. From the matatu station she has to take a motor bike to her flat after night fall because of security reasons. For this ride on the moto bike she pays another 0.50 USD. Her employer secures her with a medical cover and paid leave.

Willingness and Ability to Pay for W+S Services

There is a heated debate about how much the poor are able and willing to pay for water and sanitation services.[112] Utilities are always under scrutiny if their tariffs are socially acceptable. But are the questions of how much the utility should demand for minimum consumption or provide free water to the poor the most relevant and tenacious in the sector? For Betty in the slum and most of the poor who depend on informal service provision or their neighbours who sell water for exorbitant prices (up to 0.5 USD for 25 l), the discussion whether the utility should demand 0.01 or 0.03 USD at the kiosk seems almost meaningless. This is one of the indications that making people pay for utility services should not be the key concern in the sector.

First, when paying around 0.02 USD (sometimes 0.01) per 25 l container for utility water the monthly spending for a family of four consuming 25 l/c/d each will accumulate to 2.40 USD which is an amount below the 5% spending from the household budget of Betty as well as Agnes. Contrary to this, spending for water is far above the 5% when depending on informal providers and neighbours. Hence, for non-served people it is much more important to gain direct access to a utility outlet than worry about the tariffs from regulated utilities. Consequently, the sector has to concentrate first and foremost on extending utility services to the poor with all means possible.

Second, the development in Burkina Faso, one of the poorest countries in the world, shows that all consumers are willing to pay for water and sanitation services to such an extent that the utility can reach economic and social goals simultaneously. Coverage in the urban setting is around 90% and a tariff structure with cross-

[112]However, there are little indications in the literature on how much the better off households should pay for services. A number of contribution defend a unique tariff across all customer groups.

subsidisation ensures that everyone can consume and contribute according to its needs and ability. ONEA, the service provider, manages to cover around 140% of O+M costs which allowed a self-financing ratio for infrastructure development of approximately 40% (see Sect. 5.6). This indicates that even in the low-income countries in Sub-Saharan Africa there are little signs of refusal to pay for water and sanitation services when people can benefit from well managed utility systems. Before the reforms it was the big consumers and state institutions which refused to pay for water and not the poor. All experts from the LIAs interviewed in the four target countries confirmed their preference for utility water and their willingness to pay for services. This finding is supported by the statement of Fankhauser and Tepic[113] that delaying reforms by keeping the average tariff below cost coverage in the name of the disadvantaged is not the solution to ensure quality services for all:

> …this [delaying tariff reforms] may not be the best policy. The delay in rehabilitation would also affect poor consumers, which often suffer disproportionately from the poor service quality.

'*Socially Responsible Commercialisation*' Is an Opportunity to Reach the Poor
In the past, most of the utilities were able to ignore the LIAs or decide on their own if and how to serve the poor. The half-hearted attention of utilities to serve such areas was the reason that access declined in the sector. There was little interest by the sector institutions to learn how such difficult areas can be reached sustainably. Reforms tried to change this by establishing a new pro-poor oriented framework, allowing the professionalization of service provision and introducing a regulatory regime. The *'socially responsible commercialisation'* in Sub-Saharan Africa was born. The increase of access to formalised services became one of the overarching goals in the sector. Decision makers realised that this goal was only achievable by extending utility services to the non-served LIAs because the residential areas of the middle and high income classes were mainly covered. Hence, the utility were forced to integrate a pro-poor element in their corporate policies and had to learn how to up-scale services in fast growing LIAs.

However, this was a new field of activities for the utilities. Accordingly, the results are quite modest in many low-income countries until today. It is to note that PSP and other measures to improve utility performance have in general not lead to the development of adequate solutions and building of relevant capacity at utilities. This is not surprising when top managers are recruited among expatriates from the private sector missing experience in development cooperation and the sector is left without effective regulation. Nevertheless, there are good examples emerging where up-scaling of access for the urban poor is taking place with a combination of improvements in utility performance, enforcement of regulation to pressure utilities to extent services to the poor and last mile financing which helps to make utilities understand how to serve LIAs with the implementation of a national up-scaling concept (see also Chap. 7).

[113] Fankhauser and Tepic (2005: 25, 26).

Acceptance of Shared Facilities Due to Reforms

Some utilities, being obliged by the state and supported by donors to serve LIAs, realised that social connections alone could not lead to universal coverage. A number of such attempts, starting as early as the late 1980s (during the first water decade), indicated the limits of social connections.[114] It is interesting to see that many utilities which increase coverage substantially increase also the number of water kiosk when promoting at the same time social connections (e.g. Burkina Faso). Progressing like this is also in line with the recommendation by the SDGs to promote first basic access in non-served areas. Despite these facts, a number of stakeholders keep rejecting shared facilities such as water kiosks from utilities as an acceptable service level (GWOPA 2010)[115] or classify them as an inferior service level compared to open wells on the premise. The argument is that households have to fetch water from public outlets in containers and thereby, risking contamination of the water during transport and storage at home.

It can be said that it is a contradiction to reject on one hand public outlets from utilities which provide safe water as adequate access with the argument that there is a risk of water being contaminated during transport and storage and at the other hand accept uncontrolled informal service provision and traditional water sources in the urban environment where the risk of consuming contaminated water is much more elevated. It is not the place of the water source or the technical construction of (improved) traditional water sources which should be more important than water quality.

Many professionals on water service provision tend to agree that the responsibility of utilities for water starts at the abstraction points of raw water and ends at the outlets of the distribution network. For plot and household connections, this is the water meter, and for water kiosks[116] this is the water tap, both are the connection or collection points for the consumers. The connection to the utility water pipe at the plot and the handling of the utility water during transport and in the household is widely considered the responsibility of the household. For sanitation, the responsibility of the utility begins at the inlet of the sewer network as discharge points for customers and ends at the safe discharge of effluent at the outlet of the treatment plant.

It seems to be acceptable by everyone that a blockage in the sewer pipe on the plot of the customer is the responsibility of the household, so why should then this be different for water one could argue by pushing the responsibility of handling water by the household to the utility. The challenge of water quality at the point of consumption in the low-income countries is not only linked to water kiosks.

[114] E.g. WB program in Cotonou, largest city in Benin, 1989. The replacements of water kiosks by household connections for all in LIAs after 2 years left around a third of households stranded because they were permanently disconnected and could no longer use water kiosks form the utilities as alternative.

[115] http://access-to-water-in-nairobi.gwopa.org (last visited 04.2015).

[116] The expression 'water kiosk' in this work is used for a public outlet (shared water facility) of a utility as formalized service provider.

Households also fetch water at yard taps with containers and store it in the house, which in contrast to water kiosks is recognised by the same critical stakeholders as an acceptable service level. Storing water is also needed when the outlet is placed in the house of the consumer and where frequent and long lasting interruptions of utility services occur.[117]

Many different solutions for avoiding contamination while transporting and storing water by households have been proposed. Nevertheless, there is still not sufficient detailed knowledge on how far reaching this contamination is, at what particular stage the biggest risks occur and what are the most effective measures to reduce contamination. Some data indicate that the mixing of water from different sources in the household is one of the main reasons why utility water is contaminated. Another reason is the use of second hand containers used for fetching or storing water, which are not clean.

Surveys in Kenya[118] show that the contamination of water fetched at utility kiosks at the point of consumption in the households might be largely overestimated. Another study carried out in Morocco[119] indicated that the use of water kiosks must not necessarily increase the health risks. This might show that most of the consumers are well aware of the risks of contamination when transporting and storing water. There is a lively discussion of how to ensure that water is safe at the time and point of consumption. It can be argued that there is a shared responsibility between the utility and the household. Therefore, the question is, what should the utility do in addition to providing water according to standards at their outlets in securing a good water quality at the time of consumption in the household.

Mitigation measures tested during the surveys in Kenya[120] showed that contamination can be largely contained with a higher level of residuum chlorine in water sold at the utility outlets. A residuum chlorine level of e.g. 1 mg/l seems to be sufficient to protect the water quality for around 24 h in the chain of collection storage and consumption.[121] The surveys also suggest that for organisational reasons post-chlorination (after the centralised treatment plant) should best be carried out at the utility storage tanks covering a specific service area and not at each of the utility kiosks. This facilitates chlorination measures and at the same time covers not only customers at water kiosks but also at yard taps and household connections. In addition, utilities should receive incentives to reduce interruption in water supply especially in the LIAs in order to reduce the time of water storage by the households.

Concerning sanitation, there is no reason for not accepting shared facilities as long as the number of users is limited and they have common interest in maintaining

[117] This problem must have been existing since the colonial time. E.g. in Kenya standards for construction, introduced by the British, require that the tap in the kitchen needs to be directly linked to the water network and cannot draw water from the storage tank in the household.

[118] GFA (2016)

[119] Refer also to Devoto et al. (2012: 68–99).

[120] Final combined report by GIZ (2017), including the GFA reports

[121] The GFA study (2016: 19) finds that the free chlorine level in a transport canister dropped within 24 h from 1.08 mg/l to 0.34 mg/l.

the facility (e.g. ten people in a yard). It is a fact that in densely populated LIAs there is not enough space that every household can have its own toilet but that people living in the same house or on the same plot will come to arrangements to keep shared toilets clean.

Advantages and Disadvantages When Moving Up the Utility Service Ladder Improving access by shifting from 'improved' water sources and informal service provision in towns to utility water kiosks and thereafter to superior service levels within the controlled system, there are advantages but also some disadvantages for consumers compared to the previous situation or steps in the ladder. It can be observed that moving up the service ladder reduces inequality but not all positive elements of a specific level on the ladder are necessarily carried forward.

For instance, moving from a household connection under informal service provision to the water kiosk of the formal provider (utility) means that water is now only available outside the household plot. This disadvantage in terms of time and the more cumbersome handling of water by the household can be considered as being more than compensated by a guaranteed water quality and price and by having access to a formalised complaint system. Furthermore, the consumers move from a system which could be ended at any time (informal provider) to a service which guarantees sustainability. It can be argued that moving from informal to the basic service of utilities brings the biggest advantages to the consumers when climbing up a service ladder. All of the experts interviewed in the four target countries, from decision makers to the poor in the LIAs, have confirmed this statement.

The next step within formalised service provision is moving from a water kiosk to a yard tap. The disadvantage is that the cubic meter of water could be more expensive as several households have access to the yard tap and share one water bill as consumption might move out of the social tariff bracket. This is more than compensated by the time saved, since the tap has moved closer to the household. The increase in spending on water consumption in this case is limited, when considering the water service provider only, because the price-per-cubic-metre water at the kiosk is usually higher than the social bracket for yard taps and household connections due to the additional costs passed onto users by the engaged kiosk operators. However, the tariff at the yard taps even in the higher consumption brackets is generally much lower than the informal service providers are demanding.

Nevertheless, yard taps can have one significant disadvantage to water kiosks when it serves several households renting rooms and the land lord, controlling the yard tap, is trying to generate additional income. He has the power to add to the utility price and decide on the way the water bill is split among the tenants. Regulation can hardly influence such practices which inflates the water bill for the consumers beyond the regulated tariffs. However, tenants can move to other locations where the land lord is not restricting access or inflating prices for water and sanitation services.

Referring to the case of Agnes in Zimmerman, yard taps in multistore buildings must also be considered as a basic service although it is placed on the plot. To move away from this basic service level, owners of such buildings should be encouraged to convert the yard tap into outlets on each floor of the building like it was the case in Europe (e.g. Basena in Vienna, refer to Sect. 2.4). For Agnes in Nairobi it would be a significant improvement because unsafe borehole water would be replaced by utility water which would be moved from the yard to the fifth floor where she is living. In the case where people can move from a yard tap to a connection in the dwelling (e.g. flat) has the advantage of total control over the utility outlet by the single household or in the case of a tap in the corridor a limited number of households. However, in most of these cases the consumer will have to make an upfront payment to the utility or face an increase in the rent and have to pay standing charges for the utility connection.

Delegated Utility Management: A Viable Solution?

This issue is included in this book because there are increasing tendencies by donor institutions to convince utilities to adopt a delegated management approach for LIAs. There is also a number of literature on the delegation of service provision by the utilities, which means transferring the responsibility of the distribution of water or the collection of effluent to a third party on local level.[122] The supporters of the delegation of utility services to community groups, women groups, small-scale enterprises, etc. or the simple acceptances of neighbourhood sale by connected customers to the poor, who cannot afford a connection or monthly bill, generally ignore a number of difficulties. The risks are multiple and often seem costly to contain for the utility and for the poor alike. The utilities have to deal with the risk of insufficient maintenance, increased water theft (rise in NRW) and revenue losses. The poor risk higher water prices and lower services than utilities can offer to the other consumers with direct service provision in similar circumstances.[123]

Delegation of water and sanitation services by utilities is unlikely to obtain its objective of better service provision for the poor when utilities are not thoroughly in control of their service areas hence, underperforming. Thus, a successful delegation (benefits for consumers and utilities) needs high performing utilities to be present on the ground as a precondition. Even the best performers like ONEA in Burkina Faso have difficulties keeping control when delegating service provision. According to some management staff, it seems that ONEA loses money with the delegation of services and consumers complain of degrading customer care and uncontrolled tariffs.

In Burkina Faso, although delegated management is implemented since more than 7 years a viable analysis on costs and income and the effect on the poor are still not available. With the present information on hand, it seems unlikely that the delegation of service provision by utilities to small-scale units add value to the system.

[122] Delegated management to small-scale sub-contractors such as practiced in Burkina Faso or Kisumu in Kenya.

[123] The customer Identification Survey Pilot Report, Kericho Water and Sanitation Company Ltd (KEWASCO), elaborated by Mboya (2015), Kisumu, Kenya (2015) documents that especially in the areas where neighbourhood sales where common, water theft and anomalies of water counting and billing were prevalent. In addition, the poor had to pay 3.5 times the price for water than the large water users supplied directly by the utility.

There are some enthusiastic narratives produced on delegated management soon after its establishment (e.g. Kisumu in Kenya) which relevance is questioned later in time by the same authors. Comparative long term studies are missing. There is need for more research before we can support or reject this solution as a mode of delivery with greater advantages for both, the utility and the poor people to be served.

2.6 The Two Monitoring Systems for Counting Access

The most prominent system for monitoring access to water and sanitation is the Joint Monitoring Programme (JMP – global monitoring)[124] hosted at the UN Institutions WHO and Unicef. It monitors access worldwide and publishes its findings annually, also for each country and some regions. Initially it reported on progress towards the MDGs until 2015 and from then on towards the SDGs (2030). The global monitoring was, for many years, regarded as a reference system in many low-income countries, as there was no or insufficient reliable information on national level. However, detailed sector information systems, anchored at sector institutions such as regulators, emerged in countries in Sub-Saharan Africa where comprehensive water sector reforms were carried out. Data from both systems express access as a percentage of population living in a defined area. Nevertheless, the figures for access to water and sanitation/sewerage from the two systems can differ substantially because of various reasons. [125]

Pro and Contra of Global Monitoring

The strength of the global monitoring is the collection of data through surveys. Nevertheless, this monitoring system has a number of weaknesses which are known since long but have not been addressed sufficiently. The most relevant weakness is the use of the proxy 'improved sources'[126] for safe water because many are traditional sources and contaminated by ground water pollution in the urban setting. There is more than sufficient evidence for this in the industrialised but also in the developing world. Among the many examples Uganda was already mentioned

[124] Established to monitor progress of MDGs concerning access to water and sanitation.

[125] Refer also to Fig. 2.2.

[126] Improved sources as defined by JMP are piped systems (on plot and shared facilities), regardless if the water quality is regularly controlled (by an authority) or not, but also single water point installations. Although SDG monitoring has defined different service levels with the highest being safely managed water, the counting of people having access to 'improved sources' has not been abandoned after the monitoring of the MDG came to an end in 2015. The use of some improved sources might be appropriate for the rural setting but it is not acceptable in urban areas for a number of reasons lined out in this work. The use of a proxy for the technical design of water installations for water quality stems from an intention to simplify monitoring of access to safe water. The fact that a technical design of infrastructure does not necessarily ensure water quality has been ignored by a number of 'specialists' who influenced the discourse on water supply and sanitation on the global level and in the developing countries.

(refer to Sect. 3.4). Others are Ghana and Nigeria. For the first, Karikari[127] noted that in a survey carried out in the main residential areas of New Endubiase town in Ghana

> All samples tested did not meet the GSB/WHO bacteriological standards for drinking water…. The presence of Total coliforms, Faecal coliforms and Enterococci should particularly raise serious public health concerns over the quality of the town's boreholes and hand-dug wells.[128]

And for the second case, Uhuo et al[129] come to the same conclusion:

> The statistical analysis ($p < 0.05$) showed that Ukwuachi and Umuoghara boreholes showed high bacteria load compared to others hence the quality of the borehole waters for drinking. From the standpoint of bacteriological analysis, all borehole water in these peri-urban areas did not meet the World Health Organization Standards and should be boiled before drinking to reduce or avoid infectious microorganism which constitute the coliform bacterium in water making it unsafe for drinking. Since most households in these peri-urban areas solely rely on borehole water as their main source of drinking and usage for domestic activities but, hence contaminated, the consequences could be fatal due to the public health risk and dangers associated with drinking such contaminated water.

Therefore, any household accessing a traditional water source in the urban setting, including drilled wells, without a sufficient large protection parameter and where water quality is not regularly controlled cannot be considered as having access to safe water. Such traditional water points must be abandoned for the use of drinking water. Hence, there is a need to distinguish between the urban and rural setting as well as between informal (uncontrolled) and formal (regulated utilities) service provision in towns, because most of the informal service providers obtain their water from such 'improved sources'. In the case of Agnes, borehole water on the plot has spoiled her cloths and could have had an effect on her health because the landlord had no means in place to control water quality. Unfortunately, JMP monitoring does not make these differentiations.

Another weak point of JMP monitoring is the acceptance of neighbourhood sales which breach basic human rights requirements even when selling safe utility water. The owner of such a household connection can deny access to anyone at any time which is a challenge in LIAs for several reasons such as tribal issues for instance. Women and girls risk to be exploited entering a private ground and tariffs cannot be regulated as the case of Betty documents (refer to the previous section). Furthermore, the owner of the connection can stop reselling at its own will at any time, hence sustainability or fixed collection hours are not guaranteed.

In addition, JMP does not count shared sanitation facilities even when they are well maintained and linked to a sanitation chain. Furthermore, results of the different surveys vary strongly often within a year because their main focus is generally

[127] Karikari (2013: iv).

[128] Bain et al. (2014a, b), Chakava et al. (2014), Karikari 2013.

[129] Uhuo et al. (2014: 27).

not access to water and sanitation[130] and most likely the geographical areas are varying from one survey to another. An additional weakness is the straight regression line which is drawn since the year 1990. It does not make apparent changes in trends which might occur within a relative short time. With such substantial improvements in access the line is only flattening instead of showing the trend reversal.

Pro and Contra of National Monitoring

National monitoring collects data from formalised service providers such as (licensed) utilities, as it is assumed that rights to water and sanitation in the urban setting can only be guaranteed when providers are obliged to adhere to controlled standards. It is also recognised that urban water and sanitation service provision needs professionals as operators, which can more likely secure a sustainable system of effective treatment and monitoring of drinking water quality. Thus national information systems concentrate mainly on data concerning access to drinking water and piped sewerage systems.

Utilities calculate access figures according to a norm set by the regulator or the line ministry by multiplying the different types of outlets of the distribution network (household connections, yard taps, water kiosks) with a certain number of consumers estimated to have access to these outlets. Also connections to sewer systems are multiplied with a number of people estimated to be served. The advantage of the national monitoring is that it largely follows minimum requirement and particularly water quality, sustainable service provision and unrestricted access to their outlets. Utilities are subject to standardised reporting about their service areas which does not change substantially from one to another year. Because informal service provision does not follow such requirements their services are not included in the count by national systems.

The challenge with the national monitoring systems is the numbers of consumers attributed to the outlets (household connections, yard tap, water kiosks), which can vary substantially from one to another area in a country and from the average household size given by the census. Utilities can inflate the numbers of consumers per outlet in order to achieve a higher coverage rate on paper. In addition, some utilities (e.g. in Tanzania) inflate access figures by counting neighbourhood sales as well. In order to cater to such difficulties some regulators allow the providers to choose, within a range, the number of consumers at the types of outlets to fit the specifics in their different service areas and set a maximum for people served by the outlets. Although regulators or utilities verify such standards from time to time by sampling, such counting remains a source of inaccuracies (see also Sect. 5.4 – adjusting access figures in the comparison of countries).

Another challenge faced by the national monitoring systems is the reporting of sanitation coverage by utilities, as the responsibility of utilities for sanitation is generally limited to sewer systems and, at best, sludge management from onsite

[130]Demographic and health surveys, welfare monitoring surveys, core welfare monitoring questionnaire, multiple indicator cluster survey, world health survey, integrated household budget survey, population and housing census, malaria indicator survey and aids indicator survey.

facilities. Utility data on sewer connections include the same difficulties as water when applying an average household number to connections, and the number of households per connection. The access figures can vary substantially depending on the areas (single villas or multi-storage buildings). Most of the utilities use data for household sanitation facilities from the census or other surveys available to them.

Some countries in Sub-Saharan Africa have started to complement sector-monitoring systems with a national water and sanitation database stemming from surveys such as MajiData in Kenya.[131] They often provide a detailed insight into water and sanitation related issues and in coverage. These very extensive data are established solely for the LIAs, as the residential areas of the middle and high income classes are considered as largely served. Such a database is generally the result of a pro-poor policy aiming to improve service provision in the LIAs. Some countries, like Kenya,[132] have started to link the sector information system to the national census in regards to population, its growth rate in sub-locations and household sizes in order to limit the temptation of utilities to inflate coverage rates.[133]

Comparison of Global with National Monitoring

Global monitoring has embraced principles for counting access which are very different to what many low-income countries have now adopted in the sector framework but also what is common practice in the industrialised world due to lessons learned in history. This approach of monitoring on global level breaches fundamental requirements which are now also enshrined in the human rights for (urban) water and sanitation. It does not help when global monitoring underlines that each country can adjust their (non-binding) monitoring system according to their needs and policies but at the same time publish misleading messages on the status of countries and regions for which the politicians in the UN low-income member states are held answerable. The donor countries evaluate low-income countries in the sector according to these messages (Tables 2.3 and 2.4).

Considering the importance of water quality, sustainability of services and the need to fulfil other human rights criteria such as unrestricted access, it can be argued that national systems with utility data are the better choice for counting access to water in the urban setting despite some uncertainties in the data provided. Nevertheless, the results of national monitoring should be crosschecked with data from the global monitoring system and especially from the census. For sanitation the results from the international monitoring system seem to be more appropriate[134] as the national urban water and sanitation sectors do usually not have viable sector

[131] www.majidata.go.ke (last visited 05.2017).

[132] Waris 3, information system of the regulator Wasreb in Kenya.

[133] In addition, service providers need to be given agreed upon standards to define their service areas in order to avoid monopolists excluding parts of the population from basic services because of profit orientation and other self-interests such as inflating coverage figures.

[134] JMP (2017: 4, 110) published for the first time access figures for sanitation, which included the sanitation chain. However only data from about halve of the world's population where available for 'safely managed sanitation'. Hence, the indication that 39% of the global population use a safely managed sanitation service needs to be taken with care. https://washdata.org/ (last visited 07.2017).

Table 2.3 Differences between JMP (international) and national monitoring systems (water)

Definition, data source, etc.	International monitoring system	National water sector monitoring systems
WHO water quality standards	Recognised	Recognised, built into national standards
Access to water	Counting sources (improved as proxy for safe[a]), and formal and informal providers	Counting only formal providers outlets (quality controlled/monitored)
Sustainability	No consideration	Utilities as proxy: holding a licence (generally around 10 years) and applying for cost recovering tariffs
Data source for access	Census and a multitude of household surveys	Data provided by utility based on certain standards
Population to be served	Urban population (some exclusion in census – e.g. informal settlements)	Population in service areas increasingly related to density of settlements (urban planned, peri-urban planned, informal/unplanned settlements)
Other access criteria – distances and minimum quantity	One km and minimum of 20 lts/c/d	Design of kiosks for 100–400 consumers per tap, up to 500 m distance (30 min cycle) and 20 lts/c/d[b]
Access to sanitation	Improved sanitation facilities	Population served by sewer networks and sometimes coverage by the promoted sanitation facilities at household

[a]The report also included analysis of the results from rapid assessment of drinking water quality (RADWQ) surveys in five countries, which showed that 13 to 32 per cent of improved sources were contaminated at levels exceeding WHO guideline values in four of the five countries.' (JMP report, 2015, p. 43); In addition, 'In 38% of 191 studies at least a quarter of samples from improved sources exceeded WHO recommended levels of FIB. By equating 'use of an improved source' with 'safe,' international estimates greatly overstate access to safe drinking-water (Bain et al. 2014a, b: 3).

[b]The WSTF in Kenya uses 20lt/c/d and 400 people served per tap at water kiosks (3 tap serving 1,200 people) as standardized design (own observation)

data on access to onsite sanitation facilities. Nevertheless, the data on access to sewerage systems from JMP needs to be crosschecked with the sector information provided by the utilities.

In the past, the development of urban water and sanitation systems in the nineteenth century gradually followed stringently minimum requirements for drinking water/effluent quality and sustainability of services. Modern sector reforms in low-income countries now following the same approach and accept only formalised services when counting access which is based on a service ladder within this formalised system. This is very different to how JMP defines a ladder for drinking water. It is based on a wide range of water sources whereby on each of the three service levels (limited, basic, safely managed) the consumers face different risks in

Table 2.4 Comparison of JMP categories and national policy direction regarding water quality requirements

JMP categories of improved water sources[a]	Mode and scale of supply	Water quality testing and control	Compliance with WHO standards	National policy guidance[b]
Piped water in house hold or yard	Utilityc	Enforced (by sector institutions)	Medium to High	Accepted
	Small-scale, informal[d]	Not enforced[e]	Unlikely	Not accepted
Standpipes	Utility kiosk	Enforced	Medium to High	Accepted
	Small-scale, informal	Not enforced	Unlikely	Not accepted
Borehole	Small-scale, informal	Not enforced	Unlikely	Not accepted
Protected dug well	Individual, small-scale	Not enforced	Very unlikely	Not accepted
Protected spring	Small-scale	Not enforced	Very unlikely	Not accepted
Rainwater	Individual	Not enforced	Medium to unlikely	Not accepted
Bottled water	Small scale	Not always enforced	Unlikely when sold by mobile vendors	Not included

[a]WHO/Unicef Joint Monitoring Programme: wat/san categories
[b]When reforms are undertaken
[c]Formalized and generally with treatment facilities for surface and groundwater treatment
[d]Mainly community and private owned systems with boreholes or rarely used springs as a water source
[e]Small-scale informal providers have in general no treatment facilities as part of their installations

terms of safety and sustainability.[135] Informality is accepted by JMP for all three levels in the ladder, including piped water (refer to Fig. 2.2, in this chapter)[136]. Hence, even safely managed water has little meaning. Also sustainability of access has never been a concern in JMP monitoring although it was specifically mentioned in the definition of the water and sanitation targets in the MDG declaration, but dropped out of sight with the SDGs declaration.

It is obvious that the already difficult situations of Betty and Agnes would further deteriorate when depending on a service ladder proposed by JMP where informal service providers use 'protected' springs, open wells and rain water collected from

[135]WHO and Unicef (2017: 12), Safely managed drinking water – thematic report on drinking water 2017. The service ladder is since the SDG defined as limited, basic, safely managed. Other service levels are also considered in some literature such as water trucking, selling of water in plastic bags, reselling of water by other mobile vendors using different types of containers and means of transport. Buying water from mobile resellers can be regarded as a choice of the household and be accepted as long as the household has the option to access at any time an outlet of a formalized service provider operating under the minimum requirements set by the human rights to water and sanitation (e.g. 30 min cycle) and as long as the transport does not compromise water quality to an extent where people's health is at risks

[136]It might be misleading to call access to a protected open well as service.

Table 2.5 Service levels for water supply according to JMP and the proposed definition (PD) in this work

Water sources, service levels	Rural setting	Urban setting	Comments
Protected **traditional sources**	**Access** according to JMP and PD, because of basic tolerable risk for water quality	**Access** according to JMP but **No access** according to PD, because of high risk of pollution and unlikeliness of adequate treatment by households (hh)	Water is often available at no cost for consumers. However, sustainable control by the state regarding adequate treatment and water quality monitoring is very unlikely
Piped public outlets, yard tap and hh connection from **informal providers or community systems**	**Access** according to JMP and PD because of tolerable risk for water quality and price control by community	**Access** according to JMP but **No access** according to PD, because of high risk connected to the use of polluted sources and unlikeliness of adequate treatment by households, spending often above 5% of hh income in LIAs	No regulated water quality, water tariffs and complaint resolution. Moving from informal to formal service provision is the biggest step in development because of improvements in water quality, price and sustainable access
Piped public outlets (water kiosks) from **utilities**[a]	Service provision by utilities in rural areas is rather an exception	**Access** according to JMP and PD, because of controlled basic standards, but according to JMP with less preference than improved sources at the plot, risks of pollution during transport and storage at household	Regulated water quality, tariffs and complaint resolution. Furthermore, high sustainability. One kiosks should serve 300–500 people per tap in order to comply with the 30 min cycle and be economically viable
Yard tap from **utilities**		**Access** according to JMP and PD, because of controlled basic standards and situated on the plot, but risks of pollution during transport and storage at hh	Regulated water quality, tariffs and complaint resolution. With a limited number of households (e.g. 20–40 people) to avoid higher tariff brackets for the poor
Household connection from **utilities**		**Access** according to JMP and PD, because of controlled basic standards, but risks of pollution at storage with frequent supply interruptions	Regulated water quality, tariffs and complaint resolution. This is the highest service level

This overview is mainly referring to the classification used for MDG monitoring by JMP, but has also relevance for the SDG monitoring

[a]In this work 'water kiosks' are outlets from formalized/controlled service providers, which carry out regular water testing, treatment of raw water (if needed) and in addition, reporting is controlled and standardized

the (dirty) roofs in the urban setting for packaged water, public stand posts, yard taps and sometimes household connections. Informal service provision will worsen their already unsatisfying situation already caused in some cases by neighbourhood sales of utility water.

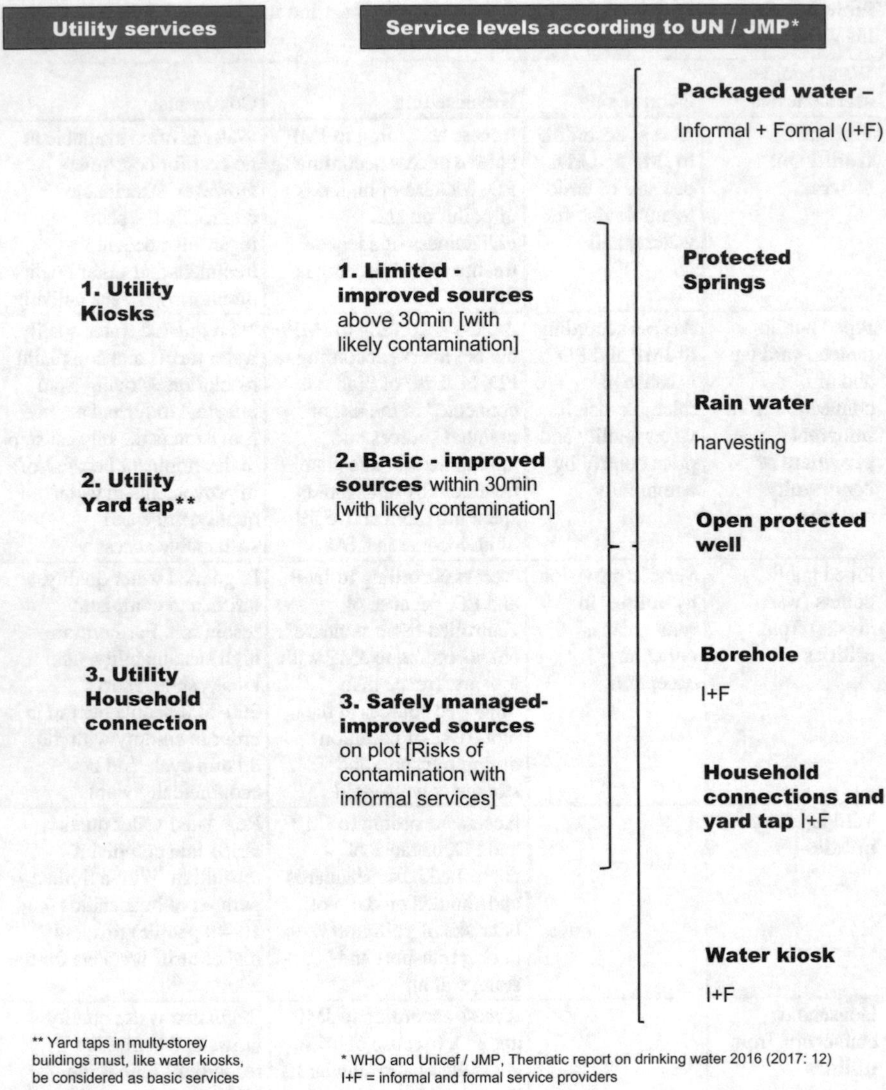

Fig. 2.2 Comparison of standards and counting of access to drinking water between national- and JMP- systems (It is to note that in the JMP document, WHO and Unicef (2017:12), there is no mentioning of 'free of faecal and priority contamination' for the service levels one (limited) and two (basic) as it is mentioned for the highest service level number three. All photos from own source or Aquapix of WSTF/Kenya)

2.7 Water Poverty and the Gender Debate

There are a number of cross-cutting issues which have gained prominence in urban water and sanitation and which became standards to be addressed by projects in the international cooperation. Some of them are mentioned in this book by other sections such as good governance others have lost importance over time such as HIV/AIDs. Two of the cross-cutting issues are worth to be discussed in more details. These are water poverty and water and gender. Both are important because by streamlining relevant issues the sector can offer a significant contribution to the countries development. Furthermore, discussions on these cross-cutting issues seem to be incomplete and therefore, the contribution in this book can help to close some gaps.

The Magnitude of Urban Poverty and Inadequate Access

According to the World Bank,[137] Sub-Saharan Africa remains the region with the highest poverty head-count ratio in the world. Fifty-seven percent of the poor in the world live in Sub-Saharan Africa, which is also the only region where the number of extreme poor is still increasing. Hence, poverty is a big issue in Sub-Saharan Africa. In general, it is understood that most of the urban poor live in LIAs, which have very different contexts. Many of these areas, including slums, are unplanned settlements and are often illegally established where housing with temporary character is made from metal sheets or wood. Others, originally planned settlements, have taken on un-planned characteristics. The 'hard core poor' are mainly found in the unplanned and illegal settlements (slums), which are composed of at least 50% shelters with temporary nature. Nevertheless, many of these settlements have existed for several decades and persist in areas which are often unattractive for housing developers because of the neighbourhood (industrial zones) or natural risks such as floods and landslides.

According to Baker,[138] 71.9% of the urban population in Sub-Sahara are slum dwellers and their settlements grow by 4.53%. Other figures show less dramatic but still impressive pictures such as MajiData[139] in Kenya, which is a very detailed survey of 276 towns documenting the water and sanitation situation in 1,964 LIAs.[140] According to this survey, the population living in the LIAs amounts to 39% of the 20.58 million people living in the service areas of the around 100 registered utilities.[141] Using the information from MajiData, around 50% of the population in the LIAs live in settlements with mainly slum characteristics.

Many documents support the estimation that between 30% and 50% of the urban population live in LIAs in Sub-Saharan Africa and that the population growth in these areas outstrips not only the growth rate in the countries but also the average growth rate in towns. According to some sources, the annual urban population growth[142] in Sub-Saharan Africa seems to remain high and increased from around 3% in the last decade to an estimated 4% in the current decade. In Kenya, for

[137] World Bank (2016: 4, 5, 38).

[138] Baker (2008: 7).

[139] www.majidata.go.ke (last visited 05.2017).

[140] First version of MajiData 2013.

[141] Wasreb Report No. 6, Kenya, http://www.wasreb.go.ke/impact-reports (last visited 05.2017).

[142] http://www.tradingeconomics.com/sub-saharan-africa/urban-population-growth-annual-percent-wb-data.html (last visited 05.2016).

instance, according to the national poverty line, 35% of the urban population is considered poor.[143] Comparing this figure with the 39% of the urban population living in LIAs within the service areas of the utilities[144] and where the poverty level stands at 55% it can be estimated that among the 8 million in these LIAs 83% are poor.[145] This represents 6.6 million out of the 8 million people reported by MajiData. Thus, this example of Kenya supports the general understanding that the biggest share of the poor live in the LIAs, where coverage by the utilities is lower than in any other areas of the towns. Therefore, it can be said that mainly the poor are deprived of sustainable access to safe water and sanitation (Fig. 2.3).

Measuring Poverty and the Link to W+S

Poverty can be defined as 'the state of one who lacks a usual or socially acceptable amount of money or material possessions'.[146] Kulindwa and Lein[147] distinguish between absolute and relative poverty by stating that 'absolute poverty means that individuals do not have the resources to meet their basic needs and relative poverty compares the individuals in terms of income and wealth standards'. Other studies on poverty establish groups of poor's in order to gain insight into the differences

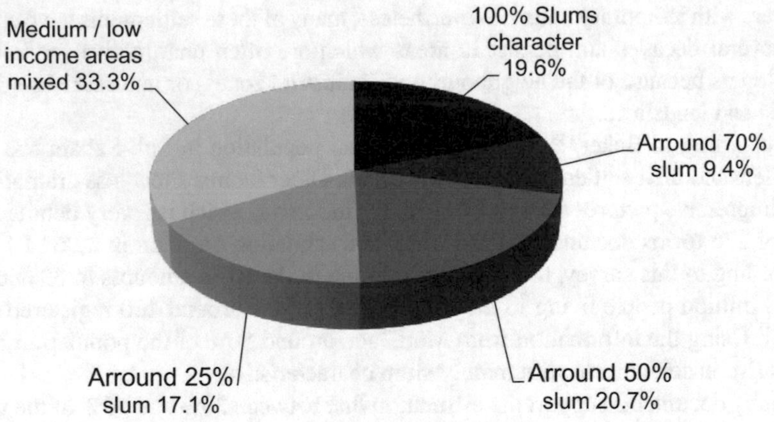

Fig. 2.3 Population in urban low-income areas in Kenya. (Source: Data from MajiData (WSTF) and figure by Werchota)

[143] Kenya economic report (2013: 158).

[144] This also includes officially labelled 'rural' but densely populated areas.

[145] In 2012: 15.3 Mio urban population (Census 2009 and 2.9% growth) from which are 35% poor = 5.3 Mio urban poor. 2.7 Mio rural population (8 Mio Majidata minus 5.3 Mio urban poor) from which are 48% poor = 1.3 Mio rural poor (peri-urban areas). The total number of estimated poor in the urban setting is therefore 5.3 plus 1.3 = 6.6 million.

[146] Merriam Webster https://www.merriam-webster.com/dictionary/poverty (last visited 05.2017).

[147] Kulindwa and Lein in Hemson et al. (2008: 1).

between countries in respect to inequality. The global monitoring of water and sanitation[148] offers a look into the different wealth groups (quintiles) for access through some country examples. Inequalities are also apparent in towns where dwellers of medium income classes can live in opulent residential areas well served by the utility and situated right next to slums across the road where dwellers depend on informal service provision or contaminated traditional water sources.

There is also much discussion on how to measure poverty. Often it is measured in monetary terms such as income available for the individual. Anyone who has to live on 1 USD or less a day, for instance, was considered poor as it was thought that this amount did not allow a person to cover basic needs.[149] Critics on defining poverty in monetary terms underline that such a measurement is insufficient because it does not reflect the multidimensional nature of poverty and does not consider the different context people live in across countries. Thus, indices reflecting different dimensions are now frequently used. However, attention should be paid to the fact that the choice of indicators for poverty and the weight attached to them influences the results and often raises the question of how reality is reflected in the specific cases. There are also a number of dimensions, which are very relevant for the poor but are difficult to be measured such as security, dependency on cartels, etc.

Nevertheless, next to income measurement, one of the standards for many studies today is the Human Poverty Index (HPI),[150] which has been used by the UNDP since 1997 in its Human Development Reports. The HPI was developed by the Oxford Poverty & Human Development Initiative for UNDP and is today used on a global and on national level in many countries with some alterations. The HPI already contains an element of access to an 'improved' water source. In the 2010 report of UNDP, for the first time, a Multidimensional Poverty Index (MPI) was used, which carried over the three dimensions of the Human Development Index (HDI): Education, Health and Living Standard, whereby the HDI uses four indicators and the MPI 10. Dimensions as well as indicators within dimensions are equally weighted. It is explained that the national poverty lines can differ as the MPI can be adjusted by each country.

The MPI can be regarded as an improvement in measuring poverty because the measurement takes on board the multidimensional nature of poverty, which provides deeper insight into poverty and also recognises the importance of water and sanitation for poverty eradication to a certain extent. However, there are still a number of questions to be raised. Next to the completeness of dimension (especially for slum dwellers) the weight attached to the indicators are disputable. Can school attendance be considered three times more important than access to drinking water and sanitation although we know that inadequate access causes millions of deaths every year among children?[151] Can nutrition be three times more important than

---

[148] JMP (e.g. report 2014: 9, 25, 32).

[149] 1 US $ a day per person was used by the World Bank in 1990.

[150] http://economicsconcepts.com/human_poverty_index_(hpi).htm (last visited 05.2017).

[151] http://www.lenntech.com/library/diseases/diseases/waterborne-diseases.htm#ixzz4lMtkr5TR 'Today we have strong evidence that water-, sanitation and hygiene-related diseases account for

access to safe sanitation considering the fact that child stunting might as much be caused by inadequate sanitation as insufficient food intake/nutrition (Fig. 2.4)?

In addition, such rankings seem to ignore the priorities the poor place on water and sanitation when they are deprived of it. Aderinwale and Ajayi[152] citing a World Bank source (2001), indicate (Nigeria for example) that the poor ranked access to water before access to education and health facilities. Equally, the expert interviews carried out for this work indicated a higher priority for access to water than for education, for instance. Furthermore, water and sanitation are indicators in the dimension 'Living Standards', but also strongly correlate with both of the indicators of the dimensions 'Health, Nutrition and Child Mortality' and to some extend with the school attendance indicator under the dimension 'Education'. This is another sign that there is a need to harmonise the priority the poor attach to water and what water and sanitation represents for the development of individuals and the society with the priority water and sanitation is given in the poverty estimations.

Olsen[153] underlines that poverty is a cycle that continuous with 'myriad hurdles' which makes it difficult for the poor to escape poverty. She uses the example of access to transport and healthcare in her analysis of poverty in the United States of America. It is obvious that for the poor in the developing world the priority is foremost access to water and sanitation before transport. However, Olsen indicates how difficult it is to escape poverty when basic services needed by humans every day are physically not accessible or expensive to obtain.

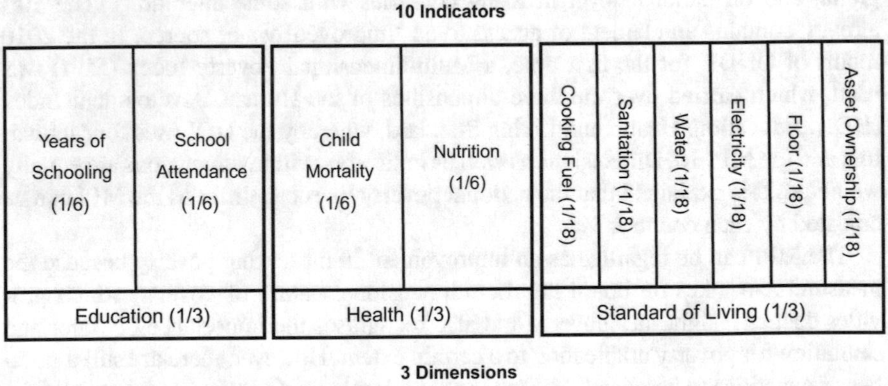

Fig. 2.4 The multidimensional poverty index. (Developed at Oxford University, MPI=Head count ratio of poverty (H) x Average Intensity of poverty (A), Poor=deprived in more of 1/3 of indicators. Refer also to http://www.ox.ac.uk/news/2018-01-23-living-poverty-global-multidimensional-poverty-index (last visited 02.2019))

some 2,213,000 deaths annually and an annual loss of 82,196,000 Disability Adjusted Life Years (DALYs) (R. Bos, Dec. 2004).

[152] Aderinwale and Ajayi in Hemson (2008: 72, 73).

[153] Olsen (2017); www.everydayfeminism.com/2017/01/escaping-poverty-is-not-easy/ *(last visited 07.2017).*

That's because the trappings of systemic poverty are woven into the fabric of daily life.' Furthermore, 'Lack of access, on multiple fronts, all add up to an uphill battle…' and in addition, 'Low-income individuals and families also suffer from interacting with a financial system that penalizes them and profits off of poverty … all disproportionally impact low-income folks, making it difficult to save money, make necessary purchases…

When these findings are translated into the situation of the *'urban water and sanitation divide'* in the developing world it becomes evident that the poor can hardly be helped with small-scale (informal) systems trying to develop into a 'business case' as many NGOs and philanthropic institutions try to do with the sale of (branded) bottled water for the poor. Such activities which focus only on LIAs are penalising the poor in the long run (for further discussions see Sect. 3.4).

Affordability for the Poor Instead of Free Services

According to Keener and previous deliberations in this book,[154] the price charged by the informal provider is in general far above the guaranteed price charged by utilities. Thus, it is obvious that the most effective measure to make water more affordable for the poor is to replace informal service provisions with utility water before discussing the affordability of the water tariffs offered by utilities. Shifting people from the informal to the formal service provision offers the biggest benefits to the poor, especially in regards to tariffs and water quality. People pay much less for water at the utilities not only because of the benefits of economy of scale, but also because most of the utilities have implemented a progressive tariff structure.

Often, when the ability to pay for services by the poor is discussed, some experts advocate that the poor cannot pay for utility services. The Burkina Faso case proves them wrong. It seems that the vast majority of the poor have the means to satisfy their basic needs for water and sanitation if services are provided by utilities (expert interviews). Also the proposal that water and sanitation services should be free for the poor is questionable because in most of the Sub-Saharan countries, the majority of the poor depend on informal service provision where the state cannot enforce free access to water anyway. Therefore, for the majority of the poor, the provision of free water has still little meaning today. Being obliged to provide water for free to the poor, utilities usually aggravate their situation in cost coverage which leads to a degradation of service provision and eventually to more of the poor being pushed back to informal service provision or traditional water sources. Thus, the demand of water for free for the poor can be seen as counterproductive for the poor.

Utility Services as a Means to Escape Poverty

The above indicates that access to water and sanitation does not automatically support the urban poor to escape poverty, especially when water is provided by informal service provision, the urban poor have to depend on traditional water points or the utility accepts neighbourhood sales (e.g. Uganda). According to Krishna, 'If a donor agency is to help the poor effectively, it must understand what keeps them

[154] Keener et al. (2010: 18).

poor.'[155] Investigating the reasons for the poor not being able to escape poverty or people to fall into poverty in rural India, Krishna[156] finds that it is foremost health and health related expenditures which keeps people under the poverty line. This is a crucial indication how important access to formalised water and sanitation services is for poverty eradications. The poor who do not have adequate access to safe water and sanitation are constantly exposed to its negative effects. They cannot benefit from the improvement in health utility water and sanitation services offers and the savings obtained due to pro-poor tariffs and therefore, risk staying entrenched in poverty.

Among their main preferences at the lower stage of poverty are sending children to school and possessing clothes to wear outside the house, which can be linked to hygiene (e.g. helping to be presentable and consequently find a job). All of these reasons and preferences outlined by Krishna are directly linked to access to water and sanitation and might be of similar importance in both settings, the urban and rural. Saving on water and sanitation and on health related expenditures when moving from informal to utility services helps the households to shift budgets to school fees and food purchases for instance. Therefore, extending utility services into the LIAs support pro-poor strategies to reduce the number of people living in absolute poverty. It also helps the people living in relative poverty because they can reduce spending on water and on the treatment of waterborne diseases and use this savings for other basic needs, which is a contribution to finally moving beyond the poverty line and maintain this status on long term. Hence, improving access to utility water and sanitation services is a contribution to reducing absolute and relative poverty.

Hemson et al[157] argue that 'the relationship of water and health, water and poverty, and water and development' is widely accepted and that the 'lack of political resolve in water and sanitation was perhaps the greatest development failure of the 20th century'. This is in line with the survey mentioned in the Preface of this book which found that experts considered the development of centralised urban water and sanitation systems in the industrialised world as the most important achievements since 1840. Hemson et al also indicate the link between poverty and water by stating that: 'Inadequate and unequal access to water [and sanitation] is thus, both a result and a cause of poverty.' Considering the sector development in the low income countries since the new Millennium and the increasing disengagement of some donors in the sector, it is very likely that the world will experience the same development failure in the new Millennium.

Understanding the Outcome of Pro-poor Measures in W+S

Utilities often do not understand the impact of their decisions on servicing the poor. A very good example for this is the Kericho case in Kenya. The utility supported the idea of neighbourhood sale because it was recognised that the poor cannot pay for a household connection. Nevertheless, Kericho ignored several factors and disadvan-

[155] Mosley (1987: 168).

[156] Krishna (2004: 123, 128, 130).

[157] Hemson et al. (2008: 13).

tages stemming from this decision, which had negative effects not only for the poor but also the utility. According to Mboya,[158] the demand at resellers was fuelled by the deliberate closure of nearby utility water kiosks. In addition, he noted that 62% of the resellers (neighbourhood sales of utility water) received very low (manipulated) water bills (under nine m3 per month and some as low as one m3) despite very high sales. It became obvious that personnel of the utility worked together with the neighbourhood resellers to increase their turnover by initiating the shutdown of utility kiosks in the area and at the same time made sure that the resellers received manipulated water bills.

Thus, the decision to allow neighbourhood sales pushed the poor who were served previously by the public outlets of the utility system back to informal service provision, led to more corruption in the utility and an increase in water theft. In addition, unrestricted physical access ended with this shift from utility kiosks to neighbourhood sales. Furthermore, this report of a pilot project for updating the customer base with the aim to reduce NRW concluded that the poor at the resellers of utility water had to pay much more than the regulated tariff:

> The implication here is that the poor in Kericho pay about three and half times more than the large water users. Large water users are mainly the well-to-do dwellers.[159]

Kericho management considered neighbourhood sales to be helpful for the poor, but they ignored the fact that it is as good as impossible for a water company to control the prices demanded from the poor for water sales on private ground and that there is no functioning market to regulate the price (refer also to the case of Betty in Kawanwgare) despite such a claim voiced in some literature. In addition, utilities often ignore the fact that in many of the LIAs universal access even with social connections cannot be achieved.[160]

The Question of Subsidies for the Poor

Evidence indicates that any tariff system where the average tariff cannot cover average costs is damaging the utility system and a cause for degradation, at least in the long term. The effect of such degradation is foremost felt by the poor because of the inability of the utility to extent services and to reach universal access in the LIAs. Therefore, holding average tariffs below average costs, often claimed to be necessary to offer access to the poor, is anything but pro-poor. The difficulties in identifying the poor and either offer them free water and sanitation services or provide direct subsidies have also been discussed previously. Demanding a little lump sum for a monthly consumption from the poor will lead to huge water wastages (refer to expert interviews, Zambia) and help (in the case of Betty) to increase the profits of resellers. Hence, decision makers in Sub-Saharan Africa should rule out such tariffs

---

[158] Customer Identification Survey Pilot Report, Kericho Water and Sanitation Company Ltd (KEWASCO), Mboya (2015: 25).

[159] Mboya (2015: 25–31).

[160] Refer also to Sect. 2.5 for the case of Betty and the issue on social connections for a few which are used to establish a business and profit from the inability of the poor to access utility services.

and decide on the two options left: the unique tariff for all or the rising block tariff system which offers the possibility to make consumers with high consumption subsidise basic consumption (quantity-based subsidies). Low consumption is thereby used as proxy for the poor and as a possible incentive for middle class consumers to save water.

Experts arguing for a unique tariff for all consumers regardless of the quantity consumed or the use of water (domestic, industrial, etc.) claim that such a tariff is more successful to balance inequalities between rich and poor than rising block tariffs. However, comparative studies with sufficient empiric evidence seem to be missing to support such a claim. In addition, a unique tariff is also questionable in a situation where raw water is not sufficiently available and higher tariffs for big consumers would offer incentives to safe water in the interest of all consumers. Furthermore, a unique tariff does not tap sufficiently into the potential generating funds from water bills. Many experts interviewed from the middle class insisted that they are willing to pay more for water and sanitation services. Tapping into this potential the sector would become more credible and it would help to close the financing gap.

There are a number of critical papers on subsidised tariffs (e.g. rising block tariffs with cross-subsidisation) which seem to ignore the desperate situation of the poor. One of the arguments is that the poor who cannot access utility services are excluded from these subsidies. However, it must be noted that everyone who cannot access utility services is excluded from its substantial benefits never mind which tariff system is in place. Therefore, this argument is not only valid for the block tariff system. Another argument against block tariff is that small-scale systems serving the LIAs do not have the possibility to cross-subsidise because of the absence of big consumers. This argument can also be used against such supply systems by indicating that block tariffs might be an incentive for decision makers to integrate small-scale systems (e.g. community system) into a utility which can offer a pro-poor strategy because of economy of scale and a diverse customer base.

Other arguments against block tariffs are that bigger consumers will leave the utility services because of high tariffs at the upper end and that the utility will neglect the collection of water bills from the poor because of insufficient tariffs at the lower end, which will have a negative effect on the collection rate. To both arguments are sometime added that domestic consumers have to subsidise state institutions like the military or the president's offices because they do not pay their water bills (e.g. Zambia until around 2006). Such arguments are not uniquely relevant for the block tariff system and such challenges are generally overcome with sector reforms without the need to change the tariff system. Increasing disconnection for non-payers, contracts with the state to ensure payment for critical institutions when they fail to pay their water bills (e.g. university, slaughter houses, etc. in Burkina Faso in the 1990s), improved water metering and other measures lead to the improvements of billing and collection during reforms. ONEA in Burkina Faso is an excellent example that utilities can reach a very high collection rate, including the LIAs, if decision makers ensure that LIAs are not neglected. In addition, there is little evidence that big consumers leave the utility system in larger numbers because of high tariffs. Often it is the erratic supply and the limits of production the utility is facing which force bigger consumers to search for alternatives and not the water price demanded by utilities. Critics also point out that mobile resellers need to buy water from the utility at the highest tariff bracket. This is not the case when water kiosks from the utility are available where generally mobile vendors access a subsidised water tariff. As water kiosks are (intermediate) pro-poor measures, selling to bigger re-venders such as water truckers should be avoided.

However, the argument that people are disadvantaged with a rising block tariff when consuming at shared connections seems to be valid.[161] This is not so much a problem for the middle class consumers than with the poor. However, regulators can oblige utilities to avoid the use of shared household connections. Regulators should also instruct utilities to undermine neighbourhood sales by offering mixed systems in LIAs (water kiosks, yard taps and household connections) wherever the security situation is permitting and insufficient water kiosk management does not push the poor to the neighbours to beg for water. In these cases the poor will have the choice between buying water from (mobile) resellers at the door step or fetch water at the kiosk which offers a subsidised tariff.

Critics of block tariffs should compare the advantages and disadvantages of both systems (block and unique tariffs) with a more differentiated analysis and decide according to the biggest aggregated benefit for the people served. Listing solely disadvantages of one system is little helpful without indicating better alternatives. On the other hand, utilities or regulators designing block tariff systems need to be aware of the possible weak points of such systems and concentrate on cost recovery and the use of rising block tariffs as pro-poor measure. This includes a cross-subsidisation which takes the existing mix of the different consumer groups, the need to make every group contribute according to their ability (including the poor) and the containment of the risks to loose cost recovery into consideration. A pro-poor tariff system must be seen as part of a set of pro-poor measures such as the promotion of low-cost technology, social connections, and etc. and as support to safeguard the environment through water savings.

Furthermore, critics of subsides should consider the deplorable situation of the poor and the fact that subsidies are already largely practiced and accepted in the water and sanitation as well as in other sectors. For instance, loans for infrastructure development (e.g. development of sewer systems, first mile for water, etc.) with preferential rates for repayment offered in the framework of international cooperation are a kind of subsidisation. Such subsidies often benefit more the upper than the lower class especially with first mile investments. If this is acceptable why the poor should then be deprived of subsidies, especially when the sector itself is generating the subsidies? It would be a milestone in the development of Betty and her family (Sect. 1.5) and the start for their escape from poverty if she or her landlord could access subsidies to build a toilet and the utility would offer a water kiosk in the

[161] The advantages and disadvantages of water kiosks and yard taps in the LIAs relative to tariffs have already been discussed.

vicinity of her home. Without doubt, a block tariff system can help many poor even when it is not perfect. The resulting benefits should be compared to the critical points raised against cross-subsidisation.

There are also systems in place, which provide free basic water, such as six cubic meters per month for the poor in some parts of South Africa,[162] or a direct subsidy system for water and sanitation to poor households like in Chile. In the latter case, the state is not subsidising the utilities but the consumers. The challenge for such systems lies in the identification of the poor and the direction of subsidies to these households. It needs a solid civil service system on local level to make such subsidisations outside the utility work according to the intentions. This seems to be absent in Sub-Saharan countries. Surveys indicate that it is not unusual that more non-poor households profit from such subsidies than the poor.[163,164]

Another tentative to avoid subsidies and make people build and maintain an acceptable toilet is the naming and shaming of households like practiced by the Community Led Total Sanitation (CLTS) approach. Not only that shaming people in a culture prevalent in Sub-Saharan is questionable; shaming poor people who cannot afford to build an adequate facility might even be considered outrageous. In the case of Betty in Kawangware for instance, there is no space and incentive for her to build a toilet in a room which she is renting while she has no means to move to another area with better installations. It seems that the way to improve sanitation for many poor is to offer incentives in the form of subsidies to the land lords. The argument that the land lord will increase the rent is not very strong because house rents are much more influenced by a functioning market than by an improved toilet.

Can a Poor Country Secure Access for the Poorest?

Alkire and Housseini[165] underline in a study that according to the MPI measured for 37 Sub-Saharan countries, Burkina Faso has a value of poverty which is more or less twice of each of the three Anglophone target countries. This result is similar to the poverty ranking of the four target countries according to income (see Sect. 5.3). The study also defines a group of the poorest among the poor (destitute/extreme poor) who 'are deprived in at least one third of the destitution indicators, which are more extreme than those used to identify the MPI poor'. The group, which is deprived in at least half of the indicators, are considered to be the very poorest of the poor. For the analysis of the extreme poor, data for 24 of the 37 Sub-Saharan African countries, including Burkina Faso and Tanzania, are available. The results indicate

[162] For instance, in Cape Town the municipality provides free toilets to the dwellers in the low-income areas and in addition a free service to empty the human waste every 2–3 days (own experience 2018). Experience shows that such extensive subsidisation is not sustainable in a fast growing environment and is of little help to develop responsibility among the beneficiaries.

[163] Henson et al. (2003: 163).

[164] Rouse (2013: 15) cites the definition of corruption by Transparency International: 'the abuse of entrusted power for private gains'.

[165] Alkire and Housseini (2014: 6–9, 15 and 17).

that over half of the poor are identified as the poorest of the poor and 38% are identified as the very poorest of the poor. From the poorest of the poor, 89% practice open defecation and 71% do not have access to safe drinking water or the water source is more than 45 min away (round trip). Burkina Faso holds second place behind Niger in the poorest of the poor at 58% compared to Swaziland with 3%.

Furthermore, the same study indicates that countries (19 of 37) that reduced MPI poverty in absolute terms, where predominantly in Eastern Africa whereby Tanzania was leading Zambia and Kenya. However, in Kenya and Zambia the population growth outpaced poverty reduction and thereby, the absolute number of poor people increased. Nevertheless, in all of these 19 countries the rate of the poorest of the poor fell in relative terms faster than the MPI rates. Hence, it can be concluded that Burkina Faso is not only the poorest country, but also has the highest rate of the poorest of the poor among the four target countries. With this information, one would expect that countries facing such pronounced poverty, struggle more than other countries to provide sustainable access to safe water and sanitation. The analysis of the four target countries (Part II) will document that a high poverty level and a big layer of the poorest of the poor does not necessarily mean that the water sector cannot achieve a high level of performance in terms of access. Therefore, the poverty levels in a country seem not a limit to move to an advanced level in water and sanitation coverage.

Donors and Utilities Need to Increase Efforts to Reach the Urban Poor

The perception that poverty is foremost a phenomena appearing in the rural setting has to be rectified because it leads to a neglect of the poor in the slums where poverty seems to be much more brutal. The poor in the slums deserve at least the same attention and should not be left out because decision makers believe that urban investments serve primarily the rich. This believe is caused because investments in water supply addresses often issues linked to insufficient services for the already connected middle and upper income groups such as insufficient service hours, water quality and pressure in the network. The concern about services to the underserved is unfortunately often enough addressed with accompanying measure of first mile infrastructure development. Hence, more attention needs to be given to extending utility services to the poor than improving an already acceptable service to the connected consumers.

Utilities and donors alike usually shy away from developing the extensions of centralised systems into the LIAs because it is difficult to implement and manage the last mile infrastructure and services (Sect. 3.4). It needs local knowledge. Reaching the urban poor cannot be adequately addressed when donors delegate the implementation of their support to construction companies and engineering consultancies. Furthermore, decision makers often do not match the need of the poor with their means available (e.g. use of low-cost technology).

Growing Awareness But Limited Progress in Closing the Gender Gap

Already in colonial times, administrators have recognised the important role women play in the provision of water for the household and their particular attention to ensure access to water and sanitation.[166] However, some argue that since then inequalities remain and in some cases have even increased because the understanding of gender inequalities was incomplete despite a growing awareness of the need to overcome bottlenecks in gender inequalities.[167]

A broader awareness of gender inequalities in the water sector commenced during the 1970s. The report from the Nairobi conference in July 1985,[168] closing the UN decade for Women (1976–1985), repeatedly referred to the role of women in the provision of safe water and sanitation. It called on governments to ensure that women are:

- consulted and involved in the planning and implementation of water and sanitation projects
- involved in the administration of water supply
- trained in the maintenance of water supply systems
- consulted on technology used in water and sanitation
- supported to reduce the burden of fetching water
- involved in improving the sanitary conditions, especially in slums

Another significant support for this cross cutting issue came with the UN Conference on Environment and Development, held in Rio de Janeiro, Brazil, in 1992, where the Agenda 21 included water and gender in Chaps. 21 and 24.[169] This agenda also emphasised on the role of women in the management of waste and again on the design of technologies for clean water and adequate sanitation facilities in order to satisfy the specific needs of women. From this point on it became accepted that gender had to be mainstreamed in the water sector. Other international conferences and UN resolutions aimed at drawing particular attention to women's participation and involvement in the water sector.

However, critics are pointing out that despite these gender mainstreaming efforts in the water and sanitation sector there was not much progress in making gender inequalities disappear in reality. One of the reasons next to the lack of distinction between rural and urban realities is most likely the fact that sector policies and strategies are not complemented with detailed documents on implementation. The document analysis carried out in the four target countries supports this observation because there were complementary detailed documents elaborated on other issues such as poverty and water by the ministries or sector institutions in two of the four target countries, but none on gender. In addition, the understanding of gender in the sector seems to be incomplete.[170]

---

166 Page in Coles and Wallace (2005: 62).

167 Joshi in Coles and Wallace (2005: 136).

168 http://www.un.org/womenwatch/daw/beijing/otherconferences/Nairobi/Nairobi%20Full%20Optimized.pdf (last visited 05.2017).

169 https://sustainabledevelopment.un.org/content/documents/Agenda21.pdf (last visited 05.2017).

170 Joshi in Coles and Wallace (2005: 136).

Urbanisation Transforms Gender Issue

General understanding is that women are responsible to fetch water for the household. According to UNDESA, 2010, in Sub-Saharan Africa, only one in four adults fetching water in the urban areas is a male.[171] Graham et al.[172] surveying households in 24 Sub-Saharan countries indicate that in a majority of the countries mainly adult women fetch water. The percentage of adult women doing so, with two exceptions, is higher in the rural than in the urban areas. However, in five of these countries urban women fetching water represent 50% or less. According to GFA,[173] the first survey in 15 towns in Kenya indicates that 62% of those who fetch water at the kiosks are male and in the second survey, the database shows that 50% are male. The same studies document that among the kiosk operator 44–50% are female. Hence, it can be assumed that urban life has changes the roles of women and men in the water sector. There are increasing cases where more male than female fetch water.

Furthermore, it is commonly argued that the time saved for fetching water allows women to engage more in economic activities or improve child care and increase school attendance of children in the household. Koolwal and van de Walle[174] studying these questions in the rural setting, find that there is no empirical proof that improved access to water has an impact on women engaging more in economic activities (other than working in the fields in rural areas). However, in the urban setting women will most likely have more opportunities to be engaged in economic activities and might use them when more time is available. Access to centralised systems under regulations serving every urban dweller will improve the situation of the women of all classes. It will safe them time, help to improve school attendance of the children and reduce the stress in the household which was present when 'chasing water'. SIPs (community and private systems) in the LIAs cannot offer all of these advantages in securing gender equality.[175]

[171] https://unstats.un.org/unsd/demographic/products/Worldswomen/Graphs/Graphs/Graph7.1.pdf (last visited 05.2017).

[172] Graham et al. (2016: 7, 8).

[173] GFA (2016: 35).

[174] Koolwal and van de Walle (2010: 35, 36).

[175] Guaranteeing access to formalized services is important because arbitrary behavior against women and girls can be better controlled and avoided. In addition, it should not be forgotten that resellers of water with a household connection from the utility (becoming informal service providers) sometimes try to prevent utilities from extending their services through water kiosks within their area of influence, which is especially punitive for women in the underserved LIAs. In 2004, the utility in Chingola, Zambia faced a number of angry resellers with household connections and non-served middle income households who wanted to prevent the opening of new water kiosks in their (underserved) area by threatening to break the kiosks down although the MD received a petition from several hundreds of underserved people (mainly women) to urgently put the kiosks into operation. Only when the utility offered, in addition to the kiosks, some household connections in the area the coalition of resellers and medium income households broke apart and the opposition to the new kiosks became manageable. To overcome the resistance by the informal service providers the utility has to make use of local knowledge and can hardly rely on the executive (police) or donors. In this case, like in many others, informality ganged up with middle income groups against the interest of the poor, whereby mainly poor women were deprived of access (own experience).

Actions to Close the Gender Gap Are Ineffective or Counterproductive

Joshi[176] indicates that the complexity of gender (its different categories) and its dependency on a specific situation (local customs and traditions) is often not sufficiently understood. Women in the developing world cannot be considered as a coherent group. Furthermore, experts in the sector often think that an increased involvement of women in water, more sensitisation and training for women and better access to water will help to overcome gender inequalities and guarantee the success of water projects. Joshi questions this. She underlines that participation by women has resulted in 'more work for those with least power' and thereby, improved access to water has not led to a reduction of gender inequality. The additional (voluntary) work women of the poor have to take over in order to obtain access takes time away from what was saved by an improved access to water and sanitation.

She also underlines that effectiveness of women raising their voices declines when it is done through committees and local leaders, often positions hijacked by men or by well off women with individual interests. This reinforces gender inequality and has a negative effect on poverty eradication because the selection of communities for new projects and the design of projects and technology are no longer significantly influenced by the needs of the poor women. According to Joshi, gender mainstreaming will only succeed when 'their [women's] position is changed and that they can meet new challenges and play new roles'. Joshi discusses a case in rural water and her findings, like the proposals of the different UN conferences, can only partly be transferred to the urban setting in the case of community management in the LIAs because urban life changes the social context even when dwellers have a rural background. Hence, conclusions on gender and water which don't distinguishing between rural and urban can be misleading.

Access to Utility Services Reduces Gender Inequality But There Is Still a Long Way Ahead

Analysing the recommendations from the Nairobi conference (1985) and the key learnings (from the Chuni project) outlined by Joshi,[177] it is obvious that certain concerns and findings from the rural areas concerning gender have no or little relevance in towns. The establishment of formalised central systems, which aim to serve everyone as well as their professionalization and regulation, has changed the situation for consumers. This has an influence on the necessary actions to end remaining gender inequalities. Connected consumers no longer need to be involved in the administration of water supply and sewerage services. They don't need to be trained in the maintenance of assets. The relationship with the management of the service provider (utility) does no longer depend on the castes, tribes, gender, economic classes, religion, and etc. The women don't need to have the chief as intermediate.

Such centralised systems serving all urban dwellers offer the same minimum services to the disadvantaged households than it does to the middle and upper

[176] Joshi in Coles and Wallace (2005: 135–151).

[177] Joshi in Coles and Wallace (2005: 142).

classes. Transfer of ownership of land has in general no influence on access to urban water and sanitation services. The cost of water for the socially disadvantaged is in general lower than for the other classes because of cross-subsidisation between consumer groups.[178] Progressive tariff structures are common in the urban sector in most Sub-Saharan countries with some exceptions like in Uganda.[179] Women of all classes are heard like men by the utility when accessing a formalised complaint system.

However, this progress in moving to gender equality is not yet achieved in all of the urban areas, especially not where the utility is half-heartedly engaged. Inequalities or insufficiencies to respond to women's needs persist in the LIAs. The voice of the many underserved people and often also from the users of the utility kiosk are not always heard by the utilities. Many formalised complaint systems do not cover consumers without household connections. Furthermore, the design and placement of shared outlets often do not meet the needs of women and girls. Job opportunities in the water industry are still not the same for women than for men. In countries where neighbourhood sales are accepted by the institutions (Tanzania) or even promoted by the utility (Uganda),[180] the underserved people are discriminated by the price for water and especially women are prone to be exploited by these re-sellers when entering the private property.

Necessary Measures to Close the Water Gender Gap in Towns

Considering the discussion so far, the following measures in the urban water and sanitation would need to be undertaken in order to end gender inequalities and satisfy the specific needs of urban women:

- Replace informal service provision and especially neighbourhood sales by utility services which can be accessed either on public ground (kiosks) or on the plot where people live (yard taps and in-house connections).
- Include women in the consultation when shared facilities for water are designed and placed in order to facilitate the use and reduce the risks for women and children when accessing (e.g. crossing main streets).∗
- Reduce the distance of shared facilities by the households without compromising the economic sustainability of kiosks (30 min cycle).∗
- Ensure women's' participation in the design of household sanitation facilities when sector institutions promote onsite sanitation in order to cater for the specific needs of women and girls.∗

[178] Especially when the standing charges of a household connection includes the real costs of mandating the connections such as meter maintenance, meter reading and billing, etc.

[179] There is no pro-poor block tariff in place which ensures a cross-subsidization from the higher consumers to the poor. There is less incentive to restrict water consumption with a unique tariff than with a rising block tariff system.

[180] The National Water Corporation allowed owners of household connections to register as a public stand post despite the outlet is placed on private ground and the tariff for the poor can no longer be controlled. The consequence is that unrestricted access can no longer be guaranteed and the poor pay a price which is much higher than the middle and high income classes pay for water.

- Fix a quota for women for water kiosks operators of 50%.∗
- Provide equal access (compared to other utility customers) to formalised complaint systems for consumers of shared facilities and underserved people within a service area.
- Provide equal access to job opportunities in the water and sanitation industry.
- Cover as many women as men with sensitisation measures for hygiene and water use.∗
- Disaggregate data in the water sector according to gender.∗
- Elaborate on detailed documents for implementers on gender issues.

The measures proposed and marked with a star (∗) are generally required standards at (national) institutions which carry out upscaling of low-cost technologies in LIA.[181] However, access to water kiosks and particularly the time spent at these outlets by the consumers remain a concern in urban water supply because of the high number of underserved people.

References

Aderinwale A, Ajayi O (2008) The link between poverty and water supply: the Nigerian example. In: Hemson et al (eds) Poverty and water – exploration of the reciprocal relationship CROP (Comparative Research Programme on Poverty). Zen Books, London/New York

Alkire A, Housseini B (2014) Multidimensional poverty in Sub-Saharan Africa: levels and trends. OPHI working paper no. 81, Oxford Poverty & Human Development Initiative (OPHI) Oxford Department of International Development, Queen Elizabeth House (QEH), University of Oxford

Bain R, Cronk R, Wright J, Yang H, Slaymaker T, Bartram J (2014a) Fecal contamination of drinking-water in low- and middle-income countries: a systematic review and meta-analysis. PLOS Med 11(5):e1001644. pp 1–18

Bain R, Cronk R, Hossain R, Bonjour S, Onda K, Wright J, Yang H, Slaymaker T, Hunter P, Pruess-Ustuen A, Bartram J (2014b) Global assessment of exposure to faecal contamination through drinking water based on a systematic review. Trop Med Intl Health 19(8):917–927

Baker JL (2008) Urban poverty: a global view. The World Bank Group, Urban Papers, UP-5, Available: http://www.worldbank.org/urban/. Last visited 05.2015

Baldwin R, Cave M, Lodge M (2012) Understanding regulation – theory, strategy, and practice. Oxford University Press, Oxford

Bohman A (2010) Framing the water and sanitation challenge – a history of urban water supply and sanitation in Ghana 1990–2005. Umea University, Doctoral dissertation in economic history, 2010, ISSN: 0347-254-X

Chakava Y, Franceys R, Parker A (2014) Private boreholes for Nairobi's urban poor: the stop-gap or the solution. Habitat Int 43:108–116

Clasen T, Boisson S, Routray P, Torontel B, Bell M, Cumming O, Ensink J, Freeman M, Jenkins M, Odagiri M, Ray S, Sinha A, Suar M, Schmidt WP (2014) Effectiveness of a rural sanitation programme on diarrhoea, soil-transmitted helminth infection, and child malnutrition in Odisha, India: a cluster-randomised trial. Lancet Global Health 2(11):645–653

Coles A, Wallace T (eds) (2005) *Gender,* water and development. Berg, Oxford/New York

[181] E.g. the WSTF in Kenya and the DTF in Zambia.

Devoto F, Duflo E, Dupas P, Pariente W, Pons V (2012) Happiness on tap: piped water adoption in urban Morocco. Am Econ J Econ Policy 4(4):68–99. MIT Open Access Articles, American Economic Association

Esrey SA, Potash JB, Roberts L, Shiff C (1991) Effects of improved water supply and sanitation on the ascariasis, diarrhoea, dracunculiasis, hookworms infection, schistosomiasis, and trachoma. Bull World Health Org 69(5):609–621

Europe-economics (2003) Scope for efficiency improvements in the water and sewerage industries. Office of Water Services, London

Exner M (2015) Regelungen zur Sicherstellung der Trinkwasserqualität in Deutschland. Lauterbach Germany, presentation at MATA/GIZ (Mitarbeiter Tagung der GIZ), July 2015

Fankhauser S, Tepic S (2005) 'Can poor consumers pay for energy and water? An affordability analysis for transitional countries' European Bank for Reconstruction and Development, Working paper no. 92

GFA (2016) Development of the water and sanitation sector – evaluation of water kiosks and water quality in the urban low income areas in Kenya, Deutsche Gesellschaft für Internationale Zusammenarbeit (GIZ) GmbH, first and second study. GFA, Hamburg

Graevingholt J, Leininger J von Haldenwang C (2012) Effective statebuilding? A review of evalutation of international statebuildings support in fragile contexts, DANIDA, (International Development Cooperation) Ministry of Foreign Affairs of Denmark

Graham, J., Hirai, M., Kim, SS. (2016) An analysis of water collection labor among women and children in 24 sub-Saharan African countries. PLOS ONE, doi: 10.137 Journal

GWOPA, UN-Habitat, INFRA Nairobi (2010) Access to water in Nairobi: mapping the inequalities beyond the statistics, Available: http://access-to-water-in-nairobi.gwopa.org/, 2013, UN Nairobi. Last visited 04.2015

Hemson D, Kulindwa K, Lein H, Mascarenhas A (2008) Poverty and water – explorations of the reciprocal relationship. Zed Books, London

Howard G, Bartram J (2003) Domestic water quantity, service level and health. WHO/SDE/WSH/03.02, World Health Organisation, Available: http://www.who.int/water_sanitation_health/diseases/WSH03.02.pdf. Last visited 04.2015

JMP-Joint Monitoring Programme on MDG (2014) Progress on drinking water and sanitation, WHO and UNICEF, Available: http://www.wssinfo.org/fileadmin/user_upload/resources/JMP_report_2014_webEng.pdf. Last visited 04.2014

JMP (2017 – WHO and Unicef) 'Progress on drinking water, Sanitation and Hygiene. https://www.who.int/mediacentre/news/releases/2017/launch-version-report-jmp-water-sanitation-hygiene.pdf

Jomo KS (2016) Public-private partnership and the 2030 agenda for sustainable development: fit for purpose? Department of Economic & Social Affairs, DESA Working Paper No. 148, February, ST/ESA/2016/DWP/148

Joshi D (2005) Misunderstanding gender in water: addressing or reproducing exclusion. In: Coles A, Wallace T (eds) Gender, water and development. Berg, Oxford

Karikari BN (2013) Physico-chemical and bacteriological assessment of selected boreholes and hand-dug wells in new Edubiase, Ashanti Region, Master Thesis for Kwame Nkrumah University of Science and Technology, Available: http://datad.aau.org/handle/123456789/1/browse?value=Nkrumah+Karikari%2C+Bernard&type=author. Last visited 4.2015

Kayaga S, Kadimba-Mwanamwambwa C (2006) Bridging Zambia's water service gap: NGO/community partnerships. Proc ICE Water Manag 159(3):155–160. Loughborogh University Institutional Repository

Keener S, Luengo M, Banerjee S (2010) Provision of water to the poor in Africa, experience with water standposts and the informal water sector, WSP 5387, Public Research Working Paper, Africa Region Sustainable Development Division, The World Bank

Koch, R. (1893) Waserfiltration und Cholera (Aus dem Institut für Infektionskrankheiten), Zeitschrift für Hygiene und Infektionskrankheiten, 1893, Bd XIV

Kolker J, Kingdom B, Tremolet S (2016) 'Financing Options for the 2030 Water Agenda' Water Global Practice, Knowledge Brief

Koolwal G, van de Walle D (2010) Access to water, women's work and child outcomes. The World Bank, May, Policy Research Working Paper, 5302, Washington, DC

Kramer K (1997) Das private Hausbad 1850-1950 und die Entwicklung des Sanitärhandwerks, Schiltach, Hansgrohe Öffentlichkeitsarbeit

Krishna A (2004) Escaping poverty and becoming poor: who gains, who loses, and why? World Dev 32(1):121–136. Elsevier Ltd. 2003

Kuks S, Kissling I (eds) (2004) The evolution of National Water Regimes in Europe – transition in water rights and water policies. Kluwer Academic Publishers, Dordrecht. ISBN 1-4020-2483-5

Mboya WO (2015) Customer identification survey pilot report, Kenya, KEWASCO CIS Pilot Report, July, Kericho Water and Sanitation Company Ltd (KEWASCO)/GIZ Deutsche Gesellschaft für Internationale Zusammenarbeit (GIZ) GmbH

Megginson W, Netter J (2001) From state to market: a survey of empirical studies on privatization. J Econ Lit XXXIX:321–389

Megginson W, Nash R, Netter J, Poulsen A (2001) The choice of private versus public capital markets: evidence from privatization, October 2, Current draft

Mosley P (1987) Foreign aid – its defence and reforms. University Press of Kentucky, Lexington

Mutono S, Kleemeier E, Nkengne C, Tumusiime F (2015) Water and sanitation for the poor and bottom 40% in Uganda: a review of strategy and practice since 2006, (Technical Assistance funded by) Water and Sanitation Program – Africa, World Bank Global Water Practice, World Bank Group

Newbery D (2004) Privatising network industries. CESifo Working Paper No. 1132, February, Category 9: Industrial Organisation

Nickson A, Franceys R (2003) Tapping the market, the challenge of institutional reform in the urban water sector. England: Palgrave MacMillan Distribution Ltd, Houndmills, Basingstoke, Hampshire RG21 6XS

Nilsson D (2006) Water for a few – a history of urban water and sanitation in East Africa. Licentiate Thesis in History of Technology, Stockholm papers in the history and philosophy of technology, pp. 35–39

Nilsson D (2010) A Paradigm of Pipes in a Society of Slums: Techno-political regime dynamics in Kenya's water sector. Paper presented at IWH conference, 16–19 June, Delft

Nilsson D (2013) Prisoners of a paradigm? What can water sector donors learn from history? In: Katko T, Juuti P, Schwartz K (eds) Water service management and governance – lessons for a sustainable future. IWA Publishing, London

OECD (2008) Service delivery in fragile situations – key concepts, findings and lessons. J Develop OECD/DAC, Discussion paper:9(3)

Oelmann M (2005) Zur Neuausrichtung der Preis- und Qualitätsregulierung in der deutschen Wasserwirtschaft, vol 2005. Kölner Wissenschaftsverlag, Köln

Olsen HB (2017) Why escaping poverty isn't nearly as easy as people think. https://everydayfeminism.com/2017/01/escaping-poverty-is-not-easy/

Page B (2005) Naked power: 'women and the social production of water in Anglophone Cameroon'. In: Coles A, Wallace T (eds) Gender, water and development. Berg, Oxford/New York

Penrose K, Caldas de Castro M, Werema J, Ryan ET (2010) Informal urban settlements and cholera risk in Dar es Salam, Tanzania. PLOS Neg Trop Des 4(3):e 631. pp. 1–11

Pollem O (2008) Regulierungsbehörden für den Wassersektor in Low-Income Countries – Eine vergleichende Untersuchung der Regulierungsbehörden in Ghana, Sambia, Mosambik und Mali. Dissertation, Carl von Ossietzky Universität Oldenburg, Fakultät II, 2008

Rouse M (2013) Insitutional governance of regulation of water services – the essential elements. IWA Publishing, London

Sabo A, Adamu H, Yuguda AU (2013) Assessment of wash-borehole water quality in Gombe Metropolis, Gombe State, Nigeria. J Environ Earth Sci 3 (1), Available: http://www.iiste.org/Journals/index.php/JEES/article/view/4007. Last visited 06.2016

Sasaki S, Suzuki H, Igarashi K, Tambatamba B, Mulenga P (2008) Spatial analysis of risk facotor of cholera outbreak for 2003–2004 in the peri-urban areas of Lusaka, Zambia. Am Soc Trop Med Hyg 79(3):414–421

Schiffler M (2015) Water, politics and money a reality check on privatisation. Springer, Cham

Schulz N, Schulz L (1977) Die limnologische Entwicklung des Ossiacher Sees (Kärnten, Österreich) seit 1931. *Carinthia II*, 167./87. Jahrgang, pp 1577–178, Klagenfurt, Naturwissenschaftlicher Verein für Kärnten, Austria, Aus dem Kärntner Institut für Seenforschung

Spears D, Ghosh A, Cumming O (2013, September) Open defecation and childhood stunting in India: an ecological analysis of new data from 112 districts. PLoS One 8(9): 1–9. e73784

Stern E, Altinger L, Feinstein O, Maranon M, Ruegenberg D, Schulz NS, Stehen Nielsen N (2008) The Paris declaration, aid effectiveness and development effectiveness – thematic study. Ministry of Foreign Affairs of Denmark. Available: https://www.oecd.org/dac/evaluation/dcdndep/41807824.pdf. Last visited 04.2017

Uhuo CA, Uneke BI, Okereke CN, Nwele DE, Ogbanshi ME (2014) The bacteriological survey of borehole waters in Peri-Urban areas of Abakaliki: Ebonyi State, Nigeria. Intl J Bacteriol Res 2(2):28–031. Available: www.internationalscholarsjournals.org (last visited 04.2015)

Van den Berg C (1997) Water privatization and regulation in England and Wales. The World Bank Group, Viewpoint, note no. 115, Available: https://openknowledge.worldbank.org/handle/10986/9427. Last visited 04.2015

Vuorinen HS, Juuti PS, Katko TS (2007) History of water and health from ancient civilisations to modern time. Water Sci Technol Water Supp 7(1):49–57

Werchota R (2013) The growing urban crises in Africa. Eschborn, Deutsche Gesellschaft für Internationale Zusammenarbeit (GIZ), Germany

World Bank Group (2016) Poverty and shared prosperity 2016 – taking on inequality. World Bank, Washington, DC

Chapter 3
Beyond the Usual Debate

Abstract W+S sector reforms often need to repair the damage of ill-conceived decentralization. Urban W+S specialists seem to have less influence in sanitation development strategies than public health officers. Unnecessary competition hinders the development of an enabling sanitation framework in the water sector and leads to contradicting definitions of sustainable access to safe W+S. Some of these adopted on the highest level breach in the urban setting minimum requirements of human rights. The shift from piped (utility) services to improved sources in towns is hailed as a success although more people have to consume contaminated water. Success is fictitious and two myths are busted: Community systems are an option for urban low-income areas and the hundreds of thousands small scale informal service providers can be regulated. A formalization of services through the involvement of utilities, a pillar of modern reforms, is the appropriate move to end the 'urban water and sanitation divide' which discriminates the poor. To up-scale adequate access for the urban poor it is necessary to distinguish between first and last mile development and operation. Because this is unknown to so many donors and national decision makers, access in the low-income areas are unsatisfying in most of the low-income countries.

This chapter debates topics, which received so far no or insufficient attention in the discussions. Thus, the elaboration on these issues can be considered as a specific contribution to the urban water and sanitation sector in the low-income countries. The chapter is organised in five sections and is winding down with a summary of the pertinent issues which contains the main messages from the Chaps. 1, 2 and 3.

R. Werchota, *Empty Buckets and Overflowing Pits*, Springer Water,
https://doi.org/10.1007/978-3-030-31383-8_3

3.1 Decentralisation and the Health Sector: Partly Cause of the W+S Crises

In Sub-Saharan Africa, decentralisation was either initiated as a national programme and thereby, concerned a number of sectors, or effected only the water sector. The latter often happened when central government institutions could no longer stop the decline in water and sanitation, even when providing substantial subsidies. The state decided to shift water and sanitation service provision to the municipalities. However, most countries in Africa have undertaken efforts to decentralise political power and to move service provision to the lowest, the so-called 'most appropriate level', which included the water and sanitation sector.

The Negative Effects of Decentralisation

The aim of decentralising water and sanitation service provision was to make the provider more accessible for customers and reduce the reaction time to technical and managerial problems. It was supposed to improve customer orientation, the maintenance of infrastructure and the overall performance of the sector. Decentralisation was also supposed to make service provision benefit from local knowledge and to direct investments to areas where the benefit for the community is highest. Thereby, services would be better aligned to demand. Another possible expectation of decentralisation was to curb the counterproductive influence of politicians from national level on water and sanitation development and thereby, improve governance. This is the theory behind decentralisation. However, reality provides another picture. Urban water sector development in countries in Sub-Saharan Africa, already in a bad shape before decentralisation, has further deteriorated in most of the cases after moving asset holding/development and operation of water and sanitation from national level to local administrations. There are several reasons for this:

- Decentralisation of the urban water and sanitation systems was often accompanied by a paradigm change. While central government structures in the past agreed that water supply and sanitation has to be subsidised, many local government structures (e.g. municipalities) regarded water supply and sanitation service provision as a 'cash cow'. With decentralisation of water and sanitation service provision, municipalities received an income source with a regular flow of cash, which often outstripped all other available income sources. Being generally short of income, the temptation to use the cash from water and sanitation income for other spending than water and sanitation is therefore very high despite the growing number of underserved people. As one governor in Kenya said: 'I need to control water and sanitation services because I have not enough funds for the health sector.'[1] Consequently, ring fencing of income from water and sanitation services for the urban sector would have been needed through national regulation for instance.

[1] Kenya 2015, own experience.

- The devolution of responsibilities to the local level meant for many central structures to retreat from its responsibilities in water and sanitation altogether. Knowledge in the sector was lost with the disengagement of central structures and the already limited collection of data on the national level was even more neglected. The national level lost largely the badly needed oversight of the sector.
- Decentralisation in the water sector was not accompanied by a simultaneous strengthening of supervision. The result was that compliance to already insufficient national standards eroded further. Service providers at a local level started to adopt their own standards. They defined, for instance, their service areas on their own. They limited it either to the areas benefiting from piped networks (as is still the case in Tanzania with a number of utilities for instance) or to the administrative borders of the municipalities. This meant that with decentralisation, the few existing national standards were gradually replaced by a number of local standards often to the disadvantage of consumers.
- Handing over the systems to different municipalities left the sector with many unviable providers. It created many small units which also made it more difficult to mobilise funds from development partners in an already chronically underfunded sector. Central governments no longer felt obliged to support municipalities in mobilising funds for investments. By getting water and sanitation off their shoulders, the already limited awareness of the importance of water and sanitation degraded further at national level. In addition, some municipalities advised their water departments not to invoice municipality institutions and municipal employees arguing that there are parts of the owners of the water services structure.[2]
- Sometimes, decentralisation followed the lines separating ethnic groups. Tribal interests were carried into the water sector on local level, which sometimes negatively influenced performance of public providers. Being from the same tribe was more important than the qualification for the jobs.[3]
- Municipalities seem to be more inclined to keep water tariffs low and use water and sanitation service provision as a vote catcher than national structures and many used the sector to start a political career. Municipalities fixed water tariffs without the involvement of professionals. Thus, insufficient tariffs, the drainage of income of the sector and the absence of subsidies resulted in the underfunding of the operations. Expert interviews confirmed that with municipalities in the driving seat service provision declined faster than under national government.
- Sometimes, politicians from the opposition at the local level used water and sanitation service provision to challenge the local power. Water and sanitation ser-

[2] Own experience in Zambia in 1999.

[3] Own observations in Kenya with devolution of water and sanitation to the counties according to the constitution 2010.

vice provision became the object of political disputes to the detriment of the underserved people as was explained by one expert in Zambia.[4]

The claim that services can be best provided on a community level does not necessarily apply to urban water and sanitation because of its dependence on costly infrastructure and the reliance on national structures for fund mobilisation, professional management and the need for strict enforcement of (national) standards. This makes water and sanitation development different from other services to be provided on a community level. The intended advantages small-scale local/community service providers are supposed to offer are often not materialising. Being closer to the consumers does not necessarily mean that the services are better[5] and if customer attendance is better it is not certain that this is a compensation for the disadvantages and risks the consumers suffer from lack of national control, economy of scale and the difficulties to mobilise funds for investments.

There are many examples of devastating deterioration of service provision after decentralisation. One is mentioned for Zambia in the 1990s (see Sect. 3.3, Luapula province). The disastrous experience of total breakdown of the formal water supply systems did later not stop local government to fight reforms in Zambia which wanted to achieve sustainability of service provision with the transfer of municipality departments to professional utilities (corporatisation) operating under national regulation. However, it is not the idea of decentralising the water sector which is to be questioned but the way decentralisation is planned and carried out. Rouse[6] indicates this for the aspect of financing:

> One major obstacle to success in decentralisation occurs when the financial aspect have not been included. In other words, decentralisation of responsibility with centralised constrains on finance and cost recovery will not work.

There are also other aspects which have to be considered with decentralisation efforts such as a minimum size of utilities, national regulation and oversight, etc. and the fact that as long as the sector and central government are not in a position to deliver the funds for asset development and donors have to provide the lion share for it, central government has to play a significant role in fund mobilisation.

---

[4] 'It was during the MMD (Chiluba) rule in 1994 and Kaunda's UNIP party was no longer in power but had strong party structures in peri-urban areas. We did not know that the people we were working with belonged to the former ruling party UNIP when we were introducing the first water kiosks. These people were political leaders (Kaunda Branch Chairmen) and their interest was to manage the kiosks. Secondly before the introduction of water kiosks in Chipata water was free and with the kiosks water was to be paid for. The political leaders' interest or money they collected was for their benefit (personal benefits) as most of them are not employed. The ruling party MMD in Chipata complained to State House and the donor was asked to explain why they were working with the opposition and not the ruling party' (expert interview, Zambia, 10.2015).

[5] According to Pauschert et al. (2012: 20): 'ISPs adapt their services and service level to the needs of their customers such as individual paying procedures, flexible supply hours, delivery to the doorstep, etc.' and 'Measured by availability and customer service orientation. Many ISPs perform better than public utility'. However, there is no evidence about these statements in the report.

[6] Rouse (2013: 6).

The Uphill Battle to Repair the Damage of Decentralisation

Reforms of the water sector frequently have to make provisions to overcome resistance in order to safeguard what is left by the damaged systems. Municipalities fight to keep the income from water bills and their influence on the management of water and sanitation services despite the downward spiral. They also fear that rising tariffs for water and sanitation will damage their political interests. To many local politicians safeguarding their own interests seemed more important than the interest of the many unserved people and the consumers receiving questionable services. This situation is aggravated when the fight is fuelled from national level. In 2000 in Zambia for instance, the Ministry of Local Government and Housing (MoLGH), holding the responsibility for water and sanitation service provision, promised more subsidies to municipalities for their water departments if they decide not to transfer their water department to the newly formed Commercial Utilities (CUs) proposed by the reforms.[7]

The fact that socially responsible commercialisation was part of the national policy did not matter to the MoLGH. It also ignored other key principles of the national framework like O+M cost coverage by consumers, for instance. The Ministry of Energy and Water, which was championing the water sector reform, could not force municipalities to move service provision to CUs. Only when the MoLGH could, in the further development, no longer provide sufficient subsidies to the resisting municipalities and cholera cases emerged in towns all over the country, the number of municipalities which joined the provincial CUs gradually increased.[8] Most of the systems were handed over by the municipalities to CUs when they were at the brink of collapse and many people especially the poor had already been pushed back to informal providers or traditional water sources. Another result of decentralisation was that more funds for premature rehabilitation of infrastructure were needed especially in the 1980s and 1990s.

In the efforts to repair the damage of decentralisation and create clustered entities, the municipalities in Zambia used the opportunity of transferring personnel other than water and sanitation when handing over the system to the CUs. In the Copper Belt among the personnel transferred were journalists, nurses, etc. which were previously not engaged in the department of water affairs but in other areas of the municipalities. The overstaffing of the newly formed CUs was breath-taking. In addition to this, the CUs often did not receive the offices and operation buildings as well as transport equipment from the out phasing municipal water department. Commencing operations under such difficult conditions was a big handicap for utilities. It is astonishing how swiftly the utilities recuperated from such an unfavourable situation when professionals for top and middle management positions

[7] The expression commercial (oriented) utility refers to public utilities, which have obtained a certain autonomy from national or local administrations (civil service) and have the potential to follow a policy of cost recovery.

[8] E.g. the number of towns under CUs started with 55 in 2005/6 and increased gradually to 72 in 2008/9, to 76 in 2011/12 and to 90 in 2014 (Sector performance reports of NWASCO, Zambia).

were employed from the open labour market.[9] Other such cases were reported in Kenya where the municipality sold property reserved for water and sanitation development before handing over the remaining assets to the newly created regional asset holders (plots designated for treatment plants and storage areas for instance).[10] These disastrous effects of decentralisation spearheaded by decision makers who seemingly had no idea about urban water and sanitation development can be observed in several countries in Sub-Saharan Africa.[11]

It should not be forgotten that such decentralisation efforts have been strongly promoted and supported by the international community. Hence, governments of partner countries should have been informed about the risks of decentralisation in the urban water and sanitation sector. It indicates how quality issues also plague international cooperation not only observed with decentralisation but also with the push for PSP. Lately it seems to repeats itself with the 'innovative' ideas to finance the sector (e.g. donor driven blended financing) and making a business case out of service provision to the poor (NGOs and philanthropic organisations). Donors should have helped partner countries to elaborate a strategy and create an appropriate framework for decentralisation. This would have avoided the dismembering or disappearance of vital structures in the sector. Hence, there was a shared responsibility of causing the damage and therefore, donors should take this as lesson learned to pay more attention in future when promoting certain approaches in the sector in partner countries.

The Questionable Role of the MoH as Champion of Sanitation

There are also other ministries opposing (intentionally or unintentionally) activities being part of urban water and sanitation reforms, which aim to improve access to services. In Burkina Faso, for instance, the MoH for years strongly opposed the engagement of the water utility (ONEA) in the development of onsite sanitation such as decentralised sludge management and promotion of household, school and public sanitation facilities. The MoH considered this as a threat to their own fund raising in the health sector for sanitation and an opposition to their policy to concentrate almost solely on hygiene education and marketing.[12] In Kenya, at the beginning of the water sector reform, the MoH even demanded that the budget for sanitation allocated to the Ministry of Water and Irrigation (MWI) for infrastructure development be transferred to its own budget.[13] This move by the MoHs to prevent

[9] Own experience in 2000 and 2001.

[10] Own experience in 2005 and 2006.

[11] It should be noted that in the meantime Zambia has transferred the responsibilities of water and sanitation service provision on national level from the MoLGH to a newly established ministry for water and sanitation.

[12] The national sanitation policy was later adopted by an inter-ministerial structure and not only by the Ministry of Health. This defused the confrontation between the health and the water sector. Own observations in 1990–1991.

[13] Own experience in 2006.

that the water sector receives funds for the development of the sanitation (infrastructure) is still ongoing.[14]

There is common understanding that education and marketing for sanitation is important. Considering the many household installation and small-scale water and sanitation projects promoted with subsidies, which failed because people and institutions did not use or operate facilities sustainably, it is perceivable that there is a strong focus on education and marketing. Therefore, many stakeholders are tempted to claim that money does not matter in water and sanitation development and that the root problem is education, sensitisation and acceptability. However, this is a very narrow view because, after all, access largely depends on first mile infrastructure and this needs money - a huge amount of money. Furthermore, support to the non-served poor in form of subsidies and infrastructure development cannot be ignored when progress in access to safe sanitation should be achieved in medium terms. It seems that often this balance of 'soft and hardware' development at the interface of infrastructure and consumers is ignored or not carefully chosen when it comes to national sanitation policies.[15]

The national sanitation policies from the health sector are generally based on the following key assumption: Hygiene education and sanitation marketing are sufficient to convince households to build toilets with their own means according to minimum standards.[16] However, many poor (and not only the poorest of the poor) do not have the means to build their own safe toilet. Hence, the expectations based on these assumptions have not materialised despite decades of sanitation education and marketing.

Analysing the national sanitation policies in several countries, numerous challenges can be identified:

- The MoH is mainly focusing on environmental health and household toilets and thereby, leaving out concerns about a sanitation chain, especially for onsite facilities. This has dire consequences for the living environment especially in the slums. It is difficult to understand how decision makers in sanitation can recognise that the majority of people depend on onsite sanitation (e.g. 90% according to the sanitation policy of 2007 in Kenya) and at the same time ignore the need for the development of a sanitation chain in the urbane setting.

---

[14] In addition, the new National Sanitation Policy in Kenya (2016–2030: 97) carries this dispute about funds forwards in demanding that the funds provided to the Water Services Trust Fund (a water sector institution promoting onsite sanitation successfully since many years through the utilities) shall be transferred to a Trust Fund to be created under the MoH: 'This policy has however, proposed the offloading of the sanitation financing function of the WSTF to the National Sanitation Fund (NASF)'.

[15] E.g. National Sanitation Policy 2007 of Kenya.

[16] E.g. the Kenyan sanitation policy of 2016 issued by the MoH indicates that the previous policy (2007, chapter 5.9.) discouraged sanitation subsidies. The orientation of the new policy has little changed in this regards of subsidies for household sanitation facilities and proposes only limited subsidies ('kept to a minimum') for the households living in extreme poverty.

- As for water, objectives are usually unrealistic and set by the health sector for activities to be carried out by other sectors, which it does not control (e.g. the sanitation policy of 2007 in Kenya aimed to reach 100% sanitation coverage of households and schools by 2015).
- The involvement of established (professional) water and sanitation utilities in onsite sanitation is not foreseen although public utilities are often named sanitation utilities (e.g. Kenya, Naivasha Water, Sewage and Sanitation Company).
- There is no indication that there are different realities in the urban and rural settings concerning sanitation.

There seems to be a strong discomfort within the health sector when the water sector becomes involved in onsite sanitation through their professionals (utilities) despite very successful cases (e.g. Burkina Faso[17] and recently Kenya). This discomfort is widespread in Sub-Saharan Africa and appears to be shared by institutions of the development cooperation.[18] Instead of regarding the activities of the water sector in sanitation as complementary to the health sector they are seen as a competition. In addition, it is very unlikely that the municipalities, especially the medium and small ones, can develop and operate assets for sanitation. Like for water, the development of urban sanitation infrastructure needs the involvement of professionals (utilities and regulators). The state administration (on central and local levels) is not the best suited to provide (urban) sanitation services. This lesson learned from water still has to trickle down to sanitation.

After all, the key drivers of providing access to water and sanitation are the line ministries and their institutions in the water and sanitation sector. They are responsible for the development of infrastructure, overseeing service provision and reporting on the progress in the fulfilment of water and sanitation rights and the water and sanitation SDGs. However, international institutions dominant in the water discourse at global level do not cooperate much with these line ministries but with the MoHs in the low-income countries. This might be one of the reasons why specialists for urban water and sanitation are not sufficiently heard. Also Hemson[19] provides an indication about this unnecessary competition in water and sanitation which also takes place within the UN system between the WHO [*and Unicef*] on one hand and UNDP [*and UN HABITAT*][20] on the other:

> 'This made explicit what was an evident tension that manifested itself to some extent in the debate over transferring water delivery out of the hands of public health engineers…'

It would be helpful to include in sector reforms a review of the roles of each stakeholder in sanitation and define a separation line between the sectors regardless

[17] ONEA in Burkina Faso promoted 10,000 to 15,000 urban sanitation facilities annually for almost 2 decades (ONEA annual reports, e.g. 2001: 17).

[18] E.g. the sanitation policy 2007 in Kenya provides special acknowledgements to the WHO, Unicef and WSP/WB for their contributions in its elaboration.

[19] Hemson et al. (2008: 25).

[20] Own observation in Kenya.

of the influence the global discourse might have today in the low-income countries.[21] Thereby, the difference has to be made between environmental health and sanitation sensitisation on one hand and the infrastructure development and service provision for sanitation on the other hand. Equally, the differences of urban and rural sanitation have to be recognised.

3.2 Translating Safe W+S into Safe Services

Water as a vehicle for the transmission of (waterborne) diseases such as cholera[22] was first made public by John Snow and others around 1849 according to Smith.[23] Snow also demonstrated that traditional raw water sources in the densely populated urban setting could no longer be assumed safe for drinking and had better be closed down for the use by households. Snow established the link between water and waterborne diseases through statistics. Only around three decades later further evidence[24] emerged which supported the understanding that drinking water needs to comply with standards.

Consequently, guidelines formulating basic requirements for drinking water were put in place in many countries, following what Grabow et al.[25] recommended '…that drinking water should rarely if ever contain total coliform bacteria, and never faecal coliforms of E. coli.' The World Health Organisation (WHO) lifted such standards for safe drinking water onto the global level, which is presently accepted worldwide with little or no deviation. JMP of the United Nation (UN) defines:

> Drinking water is water used for domestic purposes, drinking, cooking and personal hygiene' and 'Safe drinking water is water with microbial, chemical and physical characteristics that meet WHO guidelines or national standards on drinking water quality.[26]

That the UN was setting quality standards on a global level for safe water can be considered a big step forward. Safe water is therefore no longer a perception, which might vary from one person to another. With the international recognition of global standards for water quality, which is also enshrined in the rights to water and sanitation, no country today can accept much longer that people consume water from unsafe and high-risk water sources for drinking, cooking, dish washing, and etc. However, setting quality standards alone is meaningless for consumers if they are

[21] As it was done in Kenya with the executive order number 2, 2013 and previously in Burkina Faso with the sanitation policy.

[22] In this case cholera.

[23] Smith (2002: 921).

[24] Start of the science of water bacteriology with publications by von Fritsch in 1880 and von Theodor Escherich in 1886 according to Kabler et al. (1964: 58).

[25] Grabow et al. (1996: 199).

[26] http://www.who.int/water_sanitation_health/mdg1/en/ (last visited 05.2015).

not enforced. Such an enforcement mechanism must be able to control regularly water quality at any raw water sources which are used and at different points at the supply chain.

Proposed Definition for Access to Safe Water

It is not difficult to understand that an enforcement mechanism for standards will have to look different when single household are encouraged to use traditional water sources and ensure water quality or when one supplier selling water to many households is made responsible to deliver drinking water. Authorities will have enormous difficulties to oblige the hundreds of thousands of urban households to monitor the raw water quality at the sources their use and document that they adequately treat the water if needed. It can be assumed that even with large scale education programs governments will not be able in this way to achieve safe access for the majority of people. Looking at the second option (utilities as service providers), enforcement can concentrate on one institution in a town which has to control regularly water quality along the chain of production and distribution - from the water source to the delivery point. However, even such a lean control mechanism will need the involvement of professionals using adequate equipment and, when the amount of water produced is substantial, a rapid response in cases of failure to avoid harm to the health of the many people concerned. Taking enforcement and the need for professional quality insurance into consideration, the following definition for access to safely managed water for the urban setting is proposed:

> People have access to safe water when water meets WHO or equivalent national standards and there is a control system in place, which guaranties regular water testing[27] at collection sites[28] and the supply chain and if needed regular water treatment by qualified personnel.

This definition follows the lessons learned from history which have now global relevance. It corresponds to the way drinking water is supplied in the industrialised world and it is in line with the aspiration of low-income countries to formalise service provision with reforms. The requirements of water treatment to be carried out by professionals, supports also the argument for centralised supply systems (utilities). In contrast to the use of thousands of wells dispersed all over the town and situated in the yards of the numerous households, utilities use a limited numbers of water sources generally situated outside towns. This makes the strict control of water quality feasible.[29] The proposed definition is different from the definition used in the global discourse on safely managed water because it excludes many of what JMP is describing as 'improved' sources. Furthermore, the notion of safely managed as defined by JMP for the different types and use of water sources is too brought to be used for the urban setting.

[27] The method of measuring residual chlorine (free chlorine 0.2–0.3 mg per liter for instance) can partly replace bacteriological testing (colony counts, presence of Escherichia coli and coliform organisms) which is more elaborate to carry out.

[28] Water sources used by provider and distribution outlets used by consumers.

[29] Koch's remarks in 1893 indicated already that the state shall be responsible to control/ensure the safety of drinking water, refer to Exner (2015).

Nevertheless, some experts underline the need for tolerable exceptions for raw water use in the urban setting as temporary solutions. They argue that water for human uses other than drinking and cooking can be extracted from the traditional water points, such as clothes washing, house cleaning and gardening. The case of Agnes in Nairobi proves them wrong. Water from 'improved' sources can only complement drinking water in very specific cases and for much reduced uses. The households of the poorest of the poor might be forced to do so because they have difficulties paying for more than the necessary 20lt per person, per day, which is the minimum quantity considered for drinking and basic hygiene practices, according to WHO. This can also apply to households where a single person cannot carry a large amount of safe water from public outlets for many household members daily.[30]

Nevertheless, using traditional water sources in towns in addition to quality controlled piped water needs the strict separation of water from different sources in the household, which is a big challenge. The use of unsafe or high risk water sources by households is often imposed onto the households of the poor since the need for investment for the development of new water resources has long been neglected and utilities are forced to implement rationing programs when facing rising demand. If the state accepts the use of traditional water sources in the urban setting to complement the use of utility water, it has the responsibility to sensitise households on the risks. A laissez-fair attitude or a recommendation to use such 'improved' sources by authorities can be considered irresponsible.

Proposed Definition for Access to Safe Sanitation

Concerning sanitation, safety is an even bigger concern in urban settings than water as many of the health experts underline, e.g. Esrey.[31] As explained, this is less of a worry in the rural setting as Clasen et al. indicate.[32] Safety in sanitation starts with the design and use of toilets in order to avoid contact with human waste. An appropriate design also helps to encourage the use of the facility and thereby, reduces open defecation. Foul odours and places for flies and mosquitoes to thrive must be avoided. A toilet with a pit or septic tank in the urban setting must have a provision to be emptied without great health hazards and personnel involved in the sanitation chain must use protective equipment.

Sanitation facilities should not be built in areas where the soil can collapse or which are prone to inundation. The use of excess water or less compostable materials for anal cleansing should be avoided because they affects the decomposition rate of human excreta according to Tilley et al.[33] Hence, acceptability by the users and adequate operation is as important as safety for the individuals. For society it is

[30] For example, a family with five members (two parents and three children), a family member (often woman or child) would have to carry home 100 kg of water daily from a water kiosk.

[31] Esrey et al. (1991: 609).

[32] Clasen et al. (2014: 645), the improvement of coverage with toilets in the rural setting has no impact on health measured according to child mortality and 7-day diarrhoea in children younger than 5 years.

[33] Tilley et al. (2014: 21).

important to know what happens with the sludge after the human waste has been stored in the pit (on the plot). Hence, in the urban setting, safe sanitation must include a safely managed sanitation chain which needs to include the emptying of the provisional storage place as well as the transport and treatment of the sludge.

According to the WHO,

> Sanitation generally refers to the provision of facilities and services for the safe disposal of human urine and faeces. Safe sanitation can be defined as storage, transport treatment and disposal of human waste which does not harm humans and the environment.[34]

The WHO[35] has set global standards and countries, such as the USA,[36] have set national standards for the safe discharge into the environment and reuse of effluent. Access to safe sanitation as defined below is linked to urban water service provision and follows the abovementioned WHO definition. However, as argued for drinking water, it is not enough to set standards. There is need to establish a control system with which such standards can be enforced and to recognise that handling and treatment of sewerage has to be carried out by professionals within a formalised system in order to protect the living environment and the people involved in service provision.

Therefore, the present work proposes the following definition for access to safely managed sanitation for the urban setting:

> Access to safe sanitation means the availability of sanitation facilities and services with reduced risks of infecting users and people providing the services and where transport and treatment of human waste is secured and controlled by trained personnel, ensuring effluent discharge according to WHO requirements or equivalent national standards.[37]

In the industrialised world, coverage by (controlled) sewer systems is as good as universal, at least in the urban setting. This is not the case in the low-income countries. Only a small part and since decades a dwindling percentage of urban dwellers, and almost no one in rural areas, is connected to a sewer network. Therefore, the safety of sanitation involving the disposal of sludge has to depend increasingly on exhauster services (truck or smaller equipment) and manual emptier in areas where the trucks cannot reach the household pit or the poor cannot pay for such service. Both should be operated under a system of control in order to be considered safely managed. In Sub-Saharan countries human waste extracted from onsite sanitation is often emptied manually by informal ‘emptiers’, and is usually dumped illegally.[38]

---

[34] http://www.who.int/topics/sanitation/en/ (last visited 04.2015). Some definitions of safe sanitation go beyond this and include personal and domestic hygiene, dishwashing, cooking, domestic and industrial solid waste and effluents.

[35] http://www.who.int/water_sanitation_health/wastewater/wwuvol2intro.pdf (last visited 04.2015).

[36] http://water.epa.gov/polwaste/npdes/Municipalities-and-Wastewater-Treatment-Plants.cfm (last visited 04.2015).

[37] WHO (1997).

[38] According to WSTF (2012: 67, 71), Kenya, it is estimated that 75% of the latrines are emptied manually and only 14% by specialised services (local authorities, private companies and utilities).

Therefore, an effective regulation for sludge management is urgent in Sub-Saharan countries but unfortunately still rather an exception today.

3.3 Breaking Down Water Apartheid and Securing a Sustainable Development Path

In the industrialised world, service provision is organised by formalised utilities. Formalisation also covers rapid responses in case of emergencies. It is common understanding that emergency services are of a temporary nature and no stakeholders would claim that such services can replace sustainable development concepts. Efforts to restore the centralised utility system are activated in parallel to an emergency response. There is no conflict between the different stakeholders involved because they know their specific roles in the different situations and both are aligned with the sector policy of governments.

The situation in the developing world is very different because between these two types of service provisions are many others, operating within the grey area of 'informality'. There are stakeholders who offer (temporary) solutions having emergency character[39] but sell their products as solutions for sustainable access. Others promote small-scale infrastructure for water, owned and informally operated by private or the (slum) communities for instance. Such products or small scale infrastructure in towns, often trying to develop a business case, are sometimes promoted by projects of the international cooperation. This leads to what was described in Sect. 2.4 as confusion or uphill battle of governments in the low-income countries when they try to put order into the sector following a sustainable development concept and the aim of equitable access for all.

The Two Class System Within Towns

Informal small-scale service provision came into existence in most of the Sub-Saharan countries because the utilities did not want to extend services to the LIAs or were not in a position to do so. There are several reasons why the public service providers could not cope with the rising demand and allowed existing infrastructure to fall in disarray and thereby, opened the window for informal service provision. Most of these reasons are linked to an ill-designed sector framework ignoring the growing needs in the LIAs and the undue interference by politicians in the management of service provision.[40] Nickson and Franceys[41] demonstrate that there is a rela-

---

Only 8% of the sludge is disposed at the treatment plant or drained into the utility system. 37% don't know where the sludge is disposed and 55% is dumped or drained somewhere else.

[39] E.g. Peepoo toilet http://www.peepoople.com/peepoo/start-thinking-peepoo/ (last visited 05.2017).

[40] This diverts the attention of the utilities concerning accountability from the consumers and underserved towards politicians or civil servants of ministries.

[41] Nickson and Franceys (2003: 32).

tionship between autonomy of utilities and their performance. It was at the time that sector institutions ignored the sector framework and the poor, clinched to the household only paradigm, politicians practiced undue interference, utilities were gravely underperforming, tariffs were ill-designed hampering cost recovery, etc. which finally led to a permanent discrimination of the poor in access to water.

Before reforms, it was convenient for many stakeholders including decision makers from ministries and utilities to see informal providers flourish in the urban LIAs. For them, the problem of underserved people seemed to be solved and it was an easy justification for not being engaged in the 'problematic' LIAs. Maybe some of the decision makers were not aware of the discrimination it brought into the sector compared with utility services. This would indicate a lack of information. Even in sanitation there is an urban divide, which most likely goes deeper than the water divide because the connection rates to sewage systems are much lower and the costs of building and maintaining septic tanks are elevated. Furthermore, even when services for onsite sanitation are formalised they reach only a fraction of the population of non-connected households and thereby often only the middle and upper classes (e. g. emptying by septic tanks).

However, these arguments only cover a part of the explanation why informal systems gained a share of the market which mainly concerns the poor. Also the interest of some stakeholders to explore additional business opportunities and prominence might have been a reason to promote small-scale systems. It seems to be easier to mobilise funds from humanitarian channels for a support to the (self-help) communities than to support utilities to extent their formalised services into the LIAs. Helping the poor communities to help themselves seems to be more attractive than helping the utilities to extent services to the poor communities regardless of the outcome for the poor and the sector development. Furthermore, for some stakeholders the situation of the sector and the utilities before reforms might have appeared so hopeless that integrating their actions into a bigger framework, linking it to the state authorities on a national level, was just not an option. Hence, the motivation of NGOs to promote community systems in the LIAs in towns is understandable but it cannot be overlooked that promoting SIPs means supporting a development in the urban setting, which is much different to what the lessons learned from the past proposes for urban water and sanitation development and what many experts consider best practices in emerging countries.

Challenges Linked to (Small-Scale) Informal Service Provision

The dependence on informal service provision of a large part of the urban population (30–50%) should be a particular concern for governments in the low-income countries.[42] This group is growing faster than any other population group in the countries. Informal service providers have most likely developed into thousands in numbers in many countries. The main challenges of informal service provision are

[42] According to MajiData, www.majidata.go.ke (last visited 05.2017), the population in the LIAs represents around 40% of the total population to be served with urban water and sanitation systems in Kenya, as indicated by WASREB in its annual reports.

their size of operation, the price of water demanded from the (poor) consumers, the constraint to use water sources with risks, the placement of the outlets on private ground and the absence of a controlled complaint system.[43] Consequently, the water delivered by informal providers to the consumers or to mobile vendors is often of doubtful quality regardless of being provided through pipes into the dwelling, at boreholes with public stand posts or even in plastic bottles.

In addition, small scale systems meet limits when it comes to financing of infrastructure for extensions and employing qualified personnel. Furthermore, because SIPs are not controlled and in general can practically not be held accountable they can distribute water with doubtable quality and manipulate supply by creating artificial water shortages. It is very common that informal providers vary their water prices in the different seasons in order to profit from supply shortages when alternative sources have dried up (refer to the case of Betty).[44] It has already been noted that there are no market forces working and the hope that competition will regulate water prices seems wishful thinking.[45]

It should not be forgotten that informality is also linked to illegality. This is mainly water theft, hijacking of community systems by local power players (individual or groups), use of the income of water by criminal gangs or political parties and even physically elimination of competitors.[46] Therefore, informality not only fosters the discrimination of the urban poor but at the same time increases the risks of corruption and supports criminal activities.[47] This hinders sustainable development and its efforts for good governance. It can be concluded that informality delays the efforts of the state to fulfil its legal obligations.

The Figs. 3.1 and 3.2 are an indication of what the urban water divide means for the consumers in terms of costs. However, the quantities of water indicated in the

---

[43] Chakava et al. (2014: 108) state, 'A study in Nairobi's low income settlements… Results showed 42% of respondent depend on boreholes within 100 m of households…Water tariffs and quality standards were not enforced…with the average price for consumers (around US$/m^3) over ten times the national approved lifeline tariff for piped water and recorded Escherichia coli counts and fluorite levels up to 214/100 ml and 9.4 mg/l respectively, [were] significantly over WHO standards.' Investigating the long term implication, this paper concludes that '…the overall water supply demand shortfall could be greater and last longer than anticipated with consequential impact on access by the poor to adequate water quantities.'

[44] E.g. 'water mafia in Dheli', http://www.bbc.com/news/world-asia-india-33671836 (last visited 07.2016).

[45] Referring to the cases mentioned in this work for Uganda, India and Kenya.

[46] 'Officials say 30% of Pakistan city's water supply is wasted or stolen, worsening an already chronic shortage… armed gangs ... controlled part of the water supply in … Karachi… illegal water stations operate tapping into underground pipelines owned by the state... denying poor residents much needed water. Water traders with 30 to 40 tankers reportedly earn as much as $16,000 a day. Of the total amount of water stolen, over 70 percent is reportedly sold to big business… the leaders of this 'underwater world' are still operating … illegal dealers said powerful and well-connected individuals are to blame for the continued illegal practice.' (Johnston N, Aljazeera). http://www.aljazeera.com/news/2015/09/karachi-water-mafia-sucking-city-pipelines-dry-150910062202773.html (last visited 09.2015).

[47] Refer also to the Kericho case in Kenya mentioned in this work.

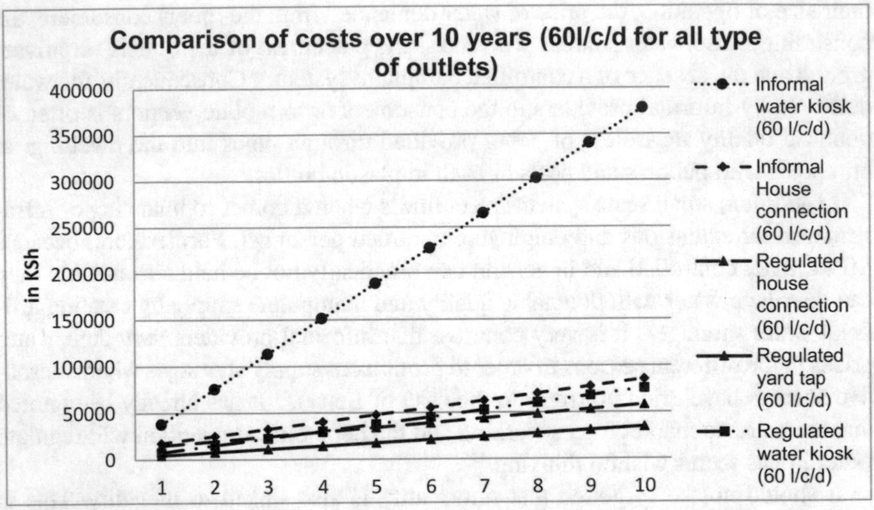

Fig. 3.1 Development of water charges at households and public outlets of informal service providers and the utility at Kericho, Kenya
Source: GIZ, Kenya Water Program, August 2016, provided with the following explanations:

- Quantity of consumption is based on empirical data per household.
- Regulated services are based on the new tariff structure for Kericho W+Sewage Company incl. connection fee.
- Prices of informal services are based on empirical data from Kericho Town.
- Only regulated services ensure drinking water quality.

Informal services: High price for low quality

- Informal services are always more expensive.
- Poor consumers pay even more at the informal kiosk than at an informal household connection.
- There is no quality control of the water. Further treatment at household level needed.

Regulated services: Fair tariff for controlled quality

- Benefits for all, but mostly for the poor who are relying on water kiosks.
- The higher the service level, the higher the cost.
- Water quality is tested. No need for further treatment.

first figure are not realistic for kiosks and yard taps as people cannot carry 60 l per capita per day from the shared facilities. Thus, the second figure indicates the spending according to the more likely quantity consumed.

Two Myths: ***Formalising Informal Service Providers*** **and the** ***Community Managed System as an Option for Urban LIAs***

Although many stakeholders demand the formalisation of informal service providers, literature research has not surfaced one example of successful formalisation on scale of SIPs which achieved the objective of standardised services for all. There are indications that it is as good as impossible to identify and register informal service providers. For instance in Kenya, the regulator asked some utilities to establish a list of informal service providers operating in their licensed service areas. This was a complementary exercise when the regulator was adjusting the service areas of ten

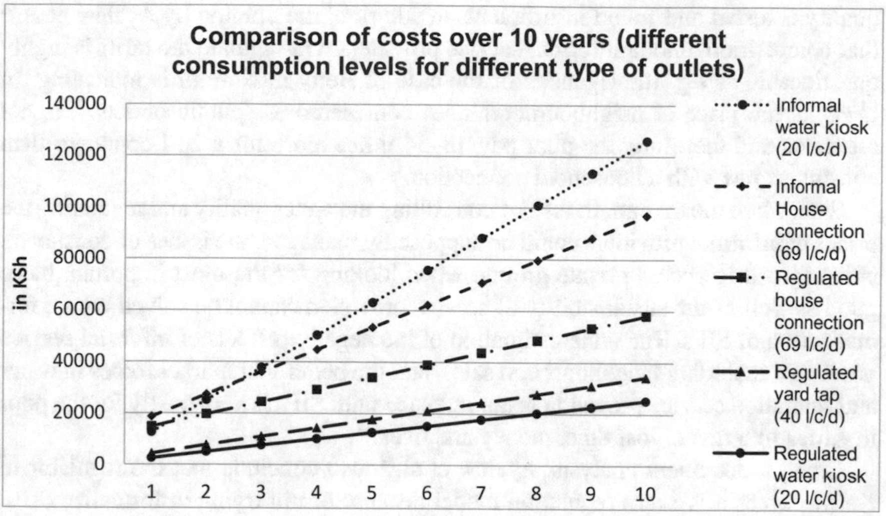

Fig. 3.2 Development of water charges at households and public outlets of informal service providers and the utility at Kericho, Kenya

utilities. None of the utilities could do so even with the support of Technical Assistance (TA) financed by a donor due to the resistance of informal service providers.[48] The below mentioned case of the 'Indian water mafia' also points into the same direction.

Research has not yet revealed how thousands of informal service providers in a country can be controlled. The proposal of Ayalew et al.[49] to supervise the many single water points in the urban areas (e.g. wells) and to formalise the many informal providers through the means of guidelines and spot checks cannot be considered feasible or goal-oriented when taking into account that a regulator with 30–40 employees already meets its limits when regulating more than 30 utilities.

> …good practice guidelines for design and operation of wells. Further guidelines could be established for water vendors, in addition to routine water quality testing and random inspection of mobile water vendors… (Ayalew et al.)

Guidelines without the possibility of enforcement on scale are of little help. Considering spot checks in the urban setting as a sufficient control is perplexing. Furthermore, it is not explained what (feasible) options are available to make SIPs comply in future or ensure that they are excluded from the market when their water

---

[48] In 2015 WASREB carried out an exercise in 15 towns to delimitate the service areas of utilities according to population density and instructed the consultant to identify the small-scale informal service providers with the utility. The result was that no one could confirm that the list of informal providers was complete and none of the informal providers returned the questionnaire to the utility (own observation 2015).

[49] Ayalew et al. (2014: 124).

quality is tested and found insufficient. In addition, the opinion by Ayalew et al.[50] that competition among informal service providers will regulate the tariff is highly questionable as e.g. the Ugandan or the case of Betty in Nairobi is indicating. In Uganda, the price at neighbourhood sales considered as 'public outlets'[51] is not regulated and therefore, the poor pay 18–34 times more for a 20 l container than consumers pay with a household connection.[52]

Even when the mammoth tusk of controlling the water quality and regulating the prices of informal providers could be supposedly managed, the issues of consumers being forced to access private ground when looking for the most important basic good as well as the sustainability of service provision cannot be solved with a formalization of SIPs. The underestimation of the negative effects of informal service provision, including neighbourhood sales, and the belief that market forces in water and sanitation service provision is taking place with SIPs is very costly for the poor in Africa in terms of lost time, money and lives.[53]

Citing a document analysis, Ayalew et al.[54] also conclude that the regulator in Kenya copies a Western regulation model because of not trying to formalize SIPs. This remark is difficult to understand. The regulator has to formalise service provision according to the sector policy and legislation. As a feasible option, regulation generally tries to makes the utilities (existing in limited numbers) extend their standardised services into the LIAs in order to replace informal service provision because it considers the formalisation of the many SIPs impossible.[55] To interpret this approach, as refusal of the regulator to recognize local reality seems therefore unjustified.[56]

[50] Ayalew et al. (2014: 123).

[51] In this case the 'public outlets' are household connections with the approval by the utility to sell water, which must be considered as neighbourhood sales.

[52] Gillian Nantume, Daily Monitor (29th December 2015), Urban poor pay expensively for cheap water, 'A 20-litre jerry can of water costs Shs 44.7 according to the NSWC tariffs for domestic consumption. For public consumption, the tariff of those who sell water to others is Shs27.5 per 20 l jerry can. At the boutique [water kiosk], I use two jerry cans of water a day, which cost Shs800 each in the rainy season, and between Shs1000 and Shs1500 in the dry season, says Nabasa.' (1997). http://www.monitor.co.ug/Business/Prosper/Urban-poor-pay-expensively-cheap-water/688616-3012400-j9uix0z/index.html (last visited 12.2016).

[53] As indicated in the GIZ (2012: 20) study on informal small-scale service providers in Tanzania, the disadvantages of ISPs are: (1) 'High Tariffs' – 'On average households in LIAs, which receive water from ISP pay 13-times the price than they would if they receive water from a household connection; and still pay three times the price than they would, if they receive water from a kiosk. (2) 'Water quality standards cannot be guaranteed' – 'Water quality delivered by ISPs is usually unknown and untested.' (3) Lack of customer rights in case the operator is not performing well. (4) 'ISPs cannot guarantee sustainable access.' (5) The operator may deny any customer access to services at any time without giving a reason or a customer an opportunity to appeal. Hence, 'ISPs do not comply with the central principles of the human right to water.'

[54] Ayalew et al. (2014: 125, 126, 127).

[55] Roughly estimated around 2.000–5.000 in the country.

[56] Ayalew et al. (2014: 125) describes a best practice example of the Phnom Penh Water Supply Authority, which achieved a '…total municipal system…' but indicates that such a system is

Furthermore, the lessons learned from the development in the industrialised world and in some emerging countries where access for all has been achieved supports the assumption that SIPs have no place next to centralised (utility) systems which are obviously the better option for the urban setting compared to a division of towns in many small areas with different service providers. Otherwise, we would see such a fractured situation within a town in the industrialised world today. It can be said that defending SIPs in a long term vision would mean to neglect academic knowledge and local feasibility.

The second myth is that community systems are the best option to ensure sustainable access to safe water and sanitation in the LIAs. Let's not forget, that in the past community systems were seen as the best solution to provide access to water in the rural areas. Considered at the time as successful, many NGOs discovered that the urban LIAs can also be a business case where such an approach could be applied and started to mobilise money to establish SIPs in towns parallel to the utility systems. Not only that in the meantime many countries realised that community managed system in the rural areas are hardly sustainable without external support and that they often do not fulfil the expected objectives, they also found out that the realities in the LIAs are very different from the rural settings. With the failure of many systems, a number of countries (e.g. Uganda, Kenya, Zambia etc.) try to integrate community systems into the utility or at least establish a link in order to maintain their operation and ensure a minimum quality of service provision. However, integrating community systems into centralised utility systems is often impossible because of the resistance by the individuals or groups, who have taken control.[57] They are threatened by the utility to lose their income generated by a system which discriminates the poor and they will defend it even it is against the interest of the public for sustainable development. In addition, the technical layout of the system might not fit into the utility network without substantial and costly adjustments to be undertaken.

The challenges community systems face have a negative impact on the people living in and around their service areas. Sometimes people living on the fringe of the area are excluded because they are not considered part of this community (expert interviews in Tanzania: 'community systems are unfair'). Therefore, unrestricted access is not always guaranteed. There are issues of water quality because of the use of local water sources and the absence of treatment facilities or a sustainable water treatment. The qualification of personnel is an issue as well as cost recovery because of the local limitation and the absence of big consumers which prevents a cross-subsidisation and therefore a preferential tariff for the poor. In addition, NGOs

hardly feasible in Africa. This is one example of how experts can underestimate the potential of sector reforms in Africa (e.g. Burkina Faso) and are willing to consider Africa as a sort of laboratory where untested and unfeasible options can be introduced.

[57] The instruction by the minister of Local Government and Housing initiated by NWASCO, the regulator, to merge the Water Trusts with the utility of Lusaka was fought off by MPs, which were mobilized by some representatives from the LIAs in 2008 (own experience).

establishing the infrastructure with the participation of the future users are hardly experts in construction. This leads often to a neglect of building standards.

The idea of community systems is to give the consumers the sense of owning the system and therefore a motivation to take care of it and satisfy the need of the people. However, it is common that community systems are controlled or hijacked by individuals or small groups, which have their particular or political interests (e.g. the case of Chipata, Zambia under Chiluba). This makes dwellers in their service area loos interest in participation. The community is also made to believe that once the system is in place their problems of access to water and sanitation are solved despite the shortcomings in water quality and sustainability and the limit to grow with rising demand.

Ayalew et al.[58] consider SIPs to be a solution for 'stopgap until municipal [centralised utility] systems catch up' and thereby underline the temporary character informal provider should have and the need to replace them as soon as possible. This is an interesting idea, but must also include a provisions when established to merge them later with the utility. This might contradict the objectives of community ownership. For instance, safeguarding the possibility of transferring the assets later to the utility would mean that the assets of the community system would need to be given to a third party instead to the community itself. Despite the well-known disadvantages of informal service provision, including community systems in LIAs, some argue that they are vital and therefore, have a reason to exist because they cover an important share of the market.[59] However, considering all the challenges SIPs come with and the disadvantages urban dwellers are submitted to, can their simple existence and, in some cases, dominance in a market be an argument for their support when having a sustainable sector development in mind?

Formalisation of Services: A Pillar of Modern Reforms

It has been argued so far that water and sanitation service provision need to be formalised not only to secure safety and sustainability for servicing the population and to support the development of a country but also because the state has given a guarantee with the signing of UN conventions to provide access for all according to minimum (human rights) standards. At least since the signing of this engagement, informality has to phase out gradually. The first step is usually to include the obligation for the formalisation in the national framework documents. However, policies, legislation and strategies are not indicating how to deal with the many informal service providers in the meantime, leaving room for the stakeholders to interpret and decide their own strategy on this issue. Formalisation of SIPs would need compliance with quality standards, regulated tariffs, regular reporting to authorities and the public, unrestricted access to private ground, etc. Trying to enforce formalisation or banning informal service providers all together without providing utility services in an area would most likely create a black market where prices for water for the poor could be well above the present high level.

[58] Ayalew et al. (2014: 110).

[59] As Ayalew et al. are pointing out in the same investigation.

In addition, sector development must be a sustainable process and it is difficult to see how informal services fit into it. Such a process of development in the sector not only needs to concentrate on a continuous functioning of infrastructure based on technical and economic sustainability but also on a constant flow of sufficient investment funds, and an enabling water sector framework, which is capable to adjust, if needed. The state has to be able to rely on the sector players which means that they cannot disappear on their own decision. Furthermore, action with emergency characters like pee-poo[60] and subsidised services financed from outside the sector are often coming to an end when external support is running out. Not even the states in Sub-Saharan Africa have managed to maintain continuously subsidies for water and sanitation service provision over decades.

Recognising these challenges, it was the only possible move for decision makers in the sector to decide on the formalisation of service provision and to promote utility services for all urban dwellers right at the commencement of the reforms. The literature on the importance of SIPs which might suggest their acceptance or integration as well as the global discourse which ignores the different realities in the rural and urban areas should not distract the decision makers in the developing world from perusing a process of sustainable sector development guided by the human rights to water and sanitation requirements.

Total Breakdown Avoided by Modern Reforms

However, even public service providers had sometimes serious sustainability challenges before sector reforms. In Zambian for instance, in Luapula, municipality departments could no longer keep infrastructure functioning and central government did not come to their rescue. The water supply systems in towns totally collapsed and were out of operation for years. The urban population had to go back to rural solutions or depended entirely on informal service providers which sold water from traditional water sources. The expectation of some local leader that the private sector would come to rescue with infrastructure rehabilitation did not materialise either.[61] Finally, a donor came to help because the only thing municipalities were able to do was waiting a decade for rescue. There are other such examples of total collapse like Yei in South Sudan.

---

[60] A well-designed sanitation concept for sterilization and composting of human faeces but less convincing when people have to use it in the slums as an everyday toilet under the conditions of sharing rooms (infringement on privacy and dignity). In addition, the challenge to dispose of urine without access to a toilet seems to remain unsolved. Also other questions seem to remain unanswered such as affordability for a household in the slums with five members in the family for instance and a comparison of long term costs of such solutions with the provision of a safe latrine. http://www.peepoople.com/peepoo/start-thinking-peepoo/ (last visited 05.2017).

[61] The contribution of funds by the private sector (in the case of PPP) in water and sanitation service provision is very limited compared to the funds needed in the sector for infrastructure development. In Senegal for instance, according to Brocklehurst and Janssen (2004: 19–21, 44, 45), the asset holder had to struggle to raise just 20 Million USD private capital and could only do so in selling moveable equipment for the same amount to the private operator.

The good news is, that reforms aim to make such cases very unlikely in future. Because authorities hold assets on behalf of the public and the national level takes a stronger lead in the oversight of sector development, even when the functions remain decentralised, any change of an operator (e.g. municipal departments to corporatized providers of PSP) can take place without compromising service delivery. In addition, regulations will blow the whistle when sustainability of operation is endangered long before the systems will totally fail.

Ending Water Apartheid in Towns

'The definition of apartheid refers to a political system where people (and their development) are clearly divided based on race, gender, class or other such factors [poverty for instance]'.[62] In urban water and sanitation in the developing world, there is a situation where access to service provision is in general divided between the residential areas of the upper and middle classes and the LIAs where the poor are living. Before reforms, this separation and a separate development approach (apartheid) was not contested by sector policies and even later still supported during the reform process by many decision makers in the ministries.[63] Breaking down this separation line is the challenge for ending water apartheid in towns.

Ayalew et al. and Schiffler[64] highlight the remarkable achievement of the Phnom Penh Water Supply Authority to move to a '…total municipal system…' ending informality in urban water and sanitation. Fortunately, modern sector reforms based on human right to water and sanitation have embraced this fundamental change with the obligation to provide services with minimum standards for everyone in towns. This makes it imperative today to question informality (SIPs) and think about its ending because the *'rescue'* by the poor through SIPs, when utility systems fall short, comes with a high price for society and the individual.

It has been explained that for several reasons SIPs cannot be controlled and utilities cannot 'catch up' by integrating SIPs into their system although such arrangements are piloted.[65] Regulating informality by contracts without making the negative effects disappear only means formalising discrimination. The strategies of SIPs include threatening of utility personnel or extending their influence into the boards of the utilities.[66] The sector should never maintain two systems composed of

---

[62] http://www.yourdictionary.com/apartheid (last visited 05.2018).

[63] Own observations in Ivory Coast 2001 and Kenya 2013. In the latter case a director in the ministry endorsed (in writing) the establishment of independent small scale community systems through an international NGO in the LIAs situated in the service area of the Nairobi water company until 2013.

[64] Ayalew et al. (2014: 125), Schiffler (2015: 161–167).

[65] E.g. Lusaka, Zambia.

[66] E.g. one of the members of the Board of Directors of the (licensed) utility, called Oloolaiser, serving Ongata Rongai, which is part of greater Nairobi, is an informal service provider active in the service area of the utility. He is supplying water from a borehole (single water source) situated in a densely populated area without water treatment facilities. He holds a water abstraction permit from the Water Resource Management Authority. However, he has no license from Wasreb, the water services regulator. The contract signed with the utility is not authorized by Wasreb. The

uncontrollable SIPs for the poor and utilities operating under regulation for the middle and upper classes because not only it foster discrimination of a large part of the urban population but also makes no economic sense. The utility expansions into the LIAs will make informal service providers obsolete. As Alayew et al.[67] indicate, the ultimate goal is to achieve '…total municipal [utility] system…'.

Consequently, the state has to instruct utilities to move more aggressively into the areas of the poor (LIAs). The representatives of the civil society and donor organisations which have promoted small scale system for the poor and carry a share of responsibility to make the poor second class citizens in the water sector should now align to sector policies and cooperate with utilities in the interest of the poor. The decline in access to piped water and the respective shift to 'improved sources' have increased the *'urban water and sanitation divide'*.[68] Therefore, it is so important to review the positive but misleading messages from the global monitoring to help overcome water apartheid in the urban setting.

Wherever new settlement for the poor emerge the extension of utility systems must have priority before the development of any community systems[69] or other SIPs. Extensions into LIAs needs local knowledge and hence, have to be spearheaded by national and local structures[70] whereby NGOs support would be crucial for utilities. Consequently, transferring assets on water and sanitation to the community or to a local private must be avoided and expectation that a utility will buy assets from SIPs discarded.[71] Stakeholders opting for support to SIPs should be asked, why not support the utilities in the first place if the state is ready to reform the sector and is convincingly concerned about serving the poor. A comprehensive view of the sector and a long term development vision moving beyond disputes of mode of delivery will help prevent the creation of SIPs.

---

informal service providers are not obliged to report or fulfil any regulatory obligations. Hence, he is operating illegally and there is no documentation on the quality of water and tariff charged to the customers. The networks of the Oloolaiser water company and this informal provider (as others) do not overlap in Ongata Rongai. There is no indication that Oloolaiser intends to extend its network to the areas of the informal service providers operating within its service area. As a member of the BoDs of the utility the informal service provider will most likely be able to secure that such extension by the utility into his area will not take place (own observation, Nairobi, 2016).

[67] Alayew et al. (2014: 125).

[68] Refer also to the Nigerian case mentioned before.

[69] Community systems in this work means a group of people in the LIAs own/operate a water supply system. This has to be distinguished from a 'municipal system' or utility.

[70] E.g. refer to the example of Chingola, Copperbelt in Zambia.

[71] Many informal small-scale providers are known for their 'spaghetti' pipes, which are pipes placed unarranged on or buried near the surface.

3.4 First Mile and Last Mile Infrastructure Development

Distinguishing first mile from last mile[72] originates most likely from the transport sector. However, certain characteristics of this separation can also be found in urban water and sanitation when systems are in a transitional development stage. A centralised management system in urban water focuses generally on the establishment of large and extended infrastructure, which can be referred to as first mile infrastructure (dams, intakes, treatment plants, distribution networks of first and second order, pumping stations and storage tanks). First mile infrastructure for water is designed to be extended with pipes of smaller diameters in order to serve households with individual connections. For sanitation it starts with the inlet of the households to the sewer collection network where the pipe diameters or the galleries are increasing in sizes towards the sewer treatment plants. Pumping stations have to be placed on the way if a collection of the effluent by gravity is not possible.

These are the systems which we see today in the industrialised world and serve everyone with connections. Presently, in the developing world, such systems serve almost exclusively the upper and middle classes in the developing world. A large proportion of the urban population, which are mainly the dwellers in the LIAs, have no access to such systems at all or only through household connections from neighbours. Some reasons for this have been explained (e.g. unplanned settlements, difficulties paying a cumulated bill). In addition, household connections for everyone in a situation of very strong growth of demand over long periods would require massive and constant upgrading of first mile infrastructure which seems impossible to finance in Sub-Saharan Africa on medium terms. Consequently, many of the poor cannot and will not benefit from household connections in the next decade. Thus, maintaining the household connection paradigm leads to a situation where the poor remain with or will be pushed back to the inconveniences of informal services or traditional water sources.[73]

To leave the poor at the mercy of the informal providers should no longer be an option, as previously explained. Hence, utilities in developing countries which are obliged to serve everyone in towns should have no other option than to focus on extensions of networks and outlets of last mile infrastructure for water supply such as water kiosks or yard taps (both shared facilities) and connect household facilities

---

[72] In this work, the expression of last mile infrastructure is used for the part of the utility systems with small diameter pipes leading into the LIAs serving primarily shared facilities/low-cost technologies (water kiosks and yard taps). Often, these small diameter networks branch off from networks which serve household connections in the middle and high income areas and are designed to accommodate a further densification of connections. However, such (last mile) extensions are also used to establish mixed systems (shared facilities and a limited number of household connections) in the LIAs.

[73] At the beginning of the 1990, a donor promoted household connections for free in a low income area in Cotonou, Benin. After around 2 years of project completion many dwellers were disconnected for non-payment of the water bill. Being asked about the present situation, they explained that for them it is worse than it was before because the public stand posts (water kiosks) have been dismantled with the promotion of household connections for all (own experience in 1992).

to their sanitation system with a decentralised chain. Such low-cost (interim) solutions offered by utilities fulfil the same basic human rights requirements for water and sanitation as consumers can enjoy with household connections. Last mile infrastructure for water in the LIAs must sometimes include small boosting stations and tanks to ensure sufficient pressure at the outlets and continuity of supply and must have meters at the outlets to measure water consumption. For sanitation, last mile infrastructure (onsite sanitation chain) is especially important because a growing majority in towns depends on onsite sanitation.

However, there is a crucial difference in the establishment of first and last mile infrastructure which has to be considered when investments are financed, planned and implemented. The creation of first mile infrastructure is mainly an affair of construction companies carrying out public works and of specialised (engineering) consultants designing the system and supervising construction work. The utility can rely on these two partners from the private sector and only has to approve the design, follow the supervision of works and get involved in the reception and takeover of the created assets. The professionals of the utility can be provided with training whenever new technologies are introduced with first mile infrastructure. The bottom line is that first mile development is mainly hardware and engineering services and its operation does not substantially go beyond the usual technical and managerial matters of utilities such as maintenance of assets, controlling water quality and customer care.

The creation and maintenance of the last mile infrastructure is much more complex. Many more stakeholders have to be involved in order to ensure acceptability and sustainable use of shared water outlets and services for the decentralised sanitation chain in order to evacuate human waste. Community and local authorities have to be consulted when decisions are taken concerning technology and the placement of shared outlets. Community plays a very important role in the protection of last mile infrastructure against vandalism once it is created. The utility needs to maintain a much closer follow up on its assets and the way they are used. In LIAs, sensitisation and community mobilisation for participation ('software'), among other things, are crucial or even more important than the infrastructure to be created ('hardware'). Conversely, there are few risks of vandalism and the utility does not need to verify very closely the use of the outlet and its pipes in the middle and high income areas served with household connections. Hence, it can be said that LIAs are contextually much different compared to the residential areas of the middle and upper income classes.

This need to differentiate the two types of service delivery with the development of assets has already been recognised by many development banks/Financial Cooperation (FC) and TA. Nevertheless, the consideration of this knowledge when funding infrastructure development is often inadequate because donors hesitate to change the 'classical' project approach for investments. Sometimes donors delegate last mile investments to NGOs[74] or include such investments as an 'accompanying

[74] E.g. the Japanese Cooperation JICA and the British Cooperation DFIT (Department of International Development), UK, in Lusaka, which channelled the funds for investments through the international NGO, CARE International.

measure' into the contracts of first mile development to be carried out by construction companies. To combine first and last mile infrastructure development under one construction project financed by donors can hardly be successful because it faces several challenges.

- NGOs are seldom specialists in infrastructure development and cooperate, in general, with the community and not with the utilities which however, should later operate the assets as part of a centralised system.
- Consultants are engaged for a determined contract period to oversee foremost the implementation of first mile investment project, for which the time of implementation can be more or less precisely estimated. Estimating the time needed for a last mile project is often difficult and sometimes impossible (e.g. LIAs where cartels are operating). Thus, first and last mile development can require very different time frames. Given a secondary importance (accompanying measure), the implementation of the 'software' measures for the last mile risks to be shortened or the anchorage at the utilities is neglected when first mile infrastructure is completed and time for last mile development is running out. Often shortcuts are taken which leads to the breakdown of last mile infrastructure soon after completion because either the utilities neglect the assets serving the poor or the ill-designed infrastructure is rejected by the beneficiaries.
- First mile investments (hardware) are planned and implemented according to international standards, generally enshrined in national building codes. For last mile investments (hard and software) international and national standards are generally not available. Thereby, the consultants invent their own standards for their projects or copy inappropriate solutions from previous projects carried out in other circumstances/contexts.
- Consultants engaged for infrastructure development focus more on technical matters and the fulfilment of the contract conditions rather than on the sustainable use of assets after the completion of the contract.
- The monitoring of the development and impact of last mile infrastructure is more complex and costly than for first mile infrastructure.

With these arguments and the fact that the number of the underserved poor is rising in the developing world, it should be evident that upscaling of last mile infrastructure is urgently needed and neither NGOs nor donors financing projects for first mile development should be in the lead. If such upscaling does not take place, the global and national goals will not be reached in the coming decades as it has been the case in the last four decades in Sub-Saharan Africa. It is time to accept that coverage fulfilling human rights requirements needs to be accelerated and how can it be done in low-income countries which face a chronical investment gap other than with up-scaling of low-cost last mile based on first mile infrastructure operated by formalised service providers.

This book explains in Chap. 7, in more detail, how the poor can be reached in the developing countries. However, three indications can already be given related to the development and operation of last mile infrastructure. Firstly, last mile activities need to be solidly anchored at national institutions. The financing and development should be organised on a national level in close collaboration with the utilities. The poverty baskets of Zambia (DTF) and Kenya (WSTF) are good examples. They have set national standards for last mile investments and operations, used funds received by donors to create competition among utilities (aid effectiveness trough calls for utility proposals) and offered training to the utilities on how to develop and operate last mile assets. These financing baskets are also able to monitor the use of pro-poor infrastructure long after the projects have been completed. It was to a large extent this contribution by the sector poverty baskets which helped the two countries to bring the long lasting decline in access to water expressed in percentage of people to be served to a hold.

Secondly, utilities have to be guided to extend services into the LIAs with interim solutions. Unfortunately, there are still utilities which resist the call in engaging in LIAs (e.g. Tanzania, Kenya). In the absence of an effective regulatory regime or political pressure, utilities are allowed to ignore LIAs, maintain the household connection only paradigm or are left to be content with third parties to deal with infrastructure development and operation of the last mile (e.g. delegated management).[75] When utilities neglect LIAs then existing infrastructure like water kiosks fall rapidly out of service. In these cases, the misleading argument is that last mile infrastructure is not accepted by the poor. In reality, it is the opposite, as the poor in the LIAs prefer shared outlets of utilities to informal service provision and traditional water points (expert interviews). Utilities need to be reminded that water and sanitation services have to reach social goals (socially responsible commercialisation) and they can no longer leave the poor in the LIAs side-lined, especially when the poor represent around 40% of the urban population. Utilities must be made aware that because water and sanitation is a ‘near public good’ their responsibility is to ensure that policy goals are achieved, regardless of their status as monopolist.

Thirdly, donors should not always follow what politicians would like to have but what people need in receiving countries, as Mosley[76] brings it to the point ‘Third world will always tend to favour the capital-intensive option’. Donors should remember their responsibility to contribute to sustainable development in the partner countries and to help in reaching global goals. Therefore, they should concentrate more on last mile investments and make use of national financing structures such as sector (poverty) baskets. The risks donors associate with baskets can be substantially reduced when TA is integrated into the national institution and it closely cooperates with the decision makers of the FC providing funds for investments. It is interesting to see that despite the importance of extending low-cost technology with last mile development, linked to first mile infrastructure, there are hardly any contributions in literature which distinguish between first and last mile infrastructure development in water and sanitation. In addition, little is explained on how formalised service provision can be up-scaled for the poor. Contrarily, there is

[75] As often promoted by WSP/WB, but hardly proven to be of advantage to the consumers and the utilities. Refer also to Sect. 3.4.

[76] Mosley (1987: 170).

a plethora of literature on how to reach the urban poor with SIPs established parallel to utility systems. Is this not another indication that the sector needs a profound rethinking on global and national levels and more knowledge?

3.5 The Fabrication of Success by Global Monitoring

The MDG declaration on water and sanitation targets[77] specifically included the notions of sustainability and safety for water and sanitation.[78] This was remarkable because it emphasised what generally is considered the most important requirements for water and sanitation service provision in order to help maintaining urban life (refer also to lessons learned from history Sect. 2.1). However, soon after the declaration, these requirements were ignored with the monitoring of the MDGs. With it the notion of improved sources, largely inappropriate for the urban setting, was introduced and neighbourhood sales was accepted in the counting of access for water. At the same time shared sanitation facilities were considered not acceptable. Sustainability of access as a requirement was dropped altogether. Hence, the two key requirements for urban water and sanitation also enshrined in the human rights were ignored which leads to a highly unsuitable service ladder proposed by JMP for the urban setting.

Fortunately, sector reforms in the low-income countries ignored these inappropriate definitions and adopted a sector orientation responding to human right to water and sanitation. However, developing countries cannot entirely ignore the misleading messages produced by the global monitoring because the donors and other stakeholders make use of it in their planning and argumentation. This puts developing countries in an awkward position because the states have to justify results of their monitoring systems against misleading messages from the global monitoring in a situation where the differences are widening in many cases. Furthermore, it influences the opinion of the decision makers and the public in the world and distracts from the problems in the sector. A very good example how people are misled, is the article on improvement of access to safe water in the world, published by the Spiegel Magazine (Germany).[79] It indicates how common public opinion is misled

[77] To halve by 2015 (from 1990 levels) the proportion of the population without sustainable access to safe drinking water and basic sanitation.

[78] Water quality, use of safe sanitation installation and safe sanitation chain.

[79] Der Spiegel, 36/2017:44, Früher war alles schlechter, No. 87: Sauberes Wasser; '1980 waren 52% der Weltbevölkerung mit Trinkwasser versorgt, 2015 sind es 92%...Uincef geht davon aus, dass 2015 weltweit etwa 600 Millionen Menschen, also jedem zwölften, nichts anderes übrigblieb, als aus unkontrollierten Quellen zu trinken…Allerdings steigt die globale Zahl derer, die Zugang zu sauberem Trinkwasser haben, weiter an. Seit dem Jahr 1990 ist sie um 2,6 Milliarden Menschen gewachsen…'It is worthwhile to remember that JMP (WHO and Unicef) published in 2014, that in the year 2013 1.8 Billion people in the world - three times the figure mentioned by the Spiegel based on the Unicef messages -, had to consume contaminated water.

when prominent media generally known for outstanding investigative journalism, relies on such sources of official, global monitoring.

The Regrettable Shift to 'Improved Sources' Is Hailed as Success
According to the JMP report 2014, access to piped water in Sub-Saharan Africa has declined since 1990 from 42% to 34% in 2012.[80] At the same time the report states that access to improved sources has increased from 83% to 85% in the same period because the sub-category 'other improved sources'[81] increased from 41% to 51%. This means that there is a shift from piped water to other improved sources which is incorrectly qualified as (impressive) progress:

> Although sub-Saharan Africa is not on track to meet the MDG drinking water target, progress has been impressive. Since 2000, almost a quarter of the current population (24%) gained access to an improved drinking water source…'[82] and 'Between 1990 and 2012, 2.3 billion people gained access to an improved drinking water source: 1.6 billion gained access to a piped supply on premises, and 700 million gained access to an improved supply, which could range from a public tap to a hand pump, protected dug well or protected spring.' according to the JPM report 2014[83] and 'In 2012 the JMP commissioned a systematic review that estimated that at least 1.8 billion people globally used a source of drinking water that was faecally contaminated.[84]

JMP reports emphasise the fulfilment of the water target of the MDGs[85] and a constant improvement of access to water in the world. But at the same time indicate that the number of people who had to consume contaminated water was around 1.8 billion in 2012 which is well above of what would have been required to fulfil the MDG water target. This figure was increasing by 2015 according to data provided by the 2017 report. JMP reporting on access shows a blatant misjudgment of the situation in the urban setting and gravely contradicts the perception of the state authorities about the urgency in the low-income countries. Such misjudgements are also supported by what the poor in the LIAs report when depending on 'improved' sources neighbourhood sales and SIPs (refer to expert interviews in dissertation from Werchota). Hence, doubting the messages produced by the global monitoring seems to have a sound foundation. Despite this, there are frequent calls by donors and their consultants that sector institutions in the low-income countries should

[80] Progress on Drinking water and sanitation, 2014 update: 65, WHO and Unicef, JMP website: http://www.wssinfo.org/ (last visited 07.2016).

[81] Tubewell/boreholes, protected dug wells, protected springs, rainwater collection, and since the SDGs bottled water, if the secondary source used by the household for cooking and personal hygiene is improved.

[82] JMP report (2014: 15).

[83] JPM report (2014: 14).

[84] JMP report (2015: 43). Refer also to Chap. 1 regarding the JMP report (2017: 3).

[85] Joint news release: UNICEF/WHO, http://www.who.int/mediacentre/news/releases/2012/drinking_water_20120306/en/ (last visited 05.2018).

reconcile their monitoring results with such misleading data instead of calling upon JMP to revise their ill-conceived definitions and monitoring system.[86]

That this success story is questionable is also supported by the findings of Bain et al.[87]:

> Access to an "improved source" provides a measure of sanitary protection but does not ensure water is free of fecal contamination nor is it consistent between source types or settings. International estimates therefore greatly overstate the use of safe drinking-water and do not fully reflect disparities in access.[88]

In the 2014 report of the global monitoring the number of people who had to consume contaminated water in 2012 was mentioned. This information was left out in the following reports. However, from the data provided by the 2017 report these figure can be calculated and shows a worsening of the situation in the world.[89]

Even the General Assembly of the UN had to recognise in 2015 that global monitoring is producing a misleading picture of the situation in the developing world.[90]

> Deeply concerned that official figures do not fully capture the dimensions of drinking water availability, safety, affordability of services and safe management of excreta…and therefore underestimate the number of those without access to safe and affordable drinking water and safely managed and affordable sanitation…need to adequately monitor the safety of drinking water and sanitation…

It becomes evident that the number of people who have to consume contaminated water is a more appropriate reality check than the number of people having access to 'improved sources' as defined by JMP. It seems that the message from the UN General Assembly in 2012 has not been sufficiently received by the global monitoring. With the SDGs,[91] replacing the MDGs, the experts were trying to solve the problems of inadequate monitoring of access by replacing the expression 'sustainable access to safe water' (MDGs) with 'safely managed water services'. However, this did not solve the issue of overstating access in the conclusions of the reports because traditional water sources, neighbourhood sales, etc. remained to be acceptable. Safely managed water services is only introduced for the highest service

[86] Own experience in Kenya and Germany 2017 and 2018.

[87] Bain et al. (2014a: 1, b) https://www.ncbi.nlm.nih.gov/pmc/articles/PMC4255778/ (last visited 05.2017).

[88] Esrey et al. (1991: 920) Faecal Indicator Bacteria (FIB) used by WHO Guidelines for Drinking Water Quality.

[89] Refer also to Chap. 2, with 1.9 billion people having to consume contaminated water in 2015 (JMP report 2017:3).

[90] The UN declaration A/C.3/70/L.55/Rev.1 from 18.11.2015, page 3/6, seventh session, Third Committee, Agenda item 72 (b) http://www.un.org/press/en/2015/gashc4160.doc.htm (last visited 11.2015).

[91] 'Ensure the availability and sustainable management of water and sanitation for all' and at the same time maintaining the acceptability of the use of basic sources in an urban environment, which include '…tube well/borehole; protected dug well, protected spring; rainwater,…' does not solve the monitoring challenge on global level JMP: http://www.wssinfo.org/definitions-methods/watsan-categories (last visited 11.2016).

level and all service levels of the ladder remain linked to the proxy 'improved sources'.

In October 2017, the Nigerian Vice-President declared a state of emergency for the urban water and sanitation sector while the JMP report indicated an access of 91.7% to improved services for the same year. This raises the following question: Would the second highest representative of a state publicly admit the failure of the state in urban water and sanitation development if the situation would not be very desperate and access stands at remarkable 91.7%? Is it not obvious that the shift from household connections (degreasing from 39.7% in 2000 to 13.5% in 2015) to non-piped (improved) water sources (increasing from 43.1% in 2000 to 78.1% in 2015)[92] is causing the water crises in Nigeria which was declared by the Vice-President?[93]

There are other critical questions to be asked: What signals do such misleading messages provide to the decision makers in developing countries and how will it influence sector development? Is it obvious to the decision makers who are not urban water and sanitation specialists that the messages derived from this monitoring approach is in contrast with the situation on the ground and also with what the human right to water requires? Concerning donors, do such messages encourage decision makers to shift support from water and sanitation development to other sectors?[94]

Global monitoring (MDGs) has also listed what it calls 'improved sanitation' as a proxy for safe sanitation, which is a mix of household facilities (flush toilet, ventilated improved pit latrine and pit latrine with slab) and the way human sludge is disposed of (piped sewer system, flush to pit latrine, septic tank, and/or composting toilet).[95] Literature complements this classification with guidelines on how sanitation facilities should be designed, constructed and operated, in order to make them safe and acceptable. The classification of improved sanitation facilities by global monitoring (MDGs until 2015) did not fully correspond with the WHO definition,

---

[92] JMP estimates (Updated 2017), Copy of JMP_2017_NGA_Nigeria – Excel, Estimates.

[93] According to world stage group, the Vice President of Nigeria declared a state of emergency on urban water and sanitation in October 2017 which can be qualified as a distress call. The article noted that: 'The Vice President in his opening remarks noted the demand-supply gap in urban water supply resulting from rapid population growth and increasing urbanization. Furthermore: 'The Minister in his keynote address highlighted the deplorable state of the Nation's urban water supply and sanitation and decried that 17 years after the approval and implementation of the National Water Supply and Sanitation Policy 2000, the objectives of the policy have not been met due mainly to lack of sufficient commitment on the part of the States.' https://www.humanitarian-response.info/system/files/documents/files/wash_sector_nigeria_emergency_technical_guidance_final20161204.pdf (last visited 11.2017). Contrary to this signal (distress call by the Nigerian government), The JMP report 2017, Copy of JMP_2017_NGA_Nigeria – Excel, Estimates, indicate that access to improved water has reached 91,7% in the urban areas, https://washdata.org/ (last visited 11.2017).

[94] Germany for instance has recently decided to close water programs in several African countries (e.g. Kenya, Uganda, South Soudan) and shift its attention to education and agriculture.

[95] http://www.wssinfo.org/definitions-methods/watsan-categories/ (last visited 04.2015).

since services for onsite sanitation are not considered for some of the installations labelled as improved (e.g. flush to pit, or disposal of sludge from septic tanks). Therefore, like for water, access to improved sanitation facilities does not necessarily mean access to safe sanitation.[96]

On the other hand, and as mentioned above, people using shared and improved toilets linked to a (controlled) sanitation chain (sewer system for instance) are rejected by global monitoring as access. This does not reflect reality either, because such toilet facilities can be considered as safe access as long as the number of users within a compound remains limited (see expert interviews). However, with the introduction of the SDGs, monitoring of access to sanitation is now linked to the sanitation chain. This leads to a challenge in monitoring as for around half of the world population there is presently no data available. To respond adequately to the concern of the United Nation General Assembly from 2015 it is obvious that the necessary adjustments in global monitoring is pending and need to go beyond the changes made from MDGs to SDGs.[97]

3.6 Summary of the Pertinent Issues

The central importance of urban water and sanitation to maintain urban life and permit individuals and society to progress is undisputable. Already in ancient times the state was aware of this and its role to secure access for everyone in towns which meant an acceptable water quality and a sustainable operation of infrastructure. The development of this infrastructure and services were too important for the cities to be left to individuals or businesses which would have excluded many people from access. Academic knowledge indicates that water and sanitation service provision is 'the natural monopoly' and both are regarded as near public goods. This justifies state intervention which should, as we know since some time, better be secured with the help of professionals for service provision and regulation. It was understood that all urban dwellers regardless of their income or status should be able to access those life supporting services because the countries development depend on it.

Thus, safe water and sound sanitation must be given at least the same importance than sufficient food supply, adequate health care, education, etc. and can be regarded as more important for securing human survival than electricity, transport and communication. Despite this overarching importance of urban water and sanitation the sector does not receive the desired priority in the international cooperation and by national decision makers. The sector is generally second in line to other sectors. This is one of the reasons why sustainable access to safe water and sanitation is declining in Sub-Saharan Africa but most likely also worldwide, as reliable information suggests.

---

[96] SDG monitoring is trying to change this by adopting the notion of 'safely managed'.

[97] Schaefer et al. (2007).

The discussion of what should be acceptable in defining urban water supply is still ongoing despite available knowledge accumulated over centuries and the widely accepted human rights to water and sanitation. The deliberations in Part I of this book underline again what safe, accessible, sustainable/reliable and affordable services must mean but also what it must exclude. The use of single/traditional water points in towns is inappropriate. Any service provision which cannot be controlled or does not guarantee a continuous treatment of raw water must be ruled out. To these requirements have to be added the guarantee of unrestricted access and the security that people can rely on the sustainability of access/supply. Consequently, neighbourhood sales and SIPs, including community owned and managed systems in the urban LIAs,[98] are unacceptable in urban water supply. There are also other (e.g. economic) arguments which support this orientation.

Because different contexts produce different realities there is a need for different water and sanitation solutions in the urban and rural settings as long as insufficient funds and other conditions do not allow everyone to have an individual connections provided by a centralised infrastructure operated by professionals. Therefore, copying rural solutions for the crowded LIAs has serious disadvantages for the poor and fosters the *'urban water and sanitation divide'*. When such discrimination comes from an approach of parallel developments, one for the poor in LIAs and the other for the better off, then it can be qualified as water apartheid.

Global monitoring of access to water and sanitation by JMP/UN defines access differently than proposed by the human rights and accepts thereby the resulting discrimination of dwellers in the urban LIAs. JMP counting of access provides the impression that the development in the sector is improving although there are multiple indications that the situation is deteriorating in Sub-Saharan Africa and most likely also in the world. The number of people forced to consume contaminated water seems to increase worldwide, politicians in low-income countries declare a state of emergency in the sector, surveys document the decline in ground water quality in towns and in sustainable access to safe water and sanitation, etc. The misleading messages from the global monitoring are disturbing because it incites donors to abandon its partners in a deteriorating situation and national decision makers don't see the need to double efforts in order to face adequately the unprecedented wave of demand.

Furthermore, the challenges in the low-income countries regarding urban water and sanitation and the growing and chronic investment gap have been underestimated since decades. One could say that missing global and national goals became a tradition in the sector. Decision makers overlook the risks stemming from a decline in urban water and sanitation which fuels social unrest and global migration in this world. It is no secret that many poor in the slums face more disastrous living conditions than refugees in overcrowded camps benefiting from emergency measures.

---

[98] Refer also to World Bank Brief on Tanzania (2017: 8), WASH Poverty Diagnostic Series, Photograph 1 on the water quality of a community system (Water from Community Source in Morogoro).

History also shows how water and sanitation service provision should be organised and that access for all need substantial infrastructure development in growing settlements. Classifying water and sanitation as social sector, like receiving countries often do, risks the neglecting of asset development. In addition, water and sanitation developments in towns have to go hand in hand and both have to comply with standards. With the acceptance of human rights to water and sanitation the states became 'officially' responsible and have to account for a development towards access for all according to minimum requirements. The poor can no longer be left to the mercy of SIPs and contaminated but 'improved' labelled traditional water sources.

Centralised systems are best suited to overcome the urban water discrimination, reduce gender inequalities and help combat poverty because they offer multiple social and economic advantages, and size matters. Modern medical advances can be used as quick fix to treat waterborne diseases but can never substitute the development of water and sanitation infrastructure. Despite all this knowledge the sector in the development world is used as laboratory to experiment with 'innovative' solutions which either do not make their promise realise or generally hamper sustainable development soon or later. Water and sanitation development is too important for trial and error experiments carried out on the underserved poor. It is too costly in terms of lost time, money and lives. Most of them do not do justice to the complexity of the sector which is today more pronounced in the developing countries as it was in the industrialised world before when countries were at the same development stage. Governments in the low-income countries have difficulties to align the many stakeholders to a path of sustainable development because they pull in different directions and pursue own interests, often under the pretext to help the poor or to fulfil their role as watch dog.

Next to the insufficient priority, use of the sector as laboratory, misleading messages and the many stakeholders working outside the national sector policies, there are other challenges which hold back progress. Politicians (mis-) use the sector to help their career and political party. The horizon of politicians is fixed at the next election while sector development needs a long term vision. Patronage and corruption is spilling into the sector. Although there is sufficient evidence that the poor are willing and able to pay for tariffs which enables utilities to achieve social and economic goals simultaneously, water tariffs are kept too low.

For the poor physical access to utility systems is so much more important than paying a regulated utility tariff and still, discussions are much more focused on utility tariff systems than on up-scaling regulated services for the poor. Despite these knowledge politicians do not allow the sector to use its self-financing potential as contribution which could match what donors are already providing. Any accumulation of funds by the utilities, necessary for investments, is branded as result of excessive tariffs by some stakeholders. This misuse of water and sanitation development by politics is counterproductive and foremost punishing the poor.

There are developments which have been massively supported by the international cooperation but have caused substantial damage to the sector. The ill-conceived decentralisation was often devastating for urban water and sanitation and needed costly repair work with reforms. Equally, the notion of the 'MoH being the mother of sanitation' in order to defend their system of public health engineers contributed to the crises in the sector. Both indicate that the international cooperation has quality issues and need to better integrate experienced experts on urban water and sanitation development in the discourse and in decision making. It also shows that the international cooperation has to take their part in the mutual responsibility more serious than in the past.

The myth that the numberless SIPs can become part of a formalised system and that community systems are of advantage to the poor in the LIAs has held back sector development for much too long. In addition, clinging on to the household connection for all paradigm has also prevented progress in access until now. Because the sector in the low-income countries is in a transitional phase different service levels are necessary to move more rapidly to a higher coverage. But basic services need to be introduced without compromising minimum requirements, especially on water quality, unrestricted access and affordability. Therefore, a service ladder has to remain within a formalised system in order to avoid the discrimination of the poor in towns.

To offer such different service levels the sector stakeholders must distinguish between first and last mile investments and recognise the different nature of the two areas of development in order to use funds (aid) more effectively. However, most of the utilities and other sector institutions still have to learn how to reach the poor with up-scaling strategies in the many LIAs, which can have very different contexts, and the international cooperation has to acquire knowledge on how to support such efforts in their partner countries through national institutions.

These issues underline that there is a substantial knowledge gap concerning the development of the sector in the low-income countries which makes it necessary to design and test a theoretical concept for development. The deliberations so far permit such an undertaking and the collected data from different sources in four Sub-Saharan countries allows the validation of the proposed model for a comprehensive sector development.

References

Ayalew M, Chenoweth J, Malcolm R, Mulugetta Y, Okotto LG, Pedley S (2014) Small independent water providers: their position in the regulatory framework for the supply of water in Kenya in Ethiopia. J Environ Law 26:105–128. https://doi.org/10.1093/jel/eqt028

Bain R, Cronk R, Wright J, Yang H, Slaymaker T, Bartram J (2014a) Fecal contamination of drinking-water in low- and middle-income countries: a systematic review and meta-analysis. PLoS Med 11(5):e1001644, pp 1–18

Bain R, Cronk R, Hossain R, Bonjour S, Onda K, Wright J, Yang H, Slaymaker T, Hunter P, Pruess-Ustuen A, Bartram J (2014b) Global assessment of exposure to faecal contamination through drinking water based on a systematic review. Trop Med Int Health 19(8):917–927

Brocklehurst C, Janssen JG (2004) Innovative contracts, sound relationships: urban water sector reform in Senegal, Water Supply and Sanitation Sector, Board Discussion Paper No. 1, 30947, The International Bank of Reconstruction and Development/The World Bank Group, January 2004, Available: http://water.worldbank.org/search/node/Innovative%20contracts%2C%20sound%20relationships%3A%20Urban%20water%20sector%20reform%20in%20Senegal (last visited 12.2015)

Chakava Y, Franceys R, Parker A (2014) Private boreholes for Nairobi's urban poor: the stop-gap or the solution. Habitat Int 43:108–116

Clasen T, Boisson S, Routray P, Torontel B, Bell M, Cumming O, Ensink J, Freeman M, Jenkins M, Odagiri M, Ray S, Sinha A, Suar M, Schmidt WP (2014) Effectiveness of a rural sanitation programme on diarrhoea, soil-transmitted helminth infection, and child malnutrition in Odisha, India: a cluster-randomised trial. Lancet Glob Health 2(11):645–653

Esrey SA, Potash JB, Roberts L, Shiff C (1991) Effects of improved water supply and sanitation on the ascariasis, diarrhoea, dracunculiasis, hookworms infection, schistosomiasis, and trachoma. Bull World Health Org 69(5):609–621

Exner M (2015) 'Regelungen zur Sicherstellung der Trinkwasserqualität in Deutschland', Lauterbach Germany, presentation at MATA/GIZ (Mitarbeiter Tagung der GIZ), July 2015

Grabow W (1996) Waterborne diseases: update on water quality assessment and control. Water SA 22(2):193–202

Hemson D, Kulindwa K, Lein H, Mascarenhas A (2008) Poverty and water – explorations of the reciprocal relationship. Zed Books, London

JMP-Joint Monitoring Programme on MDG (2014) Progress on drinking water and sanitation. WHO and UNICEF, Available: http://www.wssinfo.org/fileadmin/user_upload/resources/JMP_report_2014_webEng.pdf (last visited 04.2014)

JMP Annual Report WHO and UNICEF (2015) https://d26p6gt0m19hor.cloudfront.net/whywater/JMP-2015-Annual-Report.pdf

JMP (2017) Progress on drinking water, sanitation and hygiene. https://www.who.int/mediacentre/news/releases/2017/launch-version-report-jmp-water-sanitation-hygiene.pdf

Kabler PW, Clark HF, Geldreich EE (1964) Sanitary significance of coliform and fecal coliform organism in surface water. Public Health Report Cincinnati, Ohio 79(1):58–60. Available: http://www.ncbi.nlm.nih.gov/pmc/articles/PMC1915497/ (last visited 04.2015)

Koch R (1893) Waserfiltration und Cholera (Aus dem Institut für Infektionskrankheiten), Zeitschrift für Hygiene und Infektionskrankheiten, 1893, Bd XIV

Mosley P (1987) Foreign aid – its defence and reforms. University Press of Kentucky, USA

Nickson A, Franceys R (2003) Tapping the market, the challenge of institutional reform in the urban water sector. England: Palgrave MacMillan Distribution Ltd, Houndmills, Basingstoke, Hampshire RG21 6XS

Pauschert D, Gronemeier K, Jebens D (2012) Informal service providers in Tanzania, Eschborn, GIZ (Deutsche Gesellschaft für Internationale Zusammenarbeit GmbH), 2012, Available: http://www.giz.de/fachexpertise/downloads/giz2012-en-informal-service-provider-tanzania.pdf (last visited 06.2015)

Rouse M (2013) Insitutional governance of regulation of water services – the essential elements. IWA Publishing, London

Schaefer D, Werchota R, Dölle K (2007) MDG monitoring for urban water supply and sanitation: catching up with reality in sub-Saharan Africa. GTZ, Eschborn

Schiffler M (2015) Water, politics and money a reality check on privatisation. Springer International Publishing Switzerland, Springer Cham/Heidelberg/New York Dordrecht/London

Smith GD (2002) Commentary: behind the broad street pump: Aetiology, epidemiology and prevention of cholera in mid-19th century' Britain. Int J Epidemiol 31:920–932

Tilley E, Ulrich L, Luethi C, Reymond P, Zurbruegg C (2014) Compendium of sanitation systems and technologies, 2nd revised edn. Exeter: Swiss Federal Institute of Aquatic Science and Technology (Eawag), Duebendorf

WHO (1997) Guidelines for drinking-water quality, 2nd edn, vol 3, Surveillance and control of community supplies, 1997, Geneva

WSTF – Water Services Trust Fund (2012) Up-scaling Basic Sanitation for the Urban Poor (UBSUP) – UBSUP Preparatory Study, document 8, Water Services Trust Fund, Kenya and GIZ - Deutsche Gesellschaft für Internationale Zusammenarbeit, Germany

Part II
Design and Validation of the Sector Development Model

Part I indicated that the urban water and sanitation sector in Sub-Saharan Africa faces a serious crisis in a situation where demand for services will be growing like never before. A number of reasons have been proposed why most of the concepts for development are not appropriate or sufficiently effective. The development process in most of the countries in the past was inadequate to achieve the goals set on global or national levels. Governments in the region have realised this and started mainly in the 1990s to change directions and the way the sector is managed. After around 20 years of reforms there is an opportunity to look closer and reflect on the way forward in accelerating access to urban water and sanitation.

As outlined, success should be measured by the number of people who have access to infrastructure and services responding to minimum requirements defined by human rights to water and sanitation. With the previous discussions in mind, further reflections must concentrate on what are the basics to consider for the development of the sector. It can be said that first, the context should dictate the orientation. This means that the sector development should respond to the challenges in the country in order to be effective. Second, the most important factors influencing the sector have to become part of a development concept whereby key activities in each factor have to be identified. Third, solutions that have already prevailed in the different contexts because they have proven advantageous must be given preference to other options. This will avoid to waste time and money with trial and error tactics.

These reflections led to a model of development, which is proposed and should respond to the complexity of the sector in Sub-Saharan Africa. The model includes an appropriate sector orientation, crucial factors for development and their key intervention areas. It is then tested with empirical data in order to verify its applicability for Sub-Saharan countries and partly for low-income countries in other regions. The aim is to suggest an approach for development which finally accelerates progress towards universal access to urban water and sanitation. Part II is organised in three chapters, starting with a description of the model of development and followed by an analysis of sector reforms in four Sub-Saharan (target) countries. It closes with explanations for the different reform outcomes.

Chapter 4
An Effective Orientation and Approach for Development

Abstract Modern reforms introduce new principles which provide the sector with a more suitable orientation. The deliberation in Part I helps to identify six crucial factors for development: Framework, Regulation, Utility Performance, Information/Data, Infrastructure Development and Financing/International Cooperation. Sector principles and crucial factors form a theoretical concept for a sector development in low-income countries and especially the Sub-Saharan region. The more detailed description of the crucial factors leads to some interesting insights. Streamlining the new orientation needs the elaboration of implementation documents for institutions on all levels, which is often the missing link. Even simplified regulation can have substantial positive effects in markets with many failures. Utility performance, measured with a range of useful indicators (especially NRW), strongly depending on corporate culture and governance. For sustainability and quality of information systems, data collection requirements must be linked with reporting to the public. Not only operation but also asset development needs professionals with sufficient autonomy and the key recommendations of the Paris/Accra/Busan agenda are still detrimental. Because of the strong interdependences, a development concept is only credible if it embraces all the identified crucial factors as stringently as possible. However, concerning reforms there can be different entry points and hence, different developments during implementation.

To review the sector, it is helpful to look first at the challenges the sector is facing and the orientation under which it was functioning in the past. The challenges are very different in the developing and industrialised world. Even among low-income countries challenges can be quite different. However, there is a set of challenges which can be found in most of the low-income countries and especially among countries in the Sub-Saharan region. Comparing these challenges with the way the sector has been operating so far, gaps and ineffective approaches can be identified. Relating these gaps and approaches with the insight obtained from the deliberation in Part I of this book, principles as guide for a new and more effective sector orientation can be derived. This and the reflection on how these principles should be applied

R. Werchota, *Empty Buckets and Overflowing Pits*, Springer Water,
https://doi.org/10.1007/978-3-030-31383-8_4

in the different crucial factors for development will permit to obtain a more effective development concept.

For instance, it is known that insufficient recovery of costs[1] of Operation and Maintenance (O+M) leads to a deterioration of assets due to a lack of funds for maintenance which eventually also demotivates management to maintain activities in other areas of corporate development. This has a negative effect on the credibility of the sector and keeps potential supporters such as donors at a distance. It will eventually lead to serious interruptions or a total breakdown of service provision in towns (refer to Zambia). Hence, to avoid such a negative downward spiral the sector needs to be guided by the principles of cost recovery. Ideally for sustainability reasons, this should be achieved as far as possible by consumers paying their water bills in order to make the sector less dependent on external factors and hence, more resilient against undue interference. For this to happen and to be maintained over time, tariff adjustments have to take place regularly prepared by professionals and not decided by ill-informed politicians.

Therefore, the sector framework has to include provisions for making tariff adjustments compulsory, the utilities need to have the knowledge to forward convincing proposals to decision makers, professionals need to analyse costs and performance of utilities based on reliable information and a monitoring system will have to confirm that cost recovery is achieved and indicate when it is time to review tariffs again in order to maintain cost recovery. Such a reliable monitoring system will also support activities for the mobilisation and the use of funds in order to secure the necessary and continuous flow of finances for asset replacements and development.

This simplified example illustrates that there are several key factors for development and several intervention areas for pursuing a key principle. If one of the links in this fabric is neglected or not functioning, then the intended outcome is compromised and with it the progress in the sector. The case also underlines the importance of adopting principles to guide the different decision makers in the sector and provide them with an argument when politicians have to be advised or the resistance of stakeholders have to be overcome. The more consideration is given to all crucial factors for development in pursuing a principle, the more comprehensive the approach and the development processes will be. This will eventually strengthen the resilience of the system. Hence, to ensure good progress in the sector, decision makers must be aware of the crucial factors for development in order to be able to design and implement the necessary comprehensive set of actions.

[1] Cost recovery (by consumers) as one possible adopted sector principle.

4.1 Navigating the Sector with Appropriate Sector Principles

Urban water and sanitation development in Sub-Saharan Africa faces a number of sector external and internal challenges, which have a significant influence on the measures to be taken and how they have to be carried out. Principles should guide the decision makers in the selection and the implementation of measures. It helps to streamline ideas towards policy goals. In order to give weight to principles and the measures linked to them it is helpful to adopt principles at the highest possible political level. The following lists of sector challenges relevant for the Sub-Saharan region and most likely many other low-income countries are not necessarily complete but should captured the most important ones.

The Challenges External to the Sector

- Unprecedented population growth and urbanisation.
- High poverty level and inequality.
- Significant number of unplanned settlements where centralised water and sanitation infrastructure is difficult to be placed and managed.
- Insufficient public funds, where governments are not able to provide the sector with a significant share of funding the necessary infrastructure.
- Insufficient governance (high level of corruption and nepotism) stretching across all sectors due to insufficient political will, the absence of checks and balances and other control and follow-up systems.
- Climate change having an effect on raw water availability and quality

It can be assumed that these external challenges cannot be (substantially) influenced by a sector concept and actions carried out by the sector stakeholders. However, improvements in the water sector can motivate other sectors to commence reforms or change directions in similar ways with comparable principles. A resulting upscaling across different sectors might influence decision on the highest level of government or gradually the behaviour pattern in the country. Nevertheless, such general developments initiated by one sector cannot be taken for granted and external challenges will always limit the options and the extent to which principles can be followed in the sector.

The Challenges Internal to the Sector

The discussions in Part I of this book already provided some indication of the internal challenges the sector is confronted with:

- Inadequate framework and political will.
- Insufficient infrastructure leading to low access rates, regular water rationing and environmental pollution.
- Insufficient fund mobilisation because the sector is not credible and water and sanitation does not receive the needed priority.
- Funds are not effectively used or are diverted to other sectors.

- Unsatisfying utility performance because the utilities cannot operate with sufficient autonomy (cover costs, lack of professional personnel, etc. because of undue interference by politicians for instance).
- Mushrooming of informal service provision leading to quality challenges and preventing the state to plan and control development effectively which is increasing the '*urban water and sanitation divide*'.
- Insufficient monitoring and information to know the situation and inform adequately actions.
- Lack of basic services in LIAs.
- The unserved people are mainly the poor whereby often their situation in sanitation is worse than in water.

Internal challenges arise through the way the sector is designed, organised and managed. In addition, international cooperation has an influence and contributes to these challenges, but also to the way decision makers can behave. Knowing these challenges, it is possible to derive the necessary principles in order to accelerate sector development towards universal access according to human right requirements.

Sector Principles to Guide Development in Sub-Saharan Africa

The importance of principles to guide a sector in a process of reforms is underlined by Cullet[2]:

> On the whole, adopting water sector reforms implies significant changes because the principles proposed are different from existing principles governing the water sector.

Introducing new principles should lead to significant changes in the sector, e.g. a new institutional framework, checks and balances and a shift of the separation line between institutions. They also lead sometimes to the need to clarify borders to other sectors e.g. health sector in sanitation. It is particularly important to keep the sector development during implementation of reforms within the corridor the principles are tracing especially when personnel is rotating and knowledge is lost, new decision makers come into the sector and a new generation of politicians gain power. It is always difficult for sector personnel to explain the thinking behind certain decisions taken before and during reforms. In these situations it is very useful to remind decision makers why certain principles have been selected. It supports technocrats in ministries and implementers in sector institutions to explain the approach chosen for the reform and avoid the repeat of past mistakes.[3]

Il-conceived and counterproductive proposals are often emerging when new politicians take office and want to give the public the impression that swift actions are taken for the supposed benefit of the voters. For instance, during a press conference, the new minister of water and sanitation in Kenya in 2008 proposed that water should be accessible free of charge in LIAs.[4] Such proposals seem to reappear regu-

---

[2] Cullet (2009: 3).

[3] Able to quickly return to a previous good condition, http://dictionary.cambridge.org/dictionary/english/resilient *(last visited 05.2016)*.

[4] Own observation in Zambia, 2005.

larly in certain countries.[5] This and the proposals of the Jubilee Manifesto 2012[6] to 'Introduce a system of flat rate water charges on an area basis, with informal settlements receiving services for free' were never implemented, fortunately for the sector. Directors in the line ministry, at the regulator, etc. could convince the ministers that the proposed measures run against adopted sector principles and that they will be counterproductive for the poor pushing the sector back to the disastrous situation which existed before the reforms. Hence, despite these publicly made announcements of top politicians, the sector was sufficiently resilient to avoid a deviation from key principles.

Another example is the devolution of power from national level to the counties in Kenya when implementing the new constitution (CoK 2010). County governments questioned, among others, the national regulation for water and sanitation service provision – a key feature of the reform. Even when the new Water Bill, maintaining the provision to separate key functions, passed Parliament and thereafter the Senate in the last quarter of 2016; some counties went to court to challenge national regulation. The arguments for defending the reform and the key principle to secure checks and balances and avoid conflict of interests were that a county as the owner of the utilities providing services should not at the same time play the role of a regulator and that there should not be 47 different regulatory regimes[7] within a country. It could still not convince some of the governors, who did not want to see control from the national level, but it influenced the ministry and the members of parliament to stand firm because the top staff in the sector institutions had the convincing arguments at hand. This shows that especially in a certain development stage anything retained in the reform can be challenged at any time when governments or governance systems change.

Some of the principles might not seem to be very relevant at a first glance, e.g. the separation of sub-sectors in the framework. Nevertheless, many water experts insisted before the reforms that water resource management (WRM) is in its importance overarching water supply and sanitation (WSS) because, as the argument went (Zambia, Kenya, etc.), raw water is at the root of all water and sanitation activities. This contributed to a chronical neglect of water and sanitation service provision especially when both sub-sectors were under the same line ministry.[8]

Another example are the recent discussions on making Cape Town resilient against water shortages which concentrates much more on issues related to water resource management, invasive species, private sector participation in catchment

[5] Refer to expert interviews in Tanzania, 2015, dissertation from Werchota.

[6] Jubilee Manifesto (2012: 55); Transforming Kenya, securing Kenya's prosperity, 2013–2017, the shared manifesto of the coalition, 2012.

[7] There are 47 counties in Kenya.

[8] This neglect of water supply and sanitation and the preference given to water resource management is especially visible in the framework documents in many countries before the reform processes took off. Even recent policy documents established by governments referring to all sectors hardly cover water and sanitation with the same priority than water resource management (e.g. Vision 2030 in the different countries).

management, etc. than on the development of the framework for water supply and sanitation. Although the rationing of urban water which hit the 50 l/person/day threshold was the news in the media worldwide, water resource management and related subjects moved into the centre of the discussion to look for solutions. This suppresses the reflection of changes needed in the water supply and sanitation sub-sector such as the incorporation of the present (municipal) water and sanitation structure and the introduction of a regulatory regime in order to ensure more autonomy for professionals and for keeping water and sanitation an arm's length from (local) politics. Especially with the legacy of apartheid which is generally used as argument for avoiding necessary changes, such a separation would help to give the sub-sector the required orientation and autonomy.

Hence, to select the separation of WSS from WRM as a key principle and include it in the framework becomes obvious. This separation is also supporting the recognition that both sub-sectors are very different in their nature, especially in regards to stakeholders involved, mode of delivery, civil service involvement, etc. and that a ranking of different sub-sectors is not conducive. The introduction of such a principle helped that WSS would no longer be considered the 'little brother' of WRM and consequently, suffers from prioritised WRM.

In addition, it became accepted that certain functions in WSS should be outsourced from the ministries to professionals. This led to the establishment of new (autonomous) sector institutions. The separation of the key functions (policy making, regulation, service provision) guided reforms in a number of countries. At the time, one of the most cited examples outside Africa for such a separation was England-Wales with the changes taking place in the sector during the Thatcher government. The economic regulator of the water sector in England and Wales (Ofwat), established during reforms, was for many regulators in Africa an example to follow with adopted tools and procedures. It was also recognised that a separation of functions on the national level alone was not sufficient to improve sector performance. The separation of functions also needed to take place on local level notably between the municipalities and the service providers according to Nickson and Franceys.[9]

Recent sector reforms often had two common features: professionalised regulatory agencies and commercially oriented public service providers. Hence, separation of key functions was combined with the principle of professionalization. Both was intended to provide these new institutions with the necessary autonomy in the day-to-day operation (shield them from undue interference coming from different levels and sides) and make them functioning similar to enterprises of the private sector. For securing autonomy of sector institutions it is crucial, next to the separation of functions, to engage the right personnel, have autonomy in financial matters and control the way government can engage with the institutions by introducing effective procedures and checks and balances.

These examples indicate that the selection of sector principles are related to sector challenges and their application must be understood as a guidance, but should

[9] Nickson and Franceys (2003: 10).

not compromise other considerations such as costs for instance when the institutional framework is designed. Following the literature review and the discussions in this section the following 12 principles can be considered key for the orientation towards a sustainable sector development in the Sub-Saharan countries (Table 4.1).

The proposed 12 principles address three areas of key concerns:

- Separation of functions, areas and sub-sectors while maintaining a lean sector structure (principles 1–4).
- Professionalization of functions, giving autonomy to the management of sector institutions and formalise services (principles 5–9).
- Balance of economic and social goals with financially sound utilities following a poverty orientation (principles 10–12).

Some of these key principles were already objects of discussions in the international dialogue on water and sanitation. The declarations on global level contain such principles or are based on some. However, many decision makers being engaged in reforms are apparently not aware of the importance of principles for the sector development and make little reference to them in their policy or do not follow through with them in the process of implementation. Other countries specifically adopted some of them on highest level like Zambia with the 'Seven Sector Principles'.[10] These principles helped during the reforms to remind the minister not to give instructions to utilities concerning operational issues (e.g. regulator in Zambia). They were also used to remind stakeholders, such as NGOs, that their role has changed with the new orientation and framework.[11]

Although it is difficult to establish a hierarchy among the proposed key principles, which would also differ from one country to another because the magnitude of the challenges and urgencies might be different, there is one additional principle to follow which is rather linked to the way development should be organised: leave no crucial factor behind / aside in the development process because eventually it will become a drag for the others.

4.2 Leave no Crucial Factor Aside

Crucial factors are dimensions of urban water and sanitation on which sector development heavily depend on. Sector concepts which do not embrace these crucial factors cannot be considered comprehensive enough to ensure substantial and sustainable progress. Single interventions in one of the crucial factors need to fit into the development in each of the other factors if the overall sector progress should be

[10] E.g. full cost recovery through user charges, separation of WRM and WSS, separation of regulation and service provision, appropriate technologies for local conditions and increased spending on WSS through the government.

[11] Own experience during 1990 and 2005.

Table 4.1 Proposed sector principles for Sub-Saharan countries

Key sector principles to guide sustainable sector development	Rationale, link to challenges and reference to previous chapters
1. Separation of WRM (water resource management) from **WSS** (water services provision)[a]	Recognition of the difference in the way organisation has to take place in WSS and WRM as well as subordination of one to the other is not conducive (Sect. 4.1)[b]
2. Separation of **urban and rural** water supply and sanitation[a]	Recognising the different realities as discussed in length in Sect. 2.1
3. Separation of the **key functions**	Such as policy making, regulation and service provision in order to introduce checks and balances and to limit conflict of interests, as discussed in chapter in Sect. 2.4
4. Lean sector structure	The degree to which the sector is fractured has an influence on its functioning and therefore on efficiency, as discussed in Sects. 2.4 and 3.1
5. Professionalising funding and asset development (separated from public administration)	Recognising that the civil servants system is not the best placed to receive funding and develop infrastructure as explained in Sects. 2.4 and 2.1
6. Professionalization of **service provision**	Recognising that the civil servants system is not the best placed to regulate and operate WSS infrastructure as explained in Sects. 2.4 and 3.1
7. Sector/utility involvement in sanitation (professionalization)	Recognition that a formalised sanitation chain, developed and operated by professionals, is needed for offsite and onsite sanitation in towns, as discussed in Sect. 2.4
8. Autonomy of sector institutions and utilities	To reduce undue interference of individuals and groups, which compromises utility and sector performance (Nickson and Franceys 2003) and discussed in Sect. 2.4
9. Formalizations of service provision	To guarantee minimum service levels for all (i.e. stipulated by the human right to water and sanitation criteria) and end the discrimination of the poor in the urban setting as discussed in detail in Sect. 3.3
10. Cost recovery through users (water bills)[c]	To enhance sustainability of WSS systems and close the chronical financing gap for asset development. Avoid the 'Negativspirale' in service provision (Pollem 2008 and Sect. 4.3)
11. Ring-fencing of income in the urban water and sanitation sector	To safeguard the sector self-generated funds for operation and infrastructure development and thereby, become more credit worthy for increased fund mobilisation, as discussed in Sect. 4.3
12. Poverty orientation	To be able to reach the majority of the underserved people and move towards universal coverage, as discussed in detail in Sects. 2.7 and 3.3

[a]Without excluding a holistic view and approach and an effective coordination between these sections. Hence, the separation in in the framework documents

[b]The separation of WSS and WRM does not mean that WRM and WSS should not be managed in an integrated manner as demanded for instance by Rouse (2013: 10)

[c]Full cost recovery and autonomy of sector institutions are in the centre of a number of contributions in literature regarding regulation and reforms (refer also to Rouse 2013: 150)

Table 4.2 Linking crucial factors with the deliberations in PART I

Crucial factors	Reference to previous discussions
Framework	State has to set an enabling framework to ensure the right for access for all and avoid abuse by monopolists, refer also to the declaration of the first water decade and to Sects. 2.4, 2.7 and 3.2
Regulation	Importance to ensure minimum standards and balancing interests of stakeholders involved. Need to end informality for a basic service and control potential abuse by monopolies (refer to Sects. 2.1, 2.4, 2.7 and 3.2)
Utility performance	Permanent underperformance of utilities and neglect of LIAs hampers sector development (refer to Sects. 2.4 and 3.4)
Information / data	To know the situation on the ground, inform planning for development and ensure monitoring of progress and thereby, make the sector credible (refer to Sects. 2.6 and 3.3)
Infrastructure development	Access to water and sanitation is largely lined to infrastructure offering outlets for water and a sanitation chain for the collection and treatment of human waste (refer to Sects. 2.4, 2.7 and 3.4)
Financing/ international cooperation	The importance of financing infrastructure and the international cooperation was outlined by the declaration of the first water decade (refer to Sect. 2.4)

supported and enhanced. Often, stakeholders demonstrate their limited horizon by only concentrating on what they want to do and thereby, forget that their actions might either produce counterproductive effects in other dimensions or their success depend on developments in the other crucial factors. From the literature review, sector challenges and other previous deliberations about urban water and sanitation, the following key factors for development can be derived (Table 4.2).

Within each of these crucial factors, there are key intervention areas[12] where activities influence sector development. For instance, corporate culture is widely regarded as an important element to improve the crucial factor utility performance. Therefore, corporate culture is an area on which experts need to work if access to water and sanitation shall be improved. Corporate culture as an intervention area also depends on other intervention areas such as securing autonomy of utilities, good governance of the BoD and what Schiffler[13] describes as a 'stretch out programme' in streamlining an appropriate management style.

The absence of key activities in one of the crucial factors will limit sector development or hold it back all together. It is a combination of activities which needs to be carried out in a more or less simultaneous way in order to ensure gradual and constant progress in urban water and sanitation development. Hence the crucial factors are strongly interlinked forming a kind of fabric in the sector (Fig. 4.1).

[12] This work does not claim that the proposed list of key intervention areas are complete.

[13] Schiffler (2015: 149).

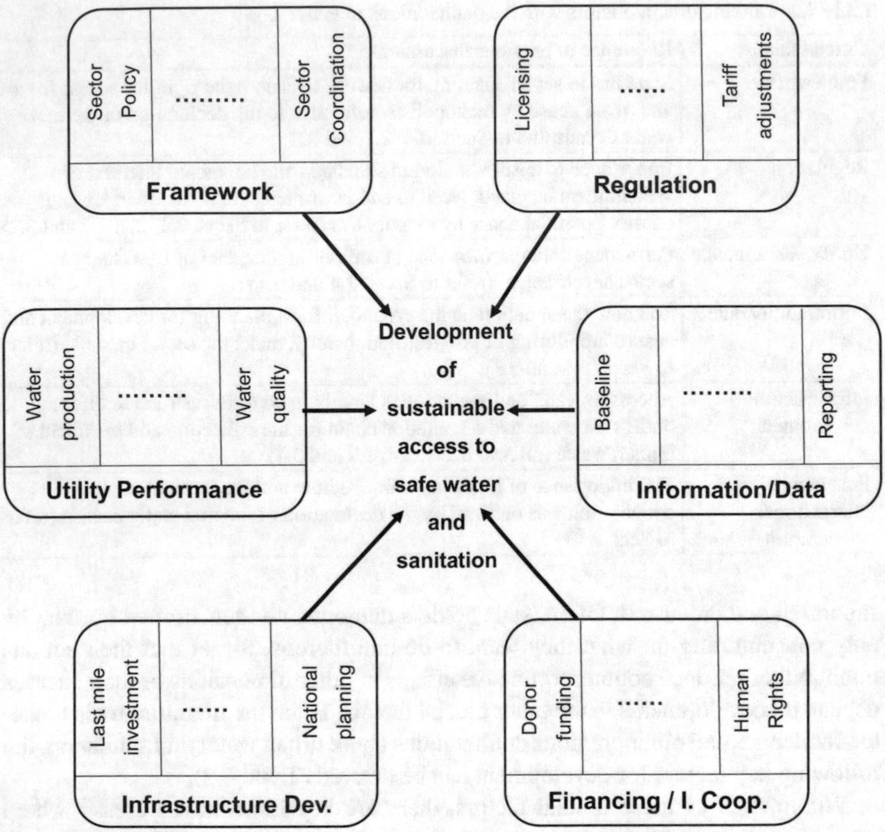

Fig. 4.1 Proposed model of crucial factors influencing sector development

4.3 Zooming in: A Development Conducive Framework, Regulation, Utility Performance, Information, Asset Development, International Cooperation

The following description of each of the crucial factors for development is intended to provide a brief overview of the interlinkages and some of the most important key intervention areas. It should also help to show how complex the sector is and that the achievement of the policy goal to guarantee sustainable access to safe water and sanitation for all depends on simultaneous advances in all of the crucial factors for development.

The Framework

The declaration of the first water decade in 1980 provided already an idea of how important an enabling framework is for a positive sector development. This importance is also underlined by Cullet[14] although he focuses mainly on the legal framework: 'In other words, water law must be at the centre of any broadly conceived strategy of development'. However, legislation is informed by the sector policy and legislation describes the institutional framework, with which the sector is to be organised. It is common practice that legislators demand the policy or at least an explanatory note on the policy when discussing proposals for new legislations (Water Bills). Hence, some of the intervention areas of the framework are the sector policy and the legislation also describing the institutional framework. In addition, the sector needs a strategy to explain the path to take in order to achieve the policy goals by following the sector principles.

A precise framework offers orientation but also security for the stakeholders as one interviewed expert in Tanzania underlined. According to this expert, the translation of the sector policy into legislation was considered a turning point in the sector development. It seems obvious that operating under a framework, which offers little orientation, it is difficult to streamline actions among stakeholders in order to overcome sector challenges. However, a comprehensive policy, legislation, institutional framework and a national strategy does not necessarily mean that stakeholders are acting within the given orientation. Coordination of stakeholders and leadership are important to make people act according to the spirit of the framework.

Referring to experts interviewed, insufficient sector coordination was one of the most cited weaknesses in the sector before reforms and has remained to be an issue despite some improvements. The following statement from an expert in Zambia after more than 15 years of reforms also reflects the opinion of experts in the other three target countries.

> We've made progress, but we still have a long way to go, the leadership on the government side is rather weak, not in a position to provide guidance. I have promised my organization that I am going to deliver, whilst trying to be in line with the rest of the donors. [But] the government unfortunately [is] not leading, that's where there is still a big gap in the coordination. Sometimes it may not be in their interest to have a donor harmonization, better for donors to be doing their own thing because then the interests in government can also do what they want to do, but in terms of really practically harmonizing with others, we haven't yet reached that stage. (Zambia 2015)

The framework documents must remain readable and therefore cannot cover every detail for implementation. Thus, framework documents require a further breakdown at each of the sector institutions responsible for implementation. This is best done by the management of the institutions with the elaboration of implementation concepts, guidelines, business plans, etc. which should be overseen by the ministry or the reform champions. The latter has to ensure that these documents for implementation are coherent with the framework documents. Closing this often missing link between the framework documents and implementation is very crucial in order to avoid that each stakeholder pulls in a different direction and thereby slows down sector development. This is also a lessons learned derived from the document analysis used by this book.

---

[14] Cullet (2009: 2).

Table 4.3 Crucial factor framework and the key intervention areas

	Key intervention areas
Framework (urban water and sanitation)	Sector policy
	Legislation
	Institutional architecture
	Sector strategy
	Documents for implementation (concepts, guidelines, business plans, etc.)
	Sector coordination/leadership
	Etc.

That all other crucial factors are closely linked to the framework does not need to be explained in much detail. The framework documents have at least to address issues such as regulation, service provision, information and reporting, financing of the sector, funding of infrastructure development and links to the international cooperation. When designing a new institutional framework, the issues of costs should not be ignored and a most appropriate balance between separation of functions and cost inflicted by the framework has to be found. However, the design of the institutional framework will also depend on how much confidence the decision makers have that certain architecture will secure sufficient checks and balances in the sector in a country. It is obvious that in Burkina Faso there is a strong confidence that sector players will respect the intended separation of functions in their behaviour which justifying a lean framework. Contrary, it seems that in Kenya decision makers are more inclined to expect that stakeholders will try to bypass certain sector principles and therefore opted for a more complex framework which offers more checks and balances as institutions will insist on their power and interests.

However, it should not be forgotten that designing an institutional framework can also be (mis-) used for nepotism and corruption. Establishing as many institutions as possible and providing them with boards composed of many directors provides an opportunity to place as many supporters as possible of a ruling party in sector institutions. Such political appointees are then useful to prepare the ground for kickbacks when for instance construction contracts are awarded. Hence, the risk to design an unnecessary complex institutional structure is especially elevated wherever a substantial amount of money is involved such as asset developer. Decentralisation / devolution of power can also be a welcome opportunity to establish such an unnecessary fractured institutional framework (Table 4.3).[15]

Regulation

The first regulatory schemes in Africa's water sector were hastily established with the engagement of multi-nationals (PSP) for the operation of public assets. In the 1990s (e.g. Guinea Conakry) regulatory functions were given to asset holders and

[15] Refer to Kenya with the establishment of regional WSBs as asset holder and developer and the devolution of power with the discussion concerning the establishment of Water Works Development Agencies (WWDAs).

developer with the intention to regulate the multi-nationals holding a contract for operation. This for obvious reasons had limited results (e.g. conflict of interests, exclusion of key regulatory functions, no flexibility in case of changes in the environment or risks, etc.).[16] There is still much discussion about the need for regulation in developing countries and particularly in the Sub-Saharan region although the benefit of regulators for the water sector is widely recognised today in the world. However, in Africa, regulation for water and sanitation supply is gaining importance.[17]

The need for state interventions due to the monopolistic nature of water and sanitation service provision and the fact that water and sanitation are near public goods was already explained in Part I of this work. Also the advantage that regulation should better be developed and implemented by professionals than by the civil service structure. Klein[18] and Pollem[19] explain a number of other reasons why a regulatory regime is needed[20] in the developing world such as opportunistic behaviour of politicians keeping the average tariff lower than the average cost. Pollem is pointing out that politicians can exercise such an opportunistic behaviour because the public perceives the utilities as a monopoly. Unfortunately, this support by the public is misused when political influence leads to clientelism and nepotism. In the industrialised world, more than hundred years of experience in sector regulation is available, but it seems that sector institutions are still exposed to undue political interference.

The differences between the two worlds are firstly, that the systems in the industrialised world are more resilient than in low-income countries and therefore can better deal with undue interference. This should not question but rather be seen as a strong argument for regulation in the low income countries because it has the potential to contribute significantly in building sector resilience. Secondly, in some low-income countries federal structure have created a fractured political and institutional environment (e.g. Kenya, Nigeria, etc.), which stretches the ability of sector institutions to a breaking point when handling the incumbent complexity. Also the urban water and sanitation sector cannot escape such situations and has to learn how to best deal with it. With the devolution of functions, risks and frictions between the two government levels are increasing which makes it more difficult to streamline

[16] The two entities (asset holders and utility with PPP) might have similar interests when it comes to keep away from serving the poor.

[17] The Conselho de Regulação de Águas (CRA) in Mozambique was one of the first water single sector regulators in Africa established by degree in 1998 in the framework of the private sector participation. Others were created or became operational: 1997 Electricity and Water Ghana; 1999 ARM, Niger; 2000 NWASCO, Zambia; 2001 EWURA, multisector, Tanzania, 2001, RURA, Rwanda, 2004 WASREB, Kenya; etc. http://www.afurnet.org/index.php/about-us/members-hdr/members (last visited 10.2016).

[18] Klein (1996: 28).

[19] Pollem (2008: 97-112).

[20] Refer also to Sect. 4.3 and Baldwin et al. (2012: 23).

sector policies and strategies and makes decision making very cumbersome and time consuming.

Such devolution of power also brings new decision makers from the lower level into the sector. Generally, they are political appointees with little knowledge about the sector and therefore, need advice in matters on water and sanitation. The regulators as professionals are best placed to guide them in decision for the sector development and advise them on their function to oversee their utilities for instance. In addition, the regulator can help to harmonise the common interests of decision makers on both government levels. Furthermore, the regulators can facilitate the exchange of these politicians between the different regions and thereby raise the level of knowledge in the country. In such a complex set-up, there is no other institution but a professional regulator which will be in a position to acquire the specific professional and local knowledge needed to play this role. There is a high probability that when the numbers of decentralised political units are elevated (e.g. counties) that decision makers (especially below the governors) will seek the advice from highly recognised professional regulators.

Hence, the responsible for regulation must be placed in a position to advise effectively politicians on the sector development or be able to influence other mechanism to keep politicians at a distance when undue interference is in the making. Klein[21] proposes that the set-up of the office of regulators must be designed to 'resist improper influence by different interest groups (companies, consumers, and government).' It follows that because of the close to politics, civil service structures are not the best suited to carry out regulation. Balance and Taylor[22] list and explain the principles for building the 'legitimacy' of a regulatory regime which also has relevance in the low-income countries.

Another main concern of regulation is that its costs ('regulatory burden') do not outstrip its benefits.[23] Because decision makers in low-income countries do not want para-statals to depend on government budgets they are open to except (utility) self-regulation which is also proposed by some experts. However, Pollem[24] discussing the different options for regulation in the sector in developing countries underlines that self-regulation might be adequate with small scale community systems in a rural setting because consumers can fulfil the functions of operation and regulation at the same time, but it will not work with bigger systems. The lively discussion of what kind of regulation is the most appropriate in the low-income countries also includes the proposal of regulation by contract[25] or pure economic regulation. The first is not flexible enough to cater for changes in a very dynamic environment and

[21] Klein (1996: summary findings).

[22] Ballance and Taylor (2001: 25–33).

[23] Which are considered to be difficult to monetarize. Parker and Kirkpatrick (2013: 42) estimate that studies on the impact of regulation do not necessarily capture the true welfare effects and thereby, underestimate the benefit of regulation.

[24] Pollem (2008: 104–109).

[25] Today it is widely recognized that no contract is sufficiently detailed to reflect the complexity of the sector and therefore, can be considered to be a full regulatory regime (situation in France).

the later does not necessarily address the substantial social issues encountered in the Sub-Saharan region.

Ayalew et al.[26] agree that the transplant of Western approaches in regulation is unlikely to solve the challenges in the developing countries. Nevertheless, sound management and sustainability of institutions are pre-conditions to achieve pro-poor goals. Unfortunately, often enough both are not achieved automatically in low-income countries just because policies emphasise on the need for autonomy of utilities. Regulation needs to implement policy decisions in order to reach policy goals. Berg[27] underlines this with the 'Third Law of Political Economy: Policy-makers who concentrate on laws and regulation that promote efficiency also promote the potential achievement of other social objectives.' Furthermore, Klein[28] emphasises that regulation needs to be based on sound information, 'The key to good regulation is the generation of information that allows the regulator to make good rules and allows interest groups to watch out for improprieties by the regulator'.

The kind of regulatory regime to put in place will also depend on the institutional framework in the sector. In the case of one national utility covering the entire market (e.g. Burkina Faso and Uganda) the establishment of a regulatory agency is generally questionable as some regulatory tools are difficult to apply such as comparative competition (benchmarking within the country). However, when a national utility becomes a 'little republic' in the sector (e.g. Uganda), the establishment of a regulator might be sometimes the only option to limit its abuse of power, despite that only one utility is in place and the fact that it will be very difficult to overcome the resistance of such a 'goliath'.

Next to the questions of what kind of regulation is needed for the situation the market is facing and the costs of regulation, decision makers want to know if and what sort of impact regulation will generate. However, it is often difficult to express benefits in monetary terms[29] which seem to help sceptics of regulation to play down its impact. Baldwin et al.[30] indicates how regulation can be assessed: '…the market and all its failings should be compared with the regulation and all its failings'. Considering the many failures in the sector, its relatively low performance in Sub-Saharan Africa and their huge negative effects on other sectors such as health, industrial production, living environment, etc., one can suggest that even simplified regulation with limited performance will have a positive impact on the sector for the countries development.

Allan et al., underlining the importance of regulation, conclude that in the UK water sector the improvements of efficiency were rather due to regulations than to

[26] Ayalew et al. (2014: 126).

[27] Berg (2001: 14).

[28] Klein (1996: 29).

[29] ceg – Competition Economists Group (2012) *Empirical studies on the impacts of economic regulation*, A report for Vector Limited, Available: www.CEG-AP.COM, (last visited 07.2016).

[30] Baldwin et al. (2012: 23).

the privatisation of utilities. Their review[31] of a number of studies emphasise that 'Efficiency improvements … are not dependent upon ownership models and are likely to be influenced more by regulatory interventions, leadership and cultural factors'.

Nonetheless, if regulation wants to be considered justified it has to provide a significant contribution to respond to the many sector challenges within a given framework.[32] Regulation in Sub Saharan Africa faces a specific complex and difficult situation which implies that regulators need to set activities beyond the range regulation have to cover in the industrialised world. One of such specific challenge is for instance the informal service provision in the LIAs. The regulator has an obligation to implement the intention of the policy makers and the requirements given by the legislators (e.g. formalisation of service provision) and can only do this with feasible concepts. As explained in Sect. 2.3, the formalisation of the many SIPs[33] seems generally impossible. Hence, the regulator has the options to either have the SIPs replaced by utility services for which he is criticised by many stakeholders or to tend the impossible; formalising the many SIPs which resist such a move. Berg[34] provides an answer to this dilemma by underlining that making a choice often means excluding other options and that policy makers need to '… translate citizen preferences into clear rankings of policy alternatives…'. Therefore, following the experts interviewed in the target countries that were unanimous in their preference for utility water and only a few accepted informal service provision merely as a fall-back position, regulation is well advised to follow this widespread desire of the underserved population and the intention of the policy makers spearheading modern reforms and push utilities to replace informal SIPs in the LIAs.

However, there is a risk (being evident in Sub Saharan Africa) that regulators focus foremost on the improvement of efficiency in utility service provision, which means to produce a service with guaranteed standards at lowest possible costs (O + M cost coverage on the possible lowest level). Thereby, regulators neglect the

---

[31] Allan et al. (2013: 470).

[32] Specific challenges in the developing countries, compared to the industrialized world, are e.g. donor dependency, stakeholders from the international community, a multitude of NGOs, rapid urbanization over many decades etc.

[33] Ayalew et al. (2014: 110, 111) argues that, 'according to the CESCR the human right to water requires the government to establish a regulatory framework that has the objective, among others, of ensuring that water services provided by third parties such as SIPs are safe and affordable. Therefore, the right to water arguably requires all governments in developing countries where SIPs operates to set up and enforce a legal framework for their regulation.' Obviously, this ignores the feasibility of controlling informal service providers or the choice of a government to only allow licensed utility for service provision and move gradually towards replacing informal service providers. In addition, Ayalew et all consider water utility kiosks operators selling utility water as (informal) SIPs, which is misleading because kiosks operators hold a contract with the licensed utility, which own the assets. Furthermore, utilities include kiosk operations in their report to the regulator. Hence, such operators of utility kiosks must be considered as part of the formalised system.

[34] Berg (2001: 1).

need for investments in first and last mile infrastructure in order to reach new underserved settlements. The sector can no longer rely as much as today on donors for investment and need a certain self-financing to make the sector attractive for investors. This is important to keep in mind when the regulatory regime is designed and gradually developed. Regulation has to be linked to investments not only to help in the provision of sufficient funds for investments but also to enhance the impact of regulation. Linking the power of funding with regulation is a very effective approach in moving the sector forward by ensuring good governance and a conducive behaviour of utilities and politicians. Regulation also has to help to cover new ground if needed such as the inclusion of the chain for onsite sanitation in service provision by professionals.

For the realisation of its functions, the regulator has a number of tools at hand, which are usually indicated by legislation or are derived from common practices:

- Licensing service providers (indicating minimum requirements, service areas, etc.).
- Standard setting.
- Tariff setting (methodology, negotiations, etc.).
- Benchmarking and reporting for comparative competition and transparency.
- Issuing guidelines on areas for improvements and comparability such as good governance, customer care, complaint resolution, cost accounting, pro-poor orientation, business planning, etc.
- Inspection as part of enforcement efforts and verification of data.
- Enforcement notices, penalties and imposition of a special regulatory regime.
- Consumer representation structures.

The use of these tools is already widespread among regulators in Africa helping to move the sector to a higher performance level and protect consumers but also to provide incentives and take enforcement measures, e.g. when approving tariffs. Generally, the effectiveness of regulation depends equally on the link of regulation and consumer representation because regulators need to know what is happening on the ground and also have an extended arm to the ground. Rouse[35] explains that consumers in industrialised countries are concerned about transparency. Hence, consumer participation in urban water and sanitation can be considered more a matter of transparency in service provision and in regulation as well as in enforcing rights than consumers digging trenches or repairing pumps like it is often demanded in the global discourse on water and sanitation. Rouse is referring to the UK Consumer Councils for Water as a best practice example.

[35] Rouse (2013: 120–123).

Table 4.4 The crucial factor regulation and the key intervention area

	Key intervention areas
Regulation	Licensing (formalisation of service provision)
	Tariff setting
	Consumer representation (sub-structure)
	Benchmarking based on reporting system for utilities
	Standards setting/guidelines (complains, governance, accounting, water quality, service areas, etc.)
	Inspection/enforcement measures
	Services for the poor
	Public reporting
	Etc.

However, regulation seems to have its limits despite the positive effects outlined by Allan et al.[36] Parker and Kirkpatrick[37] referring to a high number of studies, suggested:

> …that the effectiveness of independent regulation is influenced by government through a continued share ownership in the regulated company or by the threat of continuing political interference in regulatory decisions.

While undue political interference limits the effectiveness of regulation, the regulator needs the support from the ministry level in several cases, for instance when public institutions and other big consumers don't pay their water bill, tariff adjustments are contested by other stakeholders, BoDs of utilities undermine good corporate governance, utilities need to build-up funds for investments, etc. It is obvious that the line ministry alone will hardly guarantee support in all of these areas because it might have conflicting interests. That's why it is helpful to involve other state institutions such as the office of the Prime Minister or the Ministry of Finance in the water sector in order to ensure that utility becomes financially sound. Utilities must not become a burden to the tax payers due to underperformance, failure to repay loans, etc. (Table 4.4).

Utility Performance

That accesses to water and sanitation in towns needs someone to provide the services has been lengthily explained in Sect. 2.4. Equally, the fact that services have to be provided sustainably and meet minimum requirements (refer to Sect. 3.2). However, it has also been argued that improvements in utility performance alone

[36] Allan et al. (2013: 463). Regulation in three of the four target countries helped to change corporate culture of utilities (autonomy, behaviour of BoD, undue interference of politician to some extent). With the establishment of a regulator, at least a watchdog was established, which did not exist before (own observation).

[37] Parker and Kirkpatrick (2012: 39).

will not necessarily lead to a continuous and sustainable increase in access and that the mode of delivery is not crucial either. Allan et al.[38] citing a number of studies:

> …conclude that although private companies should be more efficient, evidence suggests this may not necessarily be the case.' Furthermore, 'The literature reviewed…provides a compelling argument that public and private companies can deliver comparable economic performance...[39]

The bottom line seems to be professionalism combined with a supporting development in the other crucial factors. However, professionals can only be effective if sector institutions / utilities are given sufficient autonomy in the day-to-day operation and are overseen by a BoD, which practice good corporate governance and also shielding the company from undue political interference. Under these conditions professionalism will lead to a new corporate culture in the sector in Sub-Saharan countries and finally to a turnaround in utility performance.[40] Corporate culture is also influenced by the way management is organised. A stiff hierarchy, which is based on outdated traditional values such as total obedience, age, male dominance, etc., is hampering performance improvements. Therefore, next to autonomy for professional staff and behaviour of the BoDs, the 'Stretch Out'[41] of modern leadership is important.

Keeping the utility an arm length away from politics (both national and local[42]) and other stakeholders and the need to engage professionals means that not only civil service structures but also community owned and managed systems cannot be regarded as best suited to provide services in the urban setting (refer to Sect. 2.4 and expert interviews).

> Now community systems, the challenge is professionalism, nobody was checking the quality of the water, they did not have any laboratories, provided water from the borehole. You know, everyone is just looking at how can I make small amount of money so the issue of governance in these areas is another challenge. They are not sustainable. Why we normally go into community involvement is, because we have challenges, issues of vandalism, non-payment. (Zambia 2015)

Considering the high number of underserved people who are mainly the poor (expert interviews, all countries) and the need for the involvement of professionals, service provision in the LIAs can be considered as another key intervention area for utilities in the low-income countries. Other important areas which need special attention of utility management are water quality, customer care which includes billing and col-

---

[38] Allan et al. (2013: 13, 14, chapter ownership and risks).

[39] Schiffler (2015: 1) reports, 'About 90% of water and sanitation utilities in the world are publicly owned and managed'.

[40] Schiffler (2015: 182); Allan et al. (2013: 472).

[41] Schiffler (2015: 159) is referring to NWSC in Uganda when using the term 'Stretch Out'; 'Gone was the distance that had previously kept the management apart from their subordinates'.

[42] There is always a risk that community owned and managed systems are captured by local politicians, traditional leaders, etc. There are several cases where local politicians build their political career by using water and sanitation issues in their election campaign – politicizing water and sanitation service provision (e.g. expert interviews, Kenya).

lection, service hours and maintenance of infrastructure because the perception of consumers concerning the service quality depend largely on them. Progress in these areas need to be measured with regional standards and should be sufficiently ambitious. Qualifying an improvement of water supply from 7 to 10 h as a successful turnaround of a reform, as Schiffler[43] proposes in the case of Aguas de Barcelona in Havana, seems overenthusiastic. Furthermore, efficiency in personnel and production are also important key intervention areas.

It is particularly noteworthy to underline the importance of NRW because of its negative effects on costs, assets, environment and especially on the behaviour of consumers and staff of the utility. Water losses around 60% of production are not rare in Africa. NRW can be considered the most difficult utility indicator to work on because of its complexity. A comprehensive set of commercial and technical measures and changes in the behaviour and integrity of personnel within the company is needed. The report on expert interviews, lists the measures to be undertaken to bring down and maintain NRW at an acceptable level of around 20%. Reaching this level in curbing water losses is an indication that a utility has obtained a new level of management quality and has become a player in a superior league, among very few members (utilities) in Africa. It also indicates that corruption in the utility has been contained, something extremely difficult to root out. No other indicator has this important meaning when it comes to management quality.

Contrary to water supply, most of the utilities still neglect responsibilities in sanitation development. Like water, human waste management needs the involvement of professionals in services and asset development. Considering the dire living conditions in slums (the case of Betty) and in planned LIAs (the case of Agnes) and the negative impact on society due to a missing or underdeveloped effluent management system, it is difficult not to regard sanitation as a key intervention area for utilities. Especially infrastructure development is important in a dynamic environment like in Sub Saharan Africa in order to make use of an improved utility performance. As one expert in Zambia brought it to the point (Table 4.5):

> And so you had the institutions established, equipped with good personnel running services but they needed investments…. (Zambia 2015)

Table 4.5 The crucial factor utility performance and the key intervention areas

	Key intervention areas
Utility performance	Services in LIAs
	Water quality (and other standards)
	Corporate governance/culture
	NRW (metering, corruption control, customer data base, billing, disconnection, leak repair, etc.
	Customer care
	Asset management
	Sanitation development (chain)
	Etc.

[43] Schiffler (2015: 29).

Information/Data

The following statement from Stonecash[44] can be considered as common knowledge: '[Information] is the raw material for making decisions for creating knowledge and fuelling the modern organization'. UKessays underlines further that information should be precise, readily available and easily accessible for better planning, decision-making and achieving planned results. Organisations depend on costly 'trial and error' activities when reliable information is missing. According to Rouse[45] the delegation of control over information is difficult to achieve. Therefore, there has to be a provision in the framework documents to indicate which institutions are responsible to control and which are to collect and update information and this for which areas of interventions. Information empowers people and institutions under the condition that it is effectively used and embedded in a certain management culture.

In many Sub-Saharan countries before reforms, it was impossible for decision makers to find reliable data for the sector. In general, sector information systems were housed in ministries and seldom functioned sustainably or produced reliable data. Often enough once installed, the systems did not operate for long and donors had to relaunch their support for information systems over and over again. Sustainability and use of the information systems were key challenges in the sector. The only option to know the situation of access before reforms was often enough the JMP data which however, could not replace sector information systems in the countries.

Although decision makers in the partner countries recognised the importance of information, the introduction of new information systems often failed to produce the desired results. There are several reasons for this. There was no pressure on civil servants (especially in ministries) to frequently update information systems. The institutions were not obliged to make data available to the public or report regularly to stakeholders by using the information system. In addition, civil servants were not regularly exposed to the critical review of the data and information provided to third parties and had therefore, little incentive to improve quality.

Software from the 'shelves', even when planned to be adapted to the expected needs, often did not meet the requirement of the institutions because the contracted consultants did not receive the necessary cooperation from the future users during the development of the systems. In addition, the middle and top management, which should have been the users of the systems, delegated the collection of data, the data entrance into the system and the validation of data to IT specialists, librarian,[46] etc. who did not understand the concept and aim of the systems. There was little under-

[44] Stonecash (1981) in www.ukessays.com (2015: 1; last visited 02.2017).

[45] Rouse (2013: 4).

[46] In the case of the Reform Support Unit in Zambia (own observation, 1999).

standing that there was a need for '…an equal contribution from both the…management and the information professionals…' as UKessays is pointing out.[47]

It took some time until the decision makers realised that the failures '… were not technical but were mainly related to employees and organisational culture…' (UKessays). The failure to ensure sustainable use of the information systems on a national level was a particular concern because it is at this level where strategic decisions are made and strategic planning for the development of the sector should take place. According to Akram,[48] strategic planning based on an information system '…has a clear effect on the Decision Making Effectiveness…'.

However, reforms have also changed the situation in this crucial factor. In many partner countries in Sub-Saharan Africa, donors supported the establishment of information systems anchored in the newly established (professional) institutions, which understood better than ministries that they cannot fulfil their functions without developing and using sustainable information systems and data-bases especially when they are obliged to report regularly to stakeholders or the public. This link between information systems and reporting obligation made a big difference. In the urban water sector, information systems are especially needed for the monitoring of utility performance, tariff adjustments and investments in order to see progress in the sector and the impact of decisions and measures taken.

How important it is to provide data for investments is indicated by Mosley[49] on the example of aid allocation by donors:

> The greater the need for aid, in other words, the greater is the risk of spending it badly out of sheer ignorance.' Furthermore, 'Even if data are available…they may well be completely unreliable. Sometimes this is because of 'pure' uncertainty and sometimes because figures have been deliberately falsified by a contractor who hopes to undertake an aid project.

This means that donors allocate aid to underperforming countries and sectors because the available information is poor. The consequence would be to concentrate first on building sufficient information in quality and quantity before making huge funds available as one interviewed expert in Tanzania proposed. This would mean that TA has to precede FC. However, establishing a national baseline and securing sustainability of information systems require an anchorage into the partner structure. Both need time. Once an investment project is in planning and is in reach of being agreed, donors and decision makers in developing countries might not like to accept delays in the implementation 'only' because of insufficient data.

Another area of intervention is the establishment of sector base-lines, which is particularly important in the countries in Sub-Saharan Africa where a very large part of the urban population is living in unplanned LIAs and their development is rapid. In these areas, development for water and sanitation needs to take place but planning is often hampered by missing data. What is reappearing regularly in the discus-

[47] UKessays (2015, chapter on problems with information systems), www.ukessays.com (2015: 1; last visited 02.2017).

[48] Akram (2011: 459).

[49] Mosley (1987: 63).

Table 4.6 The crucial factor information/data and the key intervention areas

	Key intervention areas
Information/data	Information systems (utility performance, investments, tariffs)
	Baselines for LIAs
	Reporting (to the public, sector wide, towns, LIAs)
	Defining service areas
	Customer satisfaction surveys
	Linking of sector information systems
	Management information systems – all institutions
	Etc.

sions about information systems is the proposal to create one (mega) system which combines several sub-sectors and satisfies all the needs of the ministry and other sector institutions. Experience shows that such tentative never succeeds because information systems must be designed for the specific needs of each institution and if needed a separate system for each key function is useful within one institution. For instance, a regulator will not mix information systems for monitoring utility performance and tariff reviews because reporting on performance is usually annually while tariff reviews are carried out only once in several years. In addition, the detailed information on costs, financing, etc. demanded for tariff reviews is not needed for performance monitoring. Furthermore, sourcing out functions to professionals includes the collection and analysis of data. Consequently, the ministry should make use of the output of such systems handled by professionals and not try to develop parallel systems in addition. Hence, information systems should be interlinked instead of trying to create one mega system (Table 4.6).[50]

Infrastructure Development

To obtain physical access to water and sanitation in towns, outlets for water and inlets for piped sewer or a link to a decentralised sanitation chain are needed. In addition, continuous water services depend on two distinctive part of the necessary infrastructure: raw water capture / storage, intake facilities and (bulk water) transfer pipelines to the towns on one hand and the production and distribution of drinking water up to the outlets for consumers on the other hand. It is helpful to make this differentiation because of the source of financing and sometimes of their operation when both parts are institutionally separated.

The huge backlog of investments for water in the low-income countries is especially worrying because it concerns the two parts of infrastructure and has two effects. First, it is the cause for water rationing due to insufficient raw water infrastructure which is felt by all consumer groups including the productive sector limiting economic growth. Second, distribution networks which do not reach all areas in

[50] Software companies offer different systems for certain key application such as text treatment, calculation software, presentation software, etc. in different components which are interlinked but not combined in one single application.

a town in a situation where the population is growing rapidly makes coverage decline or at least stagnant. Because generally the LIAs are not reached the discrimination of the poor is prolonged. For sanitation, the backlog in infrastructure development means that the percentage of urban people having access to piped sewer systems is declining all over Sub-Saharan Africa which is not yet compensated by infrastructure and services provided through a formalised chain for onsite sanitation. The result is an increase in the pollution of the living space and water borne diseases (Africa as the new homeland of cholera).[51]

Bringing more raw water to towns and increasing access to sewer networks will need huge investment projects in the near future in many low-income countries. However, the infrastructure to reach the urban LIAs, needs less investments per capita if it is carried out with low cost technology (last mile). The repayment of credits of infrastructure serving several generations cannot be impounded on the present consumers only. Considering the financial situation of the states in the low-income countries, funding will require an extended and combined effort from the sector and donors offering funds with favourable conditions of repayment. It will also need changes in the way the sector is mobilising and using funds.

Bearing in mind the recent case of water shortages in Cape Town, widely publicized, many decision makers in the partner and donor countries still do not realise that this has already started to happen on scale in the Sub-Saharan region and the potential it has to reach a significant magnitude in the future with the rapid urbanisation. The 50 l per capita per day restriction in Cape Town was seen as a major crisis and yet it is already a permanent event in many towns in low-income countries which experience a declining of raw water availability for towns.[52] When a well organised water sector and strong economy like in South Africa struggles to deal with such issues one can imagine what will happen in the low-income countries when the many signals today are carried on the be ignored. It will contribute to an acceleration of decline in living conditions for many more of the poor which might further fuel the already worrying migration movements.

There are still stakeholders using the argument that it is not so much money which makes people gain sustainable access, but the missing sensitisation of users of infrastructure. This argument, mostly used by civil society organisations, is mainly based on experience obtained in the rural setting where sometimes new infrastructure is not used and maintained by the beneficiaries. People continue to make use of traditional water points such as hand dug wells for water instead of paying for the electricity of a motor pump installed in a new borehole or go back to traditional latrines for defecation and use a new toilet facility, often given as a gift, as a storage room. However, this might be possible in the rural setting but the people in the urban areas can hardly behave like this because they know the risks stemming

[51] Gaffga et al. (2007)

[52] According to the Impact Report of Wasreb in Kenya, presently more than half of the around 90 utilities in the country provide 50 and less litres per capita per day.

from contaminated ground water and in general they don't have traditional latrines or a garden or field as alternatives for defecation.

Furthermore, in the urban setting the existing infrastructure for water and sewer is in many cases hopelessly overstretched that the need for hardware development is today overarching software concerns. The observation of Lord Milner,[53] almost 100 years ago, seems to have more relevance today for urban water and sanitation then ever: 'What these countries need … is economic equipment … Their development is a question of money – and money from outside'.

Once reforms have helped to overcome challenges such as inadequate frameworks, insufficient human resources, missing policy orientations, inadequate governance, and etc., the limiting factor is physical infrastructure and the challenge is the efficient use of funds. Hence, there are three interesting questions to be answered:

- Where and how can the sector find sufficient funds and what else can be done to ensure a development (extensions) and replacements of aging infrastructure, which matches the rapid growth of demand?
- How should funds be channelled into the sector to obtain the highest impact (fund / aid effectiveness)?
- What type of infrastructure should be put in place and how shall it be implemented considering the fact that there are many unplanned settlements?[54]

The first two questions will be answered under the next sub-heading (financing / international cooperation). Regarding the third question, which concerns rather access in the LIAs than bringing sufficient raw water into towns, the deliberation in Sects. 2.5 and 3.4 provide already some indications. Last mile infrastructure development for water and basic sanitation provided as interim solutions by utilities needs a stronger focus and there should be a mix of first and last mile investments which is appropriate for the country in order to increase rapidly access. It has also been underlined that the design and implementation of these two types of infrastructure is very different and that the up-scaling of low-cost solutions (basic service level) in the LIAs has to be carried out with a national concept and by a specialised national institution for which the donors should provide grants.

Such a (pro-poor) financing mechanism (basket) which also sets standards and build capacity among utilities can be established as a 'stand-alone' institution (e.g. Kenya and Zambia) or as a window of a sector investment basket (first and last mile). It is important to note that such baskets should be placed on national, and not on a regional level in order to ensure that they are able to support the ministries for fund mobilisation, sector wide planning, financing modelling, monitoring of investments, etc. Using cross-sector funding baskets for municipalities for water and sanitation development is often not helpful because it mixes different areas of development which need very different procedures for funding and investments.

[53] Lord Milner (1923: 40 in Mosley 1987: 16).

[54] Connecting everyone is also out of reach because many consumers are not able to pay an accumulated consumption, e.g. monthly water and sanitation bill.

Table 4.7 The crucial factor infrastructure development and the key intervention areas

	Key intervention areas
Infrastructure development	Financing mechanism (professionals and appropriate mix of first and last mile)
	National planning (bottom up)
	Sector financing model
	Standard setting for investments (especially last mile)
	Monitoring and reporting
	Capacity building among utilities for last mile extension and operation
	Coordination in asset development
	Etc.

Asset ownership and the design of projects, especially when production and distribution for water as well as collection and treatment for sanitation is concerned above a certain volume, are better left with the utilities or a specialised water and sanitation asset development fund for the sector. However, the conditions of handing over assets once completed to the utilities must be clear and the hand-over procedures strictly enforced because sometimes asset developer use their power to deny utilities new infrastructure in order to make money by starting illegally own operation.[55]

Concerning the challenge to increase the raw water availability for towns, new technologies might have to be considered. Many big and fast growing towns in the developing world are far from major rivers or ground water sources because they were founded for other strategic reasons such as Nairobi as a railway stop for the line from the Indian Ocean to Uganda. As it is now growing into a mega city (around 16 million people in 2050), raw water will have to be piped from increasing distances as demand is rising. At the same time the river runoff in the surrounding areas is substantial or even increasing. Topography might not permit to build big dams near the towns. Hence, very large open ground tanks near the cities covered with floating sheets for protection and reduction of evaporation might be a cost saving solution for investments and operations in order to bridge shortages in dry seasons (Table 4.7).

Financing/International Cooperation

This sub-heading is organised in two parts: the financing of infrastructure development and the influence of the international cooperation in the low-income countries. However, international cooperation in the water and sanitation sector is generally considered helpful but unable to unfold its full potential and therefore limiting development in the receiving countries. In short, there is ample room to improve aid effectiveness.

[55] E.g. in Kenya with the regional WSBs, own experience.

Sector Financing The huge amount of funds needed for infrastructure development in urban water and sanitation in the industrialised countries were generally made available by the public in the second halve of the 19th century. This was the time when the benefits from the colonies and industrialisation provided sufficient funds to the states for financing the necessary public works. It has already been indicated that the Sub-Saharan countries are in a very different position today in respect to the availability of funds and the persistent growth of rapid demand. National budgets in Sub-Saharan countries can barely be balanced without contributions from the international cooperation and the competition for funds between the sectors makes it unlikely that funds from general taxes will be one of the main sources for investments in urban water and sanitation in the near future. Despite this fact, national policies repeatedly insist that government has to provide more funds. This un-reflected repetition limits the search for alternatives.[56]

Schiffler[57] estimates that worldwide 100 billion Dollars are needed annually in the developing and emerging countries for investments in the urban water and sanitation sector compared to the 50 billion invested. The gap between investments needed and undertaken will most likely be much higher in Sub-Saharan Africa if one considers that all of the 10 poorest countries in the world are in Africa and urbanisation is rampant. Hutton and Varughese[58] estimate that the requirements worldwide for reaching the SDGs (target 6.1 and 6.2, water, sanitation and hygiene) is around three times the current investment level, whereby Sub-Saharan Africa would need around a third of it. This is estimated at 35.5 billion USD per year for the next 14 years.

Another indication, that the gap might be bigger than thought, is an estimation for Kenya which shows that the gap is around three times of the amount currently invested (see Chap. 5, Fig. 9) and this despite a deep-rooted sector reform helped already to increase funds three times. Investments grew from 3 to 9 billion KES (around 30–90 million USD per year) over a period of 8 years (2008–2014).[59] It is projected that the gap will further widen in future with the adopted new policy

---

[56] Even in Zambia, where one of the adopted seven sector principles is increased funding by government or in Kenya, where the Sessional Paper No. 1 proposed more fund allocation by the Treasury, government contribution for asset development remained very low (less than 10%) during the 15–20 years of reforms. It is of little help when sector policies insist repeatedly that the state has to contribute more to the sector and it remains wishful thinking.

[57] Shiffler (2015: 7).

[58] Hutton and Varughese (2016: 15).

[59] According to the *'Strategic Investment Plan for The Water and Sanitation Sector in Kenya, Final Report 2014'*, (2014: ii – Scenario 3) Ministry of Environment, Water and Natural Resources, the investment needed for full coverage (Horizon 2030) for 2014 for water was 12.8 Billion KES and are increasing according to an average of 31 billion in 2025–2029. For sanitation, 22 Billion of today's needs will increase to 88 Billion in 2025–2029. Therefore, the total investment needed for 2014 (35 billion KES) was only matched by around 26% with available investments of approximately 9 billion KES.

goals.[60] The persisting investment gap in the sector is also the result of no or little consideration regarding transitional service levels such as public outlets for water (water kiosks), shared sanitation facilities and infrastructure for decentralised sludge collection and management. Extensions with such low-cost / last mile solutions (basic service level) would not only reduce the required amount of funds for investments in LIAs but also help to delay the need to upgrade first mile investments (refer to Sect. 3.4).

The options to find funds for investments to close the financing gap in the sector are limited in Sub-Saharan countries. There are international and domestic sources available, which are provided by the public or the consumers, multi- and bilateral donors in the form of loans and sometimes in grants and to (very) limited extent the private sector. Donors generally provide grants for pro-poor investments or demand national governments for co-financing to cover the LIAs. Co-financing of donor grants and loans by national governments of receiving countries seldom exceeded 10% over a longer period but generally remain under 5%. Consumers could contribute much more than in the past to finance asset development by paying an average tariff well above 100% O+M costs with their water bills. This would mean that the middle and upper class consumers would contribute according to their ability / willingness to pay.

Unfortunately, this source of financing still receives little considerations in most of the low-income countries as politicians try to keep the tariffs as low as possible for everyone who benefits utility services. This is no help to the underserved people. Most likely it will need other substantial water and sanitation crises in the partner countries to make politicians understand that the next unavoidable steps in reforms need to be an increase in self-financing of the sector through water and sanitation bills. However, in crises where the utilities have to ration water drastically for everyone including the big consumers the potential of increasing self-financing of the sector diminishes rapidly. In addition, in such a situation big consumers might be lost for the utility for a long time because they are pushed to invest in boreholes to tap ground water resources. In such a case, the most lucrative clientele brakes away from services and usually there is little utilities can do about. The consumers who have up to this point covered the subsidies for the poor are no longer contributing which makes it as good as impossible for the utilities to maintain cost recovery. Even water resource management authorities would find it difficult to obtain in such a situation sufficient political support to stop people from sinking boreholes. To try in such a situation to introduce a reform by increasing utility tariffs for the remaining consumers including the poor is hardly an option because it will face stiff opposition from politicians and the general public. Therefore, it is always better to start reforms before critical supply situations escalate.

When donors are criticised for insufficient aid effectiveness it should not be forgotten that this is also due to the insufficient system in the receiving countries which has to change and that there is ample room for improving the effectiveness of

[60] 100% access for water and 50% access to sewerage networks in the urban setting.

national government contributions to the sector. Many sector institutions receive substantial amounts of money from the national budget for which there is little scrutiny of how the funds are used. Unfortunately, often enough it is considered an internal affair of partner countries and that donors stay out of it. This should be seriously reviewed by donors because why should public money from donor support partner countries which waste or misuse their own public funds in the sector? Donors should demand that their contributions are pooled with government funds for the sector and submitted to the same strict supervision and procedures as donor money is obliged to undergo. This would need the outsourcing of financing and asset development from civil service structures (ministries / municipalities) to corporatized professional entities. It should be an area where donors should not make compromises.

The contribution of NGOs to investments in the sector is estimated to be less than 1% worldwide according to Schiffler[61] while from donors ('official development assistance') it is around 23%.[62] The donor contribution is now shifting increasingly from grants to subsidised / concessional loans. Considering the high poverty level and other factors, such as pressure on European countries by immigration from Africa, a higher share of donor contributions for the sector can most likely be expected in future, once the full extent of the water crises is recognised. Burkina Faso, which received around 60% of total funding for asset development from donors over two decades and invested substantially more than the four target countries per capita to be served, is an example of what countries could expect as total amounts from international development cooperation once the sector becomes credible.

Schiffler[63] also indicates that financing of urban water and sanitation infrastructure by the private sector is around 3% in the developing countries. The reasons for this limited contribution are higher interest rates, shorter period of repayment and especially the currency exchange risks which will always make international private lenders stay away from the water sector in the developing world. Consequently, there is the local private capital market remaining which will only be available for investment which guarantees a significant and constant income stream to which the banks can access in case of repayment difficulties by the utilities. Large scale private funding will also lead to unacceptable high tariffs for the urban population at one moment in time.

In addition, it cannot be expected that the contribution of the private sector for infrastructure funding will increase trough more PSPs because it is not certain that the efficiency gains of this mode of delivery will outweigh the additional costs of financing hard ware in the developing world, as Schiffler observes.[64] Hence, private sector financing including blended financing in the water and sanitation sector will

---

[61] Schiffler (2015: 8, 9).

[62] 20% direct to the states and 3% through private.

[63] Schiffler (2015: 9).

[64] Schiffler (2015: 33).

remain at best a niche product. Also available empiric evidence points into this direction. The water and sanitation sector is simply to complex, too risky and further progress in access needs increased poverty orientation (the high hanging fruits are next).

It is also important to highlight that investments need to be undertaken at a certain moment in time in the development of the system. Consequently, a huge amount of funds is needed at a certain time when, for instance, the treatment plant has to be extended and its design has to cater also for the next generations. This means two things. Firstly, the utilities will have to have a tariff which allows the accumulation of funds prior to such big investments as self-financing contributions in order to attract sufficient funding from external sources. But accumulating huge funds is very difficult for a utility in an environment where (national and local) governments are generally cash strapped and politicians promise a group of voters that water will be free of charge or tariffs will not be raised in future. Secondly, cumulating funds need ring-fencing of income which must be considered a key intervention area under the crucial factor financing. The need to cover average costs has been explained. Therefore, tariff studies and the preparation for tariff negotiations are other key intervention areas. For both, there are needs for a strong framework and discipline in the enforcement of the sector principle of cost recovery.

According to the available information, sector funding in Sub-Saharan countries faces the same constrains than sector funding in the rest of the world, although with tighter limits. In all four target countries, donors are the biggest contributor to asset development. In exceptional cases, self-financing through the water bills is second in place, but usually it is the national governments with limited contributions. Private capital is only available in very specific situations and NGO contributions for assets[65] seem to be as good as inexistent in the urban setting. Bankable projects financed by local institutions are limited to extensions into high-income areas or industrial zones as well as for specific components of the systems such as treatment plants or pumping stations for instance where a block tariff with a steep progression enables the utility to recuperate sufficient benefits from the investments within a relative short time.

Despite the limited potential of generating private sector funding some experts interviewed (Zambia) still insist that it has the highest potential in the sector in the developing world.[66] This unrealistic vision is most likely routed in the messages from literature and a constant push from some donor institutions in this direction of thinking.[67] It is surprising that the unsuccessful call for more private sector funding,

[65] In all four target countries there was no information available concerning spending of NGOs for asset development in the sector. A number of government officials criticised this missing transparency of NGOs especially when considering their claim to watch behaviour and governance of the sector institutions.

[66] Also this demonstrates an unrealistic vision often observed among decision makers in the developing world, most likely because of missing information, knowledge, etc.

[67] E.g. WSP/WB (Banking the un-banked), PPIAF/WB (PPP in small scale service provision and in financing infrastructure / Meeting the Challenge of Financing Water and Sanitation) or the Dutch initiative 'Kenya Pooled Water Fund' (Embassy of the Kingdom of the Netherlands).

raised since decades, is now reappearing with the proposal to increase blended funding. According to Shiffler,[68] also ‘Equity capital and bonds play a limited role’ in developing and emerging countries worldwide and, as outlined above, most likely even less in Africa. It is also unlikely that for instance pension funds will be available on scale for the water sector considering the different opportunities in and outside countries and the limited performance in the urban water sector. As one expert interviewed stated, ‘You cannot sell a bond to someone who is dying [when you are dying].’ (Zambia 2015). The concern about costs of capital investments is justified, like Klein[69] underlines with the suggestion that regulatory arrangements should play a role in ensuring that funds for investments are obtained at acceptable costs and thereby make sure that tariffs remain affordable.

Now the first question raised in the previous section can be answered: Where and how can the sector find sufficient funds and what else can be done to ensure a development (extensions) and replacements of aging infrastructure, which matches the rapid growth of demand? First, the sector should promote more low cost technology (basic service level) in order to close the financing gap by reducing the need for more first mile investments and reach as many people as possible. Concerning the funding of first mile infrastructure, the sector in Sub-Saharan Africa remains mainly with the (three) ‘*conventional*’ options. The self-financing of the sector, funding through international cooperation, which can provide loans with favourable conditions in terms of repayment time and interest rates (concessional loans), and to some limited extent contributions by the national governments. The latter should primarily concentrate on unlocking the raw water shortages in towns.

Concerning the second question (How should funds be channelled into the sector to obtain the highest impact?), donors need to take decisions and provide funds in a way which increases aid and government funds (receiving countries) effectiveness. In short: make better use of what is already available. The luxury of wasting funds because of the relaxed attitudes of donor staff and politicians who allocate funds for other reasons than sustainable development should now phase out because the situation in urban water and sanitation is becoming too critical. The time to support countries which accept underperforming utilities, hold on to ill-conceived frameworks, have a laissez-fair attitude in corruption and towards the household connection only paradigm, accept that decision makers use the water sector for political careers, ignore the plight of the poor, etc. must be over. There are too many low-income countries ready for sweeping reforms where a high level of aid effectiveness can be secured and where donors can pour money in a tight instead of a leaking bucket. Wherever it is possible to make utilities compete for funds a high level of aid effectiveness can be reached (e.g. basket funding). Unfortunately, the use of this mechanism is today an exception because it would mean to control political interference and ensure a close and sometimes cumbersome follow-up by donors (TA and FC) on good governance.

---

[68] Shiffler (2015: 7).

[69] Klein (1996: 27).

In addition, donors need to give the sector the same attention in the partner countries as it was given in their own countries in the past. This implies that donor contributions have to increase by shifting funds from other less important sectors to urban water and sanitation development. All other financing option often praised as the solution for closing the financing gap must be viewed with caution such as commercial or blended financing, mobilisation of pension funds, etc. because they have not been used on scale so far. There are pilot projects which have not been sufficiently tested yet and never up-scaled in the developing world. Therefore, they can in the best case be regarded as niche products. The disappointing Kenyan example on PSP, outlined further below, with the unrealistic promise of generating private sector funds for investments will demonstrate how devastating it is for a low-income country to rely on 'innovative' but untested solutions and on top of it establish a tailor made framework for it (see Chap. 5).

There is also need to distinguish between sector and utility performance because government contribution has an effect on the sector beyond utilities. National governments receiving concessional loans from development partners and are deciding to pass them on to the utility as grants must not necessarily help the sectors' and countries sustainable development. Firstly, it increases national debts which have to be repaid by general taxes and secondly it might not help to put enough pressure on the utilities to increase performance as much as when the utility is obliged to repay the loans and at the same time maintain affordable tariffs. Furthermore, there is the risk that governments use this as an argument to keep on unduly interfering in the sector and deny progress in its build-up of resilience.

International Cooperation There are two major external influences in the water and sanitation sector in low-income countries. These are the UN system and the donors providing development aid in the form of financial cooperation and technical assistance (FC and TA). Philanthropic organisations and NGOs are part of the international cooperation concentrating mainly on research, small scale systems and provision of TA. Because donors finance the lion share of asset development in the low-income countries they can have a significant influence on the sector policy and development if Technical and Financial Cooperation work closely together. There are good examples how effective cooperation between autonomous acting Technical and Financial Cooperation can be (e.g. Kenya).[70] The opinion on the role of the donors in the partner countries differ often substantially from the judgements made by stakeholders from the international cooperation. Experts in low-income countries seem to have in general a positive opinion about donors and their input (expert interviews). But this does not mean that all is fine with the contribution of donors.

Not long after its commencement (1949), the International Development Cooperation was criticised because the way it was functioning and the effects it was creating. It had to go through a remarkable learning process, which is still ongoing

[70] Mwega (2009: working paper 8), regarding German Development Cooperation in the water sector, cooperation of GTZ and KfW.

and there are reasons to assume that it will have to carry on improving in future. It started with the lesson learned that development support needs a different and more complex input than the reconstruction of the industrialised countries after the World War II with the Marshal plan. Thereafter, during the Cold War, aid became an instrument of power for the East and West competing for alliances in the developing world. Mosley[71] labelled the use of development aid during this time as an 'instrument of persuasion'.

Also the report of the Brand Commission[72] labelled aid as an instrument used in the interest of the donors instead helping the development in the partner countries. The post-colonial discussion underlined that the one-sided and discriminatory relationship between the colonial powers and the 'rest' of the world was carried forward after independence.[73] Local knowledge'[74] did not count much in the discourse of development.[75] Moyo,[76] a fierce critic of traditional aid, argues that aid is not working because it brings no improvements in the development countries. Her arguments are that in the development world, public institutions (civil service, police, and judiciary) are ineffective, corrupt and have no real power next to the president, who is often acting as dictator. She claims that aid is fostering corruption and conditionality's do not work.[77] The latter is only partly shared by Schiffler[78] who stated that external influences cannot force decisions onto politicians in the partner countries. However, he recognised that donors can 'convince' partners to choose a certain development path and can help to amplify efforts considered as purposeful. He[79] linked the success of Phnom Penh Water Supply Authority (PPWSA) in the Cambodian utility to the fact that 'Foreign donors also helped to push for the creation of the right institutional framework: the creation of an autonomous utility.' Nevertheless, as explained, the intervention in the framework development goes beyond the intention to secure autonomy for sector institutions and sector development depends on more than utility performance.

[71] Mosley (1987: 24).

[72] Brand Commission (1980: chapter 3), '…argued the case for international aid as a self-interested rather than a moral or redistributive act.'

[73] Refer to Hall (1994), post colonialism.

[74] Refer to Mignolo (2008) and Escobar (1994).

[75] However, critics of the post-development discussions pointed out that in order to guarantee sustainability in services, the state needs centralized structures on national level as support for local systems and as insurance that they comply with minimum requirements in the interest of the one to be served. Hence, it can be concluded that one needs the other in order to serve the population adequately. In addition, the critics underlined that local dominance does not necessarily support development and ensures equality (refer also to Joshi in Sect. 2.7).

[76] Moyo (2009: chapter 3).

[77] Moyo (2009: 52, 38).

[78] Schiffler (2015: 4, 183).

[79] Schiffler (2015: 168).

Furthermore, Moyo[80] indicates that aid can damage the framework in the partner countries and interventions to benefit the poor can harm sustainable development and thereby, increase poverty. 'For most countries, a direct consequence of the aid-driven interventions has been a dramatic descent into poverty.' The 1980s, called by Moyo[81] 'the lost age of development', saw many partner countries defaulting on the repayment of loans. Contrary to these critical opinions, Mosley[82] made a solid case for the defence of foreign aid without neglecting critical issues. He analysed aid as an instrument promoting development (growth), redistribution (poverty alleviation) and exports in the partner countries. According to Mosely, international cooperation as a state intervention by donor countries can be justified with several arguments such as moral reasons, promotion of growth for bilateral benefit, market substitute[83] and one can add today, as a tentative to limit mass migration. Furthermore, he documents[84] that the arguments of aid as an instrument to extent former colonial power into modern days or donors ideology are unfounded: 'it is no longer capable of doing the job' because of the considerable power the recipient countries have today.' However, conditional aid became acceptable to many experts and the World Bank was spearheading this orientation[85] because donors had to limit risks for their contributions and have the recognised obligation to influence changes in the receiving countries in order to achieve more growth and equity in the world.

Considering the unused potential of aid in the sector and the huge investment gap, conditionality's must be upheld as instrument but more effectively designed and implemented. Following Mosley's argument that aid is to be used to develop functions and tools in a market which are not operative yet, donors providing concessional aid with public funds (e.g. 1.5% interest rate) have the obligation to generate with their contributions an additional benefit compared to financing by local private banks (with e.g. an 10–25% interest rates). This benefit, next to the aim to increase access to water and sanitation in low-income countries, must come as a result of the framework development and its implementation and should be in an order of magnitude which ensures that the sector can increasingly tap into national commercial financing sources.

This would mean that over time donor contributions help to increase credibility of the sector by curbing sustainably corruption and undue political interference leaving a footprint beyond their project. It is obvious that thereby donors have to think beyond infrastructure development and utility performance. The result of such a contribution of donor funding for instance will be a gradual shift from pure donor funding to blended financing and thereafter to increasing funding by local banks, etc. One could imagine an equation where at the beginning of such a process com-

[80] Moyo (2009: 47 and 51).

[81] Moyo (2009: 17).

[82] Mosley (1987: chapters 5, 6 and 7).

[83] Mosley (1987: 3–9).

[84] Mosley (1987: 43).

[85] Mosley (1987: 29).

mercial financing with high interest rates equals concessional loans and their benefits in terms of framework improvements. Over time when the risks in the sector decline, interest rates from commercial banks will decline and concessional loans can drop as further benefits through framework improvements are generated.

Moyo's opinion about foreign aid and its effects in the low income countries seems to be an overstretched generalisation. It cannot be neglected, that foreign aid is channelled to very different types of areas, having different aims and also time horizons. Hence, indicators to measure success or failure have to differ at least in these cases. Humanitarian aid and response to disasters are very different from investments in urban water and sanitation, where infrastructure serves several generations. For instance, handing out mosquito nets to contain malaria, or food in case of large scale starvation have most likely different effects on people and the economy than funding assets for water and sanitation.[86] However, it is recognised that aid as a short-term intervention provided outside a sustainable development concept can be harmful to the partner countries and undermine effective support.[87]

Nevertheless, the discussion about the usefulness of foreign aid was very helpful and finally led by the end of the last millennium to changes guided by new principles. International cooperation strived to move to a situation where partner and donor countries could cooperate on par and merge their different knowledge in the interest of higher aid effectiveness. This process of change in development support was spearheaded by the OECD commencing with the organisation of a high-level meeting in Rome (2003). Three other high-level meetings followed, which led to important outputs: the Paris declaration (2005), the Accra action plan (2008) and the Busan agreement (2010) on the Global Partnership for Effective Development.[88] The high-level participation from developing and donor countries as well as from the civil society in this process of the aid debate secured recommendations, which became acceptable to all stakeholders. This large participation also added weight to the recommendations which introduced a new set of principles and standards for international cooperation:

[86] The rapidity of disaster relief will save many lives and is therefore a key criteria and more important than the possible negative effects on the economy. Handing out imported mosquito nets will reduce the cases of malaria. Consequently, it is more important to reach as many households as possible, than to safeguard a handful small scale local entrepreneurs. If a sustainable use of the actions can be achieved by establishing a local network for services guaranteeing regular treatment and repair of the nets, the benefits might outstrip the negative effects on the local economy when local producers of mosquito nets disappear.

[87] E.g. in Torit, South Sudan, where an NGO is operating in the same area where the utility is trying to extend standardised services (e.g. confirmed water quality through treatment and regular testing) to the poor within a framework of commercial viability. The NGO is undermining this effort of sustainable development by subsidising its water production and by holding production costs low by ignoring minimum requirements (e.g. distributing raw water from a borehole within the settlement). It can be assumed that once the NGOs will cease its support, the system will collapse as it happened in towns like Yei before (own observation, 2016).

[88] http://www.oecd.org/dac/effectiveness/parisdeclarationandaccraagendaforaction.htm (last visited 02.2017). In this work the outcome of these three high level meeting is referred to as Paris/Accra/Busan Agenda.

- Ownership of developing countries over their development process, their strong leadership in coordination and the use of country systems for aid delivery.
- Alignment of donors and other development partners to local systems.
- Harmonisation of donors to avoid duplications and simplify procedures.
- Delivering of results, which need to be sustainable and measurable.
- Mutual accountability and transparency of partners (institutions of receiving and donor countries) which are accountable to all citizens and should ensure the complementarity of their actions (inclusive partnership).
- Capacity development to enable countries to build and manage their own future.

The Accra meeting emphasised again strongly on poverty reduction, gender equality and social justice. Since the Paris declaration, progress in the implementation of these principles is regularly measured and the results are published.[89] Another high-level meeting followed in 2014 in Mexico.

Although there are signs of improvements in the role the donors play in financing infrastructure for water and sanitation according to the results of the experts interviews,[90] it seems that the process launched by the OECD has lost steam over time and monitoring of the implementation of the above mentioned principles does not necessarily document a substantial breakthrough in aid effectiveness on large scale. There are even signs that strategies of development banks have shifted and this not always in the interest of aid effectiveness. Moretti and Pestre[91] provide a critical view on this shift by analysing the language of the WB used in its reports since 1958. This analysis reveals a remarkable swing from 'a strong sense of causality links expertise, loans, investments, and material realization' to a more cryptic and less precise language which no longer addresses challenges in a direct message based on a 'critical understanding' of the situation.[92] In the last years it became obvious that the positive OECD initiative needs to be relaunch as soon as possible with more precise definitions, quality in monitoring and mechanism of enforcement.

It is understandable that development banks like any other commercial lending institution are concerned about repayments of loans and therefore, need to establish a constructive relationship with governments of the receiving countries. It is also understandable that development banks want to escape criticism regarding their contribution which is often difficult to avoid with interventions based on insufficient insider knowledge of partner institutions and their practices. However, donors too often shy away from their share of mutual responsibility and the control how funds are used. Signs of insufficient governance and inadequate management at the receiving institutions which compromises the sustainability of the investments funded by donors are regularly ignored. The decision makers of the donors do not want to see

[89] E.g. http://www.oecd-ilibrary.org/development/aid-effectiveness_9789264050877-en (last visited 02.2016).

[90] Almost all experts considered the role of the donors as generally positive, although many indicating also some flaws in the provision of funds.

[91] Moretti and Pestre (2015).

[92] Moretti and Pestre (2015: 77 and 95).

a disruption in the disbursements of funds once contracts have been signed and in addition, can still rely on Treasury of the receiving country for the repayment of loans when utilities fail.[93] However, it should be recognised that in the interest of sustainable development in the sector, utilities should not be allowed to offload their repayment responsibilities to the government for investments in production and distribution.

Donors have a crucial role to play in the sector development in receiving countries because the impact of aid must be measurable in terms of sustainable development. The difference in the missions of development and commercial banks start disappearing when the importance of aid for the promotion of soft elements of development is fading. To gain sufficient insight and use local knowledge, financial cooperation need to be combined with competent TA being integrated into the partner system (refer to Wolfensohn Centre for Development). Mutual responsibility needs to be taken more seriously by donors from the design to the completion of projects.[94]

One would think that a high demand for limited funds would automatically create healthy competition where a receiving country which can offer the best proposals in terms of aid effectiveness, transparency and progress in sustainable development obtains most of the funding / will have the best chance to be selected. This is not necessarily the case as underlined by Mosley.[95] There are other criteria for donors to allocate funds such as past history of relationships, regional and political preferences of donors, strategic importance of the country, poverty level in the receiving countries, administrative constraints of donor structure, etc. The financing of investments in the water sector in Tanzania and Uganda by donors are examples of this phenomenon. But providing funds with the sole intention of helping the poor or the interest of the decision makers in the donor country can be counterproductive in regards to sustainable development on medium and long term. Often, the 'fiscal impact' of donor contributions are not sufficiently taken on board, according to Schiffler.[96]

It would be in the interest of all parties that the development banks focus better on their mandate which is to promote sustainable development in the low-income countries and provide continuity in their support. For the water and sanitation sector

---

[93] Own observation and also indications from the expert interviews. For instance, a donor from the financial cooperation carried on investing in the Nzoia Company situated in the Lake Victory North region despite advisors from the TA warned in 2013 that progress of the water company in performance is inadequate and in some areas deteriorating due to serious governance issues. The message was clear: Without changes at Nzoia oversight and management the sustainability of the investments are compromised. Despite this warning, investment funding continued and not long after the final commissioning of the investments (2016) the regulator WASREB had to use the provision of the new Water Act (2017) to put the Nzoia Company under a special regulatory regime, an instrument for regulation to achieve substantial changes at the utility (mainly a replacement of management and BoDs).

[94] Wolfensohn Centre for Development (2009, working paper 8).

[95] Mosley (1987: chapter 3).

[96] Schiffler (2015: 174).

in the development world it would mean that basic conditions have to be fulfilled in the receiving countries such as the ability of utilities to repay their loans for instance.[97] The responsibility of the state to guarantee human rights does not mean that politicians have to dish out gifts with the intention to maintain counterproductive control. Furthermore, a sound concept for the sector means that an application for funding is based on a national investment plan (if possible with a bottom-up-planning), a sector-financing model and a monitoring system which can document that the intended use of the funds and infrastructure was fulfilled. Unfortunately, these instruments are missing in many receiving countries.

In addition, donors should not try to ignore the need for professional TA which should function autonomously from bankers. Capacity and institutional building is as vital as money and too important to be provided as side event or by accompanying measures of infrastructure funding. TA must be embedded into the institutions of the receiving country for a substantial amount of time and should be a mix of short and long-term interventions (expert interviews). Reforms with technical support requires more time (10–15 years) than a construction project (2–3 years).

Next to donors, the UN system has a significant influence on the sector in the developing and donor countries alike. With the first UN Development Decade (1960), economic goals for partner and donor countries were established. The aim was that the GDP in partner countries should grow annually by 5% and that the contributions to the development cooperation by the donor countries should reach 0.7% of their GDP per year. However, as Mosley[98] noted, already by the end of the 1960s, it became evident that partner countries faced limits in the absorption of aid and that aid did not reach the poor segment of the population to the intendent extent. It was recognised that it needed more than a growing GDP if everyone in a country should benefit from development. Thus, the reduction of poverty and inequality received stronger attention in the discourse of development. The perception that most of the poor live in the rural areas[99] led to a much stronger support for rural development, agriculture and social services from the 1970s on, according to Moyo.[100]

---

[97] However, there are indications that donors do not always insists on the repayment by utilities as long as there is a high possibility that the national treasury will pay back the loans. This might suggest that for some desk officers of donors the disbursement of funds has a higher priority than to reach the objectives of sustainable development. In additions, donors use the argument that once the contract has been signed and money has been disbursed it is impossible to put a project on hold during implementation just because the receiving partner does not fulfil certain conditions (own experience in several Sub-Saharan countries). This would suggest that it is in the interest of both partners (donors and receiving countries) that long term arrangements are signed and implemented with a series of projects carried out in steps whereby each party can verify the engagements of the other and the fulfilment of condition agreed upon together. Everyone can also pull out of the arrangement at any time with justified reasons when made public. Such arrangements would help to improve governance and sustainable development in the sector.

[98] Mosley (1987: 27).

[99] The outcome of poverty surveys depends on the selection of indicators – refer to Sect. 2.7.

[100] Moyo (2009: 15).

The selections of key issues for development, declarations of decades, goal setting and monitoring (MDGs/SDGs), human rights, and etc. had a tremendous influence on the developing and donor countries. As signatories to the different declaration and conventions,[101] countries are obliged to work towards the targets and intentions set by the UN. This is reflected in the framework documents in a number of countries, for instance. Kenya has included access to water and sanitation as a right in the Bill of Rights in the new constitution (2010). Hence, many UN declarations and conventions are very helpful to provide new orientations in the partner countries which can be used by civil society organisations and donors to align their contribution and convince national government to work towards such aims.

However, as already outlined in Chaps. 1 and 2, there are flaws in some of the declarations and in the mechanism for implementation regarding urban water and sanitation development, which seems to delay instead of accelerate development in the partner countries. Furthermore, because global goals and targets have never been reached in the last decades and messages based on the monitoring are misleading they largely lost relevance for the sector in the developing world. It is counterproductive to produce success *'on paper'* which is in stark contrast to the reality on the ground and therefore, it is time that the critical remarks of the General Assembly of the UN on the weaknesses of the JMP monitoring leads to a better scrutiny of messages concerning achievements of water and sanitation related MDGs and now SDGs.[102] One expert in Tanzania explained this almost dogmatic believe in unachievable global targets / goals using an example for sanitation in Africa (Table 4.8):

> I feel, we are working very strategically to destabilize this government. The government should put 0.05 of GDP in sanitation [AU proposal]. 10 years down the line, no government has honoured it [in Sub-Saharan Africa]. Now, you are coming with the idea 0.10 GDP. Yes, global world, which comes with us partners, at the dialog table. We go with our own strategies before all expertise….

Table 4.8 The crucial factor financing/international cooperation and the key intervention areas

	Key intervention areas
Financing/international cooperation	Donor mobilisation/coordination
	Feedback/reporting to global level
	Ring-fencing of income in the sector
	Ownership/leadership
	National funding mechanism
	Financing modelling
	Etc.

[101] Such as human rights, water decades, sustainable development, MDG/SDG with its monitoring by JMP, etc.

[102] Tortajada and Biswas (2018), indicating inaccurate figures used for water and sanitation MDG monitoring and the missing questioning of it by the stakeholders, including the academic world.

References

Akram JK (2011) The significance of management information systems for enhancing strategic and tactical planning. J Inf Syst Technol Manag 8(2):459–470

Allan, R., Jeffrey, P., Clarke, M. and Pollard, S. (2013) 'The impact of regulation, ownership and business culture on managing corporate risk within the water industry, School Appl Sci School Manag, Cranfield Water Sci Inst, Water Policy 15 (3), pp. 458–478

Ayalew M, Chenoweth J, Malcolm R, Mulugetta Y, Okotto LG, Pedley S (2014) Small independent water providers: their position in the regulatory framework for the supply of water in Kenya in Ethiopia. J Environ Law 26:105–128. https://doi.org/10.1093/jel/eqt028

Balance T, Taylor A (2001) The principles of best practice regulation. A report commissioned by Water UK, Stone and Webster Consultants

Baldwin R, Cave M, Lodge M (2012) Understanding regulation – theory, strategy, and practice. Oxford University Press, Oxford

Berg SV (2001). Fundamentals of economic regulation, international training program on utility regulation and strategy. University of Florida, June 2001

Cullet P (2009) Water law, poverty, and development – water sector reforms in India. Oxford University Press, Oxford

Escobar A (1994) Introduction: development and the anthropology of modernity. Ders.: encountering development. The making and unmaking of the third world. Princeton University Press, Princeton. 2011

Gaffga, N.H., Tauxe, R.V. and Mintz, E.D. (2007) Cholera: a new homeland in Africa?, Am J Med Hyg, 77(4):705–713

Hall S (1994) Der Westen und der Rest. Diskurs und Macht. Ders.: Rassismus und kulturelle Identität. Ausgewählte Schriften 2, Hamburg, pp 137–179

Hutton G Varughese M (2016) The costs of Meeting the 2030 Sustainable Development Goal Targets on Drinking Water, Sanitation, and Hygiene, Technical Paper 103171, Water and Sanitation Program, 2016, World Bank Group

Klein M (1996) Economic regulation of water companies. World Bank Policy Research Working Paper, Nr. 1649, September, Private Sector Development Department

Mignolo WD (2008) Preamble: the historical Foundation of modernity/coloniality and the emergence of decolonial thinking. In: Castro-Klaren S (ed.) A companion to Latin American literature and culture. Malden pp 12–23

Moretti F, Pestre D (2015) Bankspeak. Language World Bank Rep New Left Rev 92:75–99

Mosley P (1987) Foreign aid – its defence and reforms. University Press of Kentucky

Moyo D (2009) Dead aid – Why Aid Makes Things Worse and How There is Another Way for Africa. Penguin Books

Mwega FM (2009) A case study of aid effectiveness in Kenya. Wohlfensohn Centre for Development, Working Paper 8

Nickson A Franceys R (2003) *Tapping the market, the challenge of institutional reform in the urban water sector*. England: Palgrave MacMillan Distribution Ltd, Houndmills, Basingstoke, Hampshire RG21 6XS

Parker D, Kirkpatrick C (2012) Measuring regulatory performance, the economic impact of regulatory policy: a literature review of quantitative evidence. OECD, August, Expert Paper No. 3

Pollem O (2008) Regulierungsbehörden für den Wassersektor in low-income countries – Eine vergleichende Untersuchung der Regulierungsbehörden in Ghana, Sambia, Mosambik und Mali. Dissertation, Carl von Ossietzky Universität Oldenburg, Fakultät II, 2008

Rouse M (2013) Insitutional governance of regulation of water services – the essential elements. IWA Publishing, London

Schiffler M (2015) Water, politics and money a reality check on privatisation. Springer International Publishing Switzerland, Cham

Tortajada C, Biswas AK (2018) Achieving universal access to clean water and sanitation in an era of water scarcity: strengthening contributions form academia. Sci Direct 34:21–25. Current Opinion in Environmental Sustainability, 2018, www.sciencedirect.com

UKessays (2015) The importance of information systems. Available: https://www.ukessays.com/essays/information-systems/the-importance-of-information-systems.php (last visited 11.2016)

Chapter 5
What Sector Reforms in Four Countries Teach Us?

Abstract The theoretical concept for the sector development is tested with quantitative and qualitative data collected in four countries. The poorest, Burkina Faso, shows the best progress in terms of access in percentage of people (first indicator of sector performance) and reduction of the non-served (second indicator). It reached the highest coverage rate for water and invested most in infrastructure per capita to be served. This top ranking, comparable with a number of providers in Europe, is followed by Zambia which also shows progress in the two indicators, but trailing far behind Burkina Faso. Kenya has managed to reverse the long lasting negative trend in water access but could not stop the increase in non-served people while Tanzania failed in both development indicators despite a huge investment program. Wherever utilities succeeded in water, there is increasing involvement in onsite sanitation development. However, improved utility performance and more money for investments is not necessarily sufficient to ensure substantial progress in access. Data for water from national information systems seem to be more appropriate than from JMP. For sanitation (except piped systems) JMP data are very useful. However, cross-checks between the two monitoring systems are helpful and comparison of data from different countries needs adjustments. For this, a methodology is proposed.

This chapter describes and compares the reform outcomes in four countries in Sub-Saharan Africa and relates them to the proposed sector principles and the crucial factors. The quantitative and qualitative data collected during the field work of the dissertation[1] allows to trace the development in all proposed crucial factors and in each of the target countries. The time elapsed from the collection of the data to the reading of this book should not be considered a hindrance because the intention of this book is to draw conclusions on what modern reforms can achieve, how they should be designed and implemented and what development is needed to reach universal access.

[1] Submitted to the University of Vienna in 2017.

R. Werchota, *Empty Buckets and Overflowing Pits*, Springer Water,
https://doi.org/10.1007/978-3-030-31383-8_5

With the description of the development in each crucial factor the summary of Chaps. 2 and 3 (Sect. 3.6) should be kept in mind because the comparison of theory with reality permits to validate the proposed concept and helps to explain why the outcomes of reforms are so different in the target countries. It helps to pinpoint the reasons behind the recommendations of this book. Before describing the development in each of the factors for the target countries the analysis of the quantitative data focuses on access to water, utility performance and investments. The importance of qualitative data must not be underestimated in the socio-economic analysis. It is helpful to cross-check statistical with qualitative data which could indicate that adjustments are needed (e.g. comparison of access data between countries, etc.) in order to ensure that established relationships reflect reality.

5.1 Brief Recap of Methodology

The four target countries, Burkina Faso, Kenya, Tanzania and Zambia, were selected according to five criteria.[2] The aim of this selection was to be able to generalise results for the Sub-Saharan region and in certain areas also for some other developing countries. Such a generalisation is possible because the sector in most of the low-income countries face the same key challenges and sector issues, operate in a similar environment, pursue the same sector goals and can decide on similar options. In addition, overarching sector goals do not change significantly. The research was undertaken between 2014 and 2017, whereby the fieldwork had started in the second half of 2014 and was carried out until mid-2016.

The comparative study used different data sources and covers different areas such as the performance of utilities (quantitative),[3] investments in infrastructure (quantitative), a documents analysis of the framework (qualitative) and expert interviews (qualitative).[4] The mix of qualitative and quantitative data is intended to help cross-validation and triangulation. For several reasons, it became apparent that access data for water from different countries can only be compared with an adjustment for which a methodology is proposed by this work. The results produced from the national information systems are cross-checked with the data of the global monitoring (JMP). The analyses is using data covering an 8 year period for the different areas investigated.

[2] Political stability, sector reforms undertaken, stakeholders with considerable expertise, Francophone and Anglophone countries and different existence of urban features.

[3] The selection by this work of indicators to measure the performance of water and sanitation utilities is guided by a combination of what is generally regarded as key for measuring utility performance and the reliability of the available data.

[4] Considering the high number of underserved people in Sub-Sahara, the interviews included five representatives from larger LIAs in the capitals of Burkina Faso, Kenya and Zambia as well as in Arusha, Tanzania.

The analysis of the utility performance indicators and the link to access to water was carried out with cumulated utility data for the entire sector (all utilities combined) in three of the four target countries. For Tanzania, only data for regional utilities where available for the investigated time which however, represent 76% of the market. In addition to these cumulated data, disaggregated data for the different sizes of utilities were available in Kenya and Zambia. This offered the opportunity to analyse different groups of utilities according to their sizes in relation to performance and the development of access.

The analysis on water supply is focusing on formalised (utility) services only because informal service providers, generally small-scale, are not fulfilling human rights standards and in addition, data on national level are unavailable. In all four countries, the framework indicates that only formalised service provision is accepted. However, in the four target countries, formalised utilities control either the entire market (Burkina Faso) or at least the lion share of water service provision in the urban areas (Kenya, Tanzania and Zambia). Concerning access to sanitation, the analysis concentrates on access to sewer networks and the involvement of the utilities in the chain for onsite sanitation.

The quantitative analysis includes both descriptive and inferential statistics. However, statistical analysis of the available data faced some challenges. The most important limitation is that only four target countries and 8 years are under observation, limiting the sample size to $n = 4$ and the time series to $n = 8$. This limitation excluded time series as analysis. Alternatively, the correlation was calculated in a way which is called 'profile correlations' and the raw data of the time series (8 years) is transposed as a 'case' for each year. This way, the target indicator 'access to water' can be correlated with all other indicators by using a bivariate non-parametric correlation; Spearman's rho.[5] The information obtained must be interpreted as association. By carrying out correlation analysis of access to water with each of the indicators separately (profile correlation), it is possible to investigate the degree of association.

This quantitative analysis is complemented with a second method; Euclidian distances. This method treats the data in the form of profiles similar to semantic profiles and tries to find out to what extent these profiles differ from each other. This allows an interpretation if there is an association between the curves. This kind of analysis cannot describe long-term effects. These limitations of the analysis have to be kept in mind when interpreting the results. Both methods, profile correlation and correlation of Euclidean distances based on standardised dimension, obtain in this work very similar results.

The qualitative data are obtained in the four target countries by 55 expert interviews, which are carried out according to Mayring,[6] and an analysis of the framework documents, which follows a methodology proposed by the author of this book

[5] Field (2013).

[6] Mayring (2010).

in its dissertation.[7] It is to note that the rankings of the target countries according to the quantitative data and the document analysis obtain very similar results.

5.2 A Standardised Reform Process?

In the 1990s, it became obvious to development partners and to the national decision makers in many Sub-Saharan countries that sweeping measures had to be taken in order to stop a further decline in the sector.[8] This triggered sector reforms with the following main features:

- Reduce inefficiencies in the sector, especially for service provision (corporatisation) and make the sector more self-sustainable.
- Improve viability of service providers (clustering of utilities, professionalization and O+M cost coverage).
- Offer access to controlled services to all (formalisation, regulation).
- Make systems more socially just/reduce in-equalities (access for the poor, gender).
- Mobilise more funds for investments.
- Improve effectiveness in asset development (aid effectiveness).
- Protect the environment (control effluents and water wastage).

Entry Points and Time Requirements for Reforms
It was usually large crises which triggered deep rooted reforms especially dramatic supply shortages felt by all consumer groups mainly during years of droughts.[9] This put an enormous stress on national decision makers to solve the crises. In some countries (e.g. in Burkina Faso and Uganda), reforms started with a restructuring of utilities, because it was considered that utilities are at the core of the problems in the sector. This can be considered as a project induced reform and is facilitated wherever either one national or a very limited number of utilities cover the market. During this restructuring process, changes in the other crucial factors were gradually initiated such as the elaboration of new sector policies and strategies, new legislation, introduction of key information systems, pilots for mode of delivery,[10] etc.

Some donors used such difficult situations to pressure national governments in accepting the generally unpopular involvement of multi-nationals in urban water and sanitation in exchange for the provision of funds for investments (e.g. Tanzania,

---

[7] University of Vienna, International development, Werchota (2017: Volume II, Appendix 3).

[8] Sector reforms in Burkina Faso started at the beginning of the 1990s while in the Anglophone countries at the end of the 1990s.

[9] E.g. Ouagadougou, Burkina Faso in 1997 and 1998.

[10] Private sector participation was high on the agenda during the 1990s, whereby SODECI in the Ivory Coast was cited as best practice, starting as a provisional service provider in the fringe areas of Abidjan in 1959 and developing into a large scale service provider with the substantial extension program for urban water and sanitation in 1973, operating under a concession contract until today.

Senegal).[11] The expectation linked with PSP to generate funds for investments from the private sector or improve utility performance to such a level that sector self-financing of asset development would substantially increase (e.g. Mali) never materialised on a noteworthy scale, at least in the region.[12] The limited success of PSP in Sub-Saharan Africa in utility performance could be due to the fact that the approach was not designed as a multi-dimensional intervention (e.g. including a regulatory regime, national investment and financing planning, etc.). Behind this was most likely the concern that a comprehensive sector approach could hinder the PSP arrangement at some point. It might be interesting to investigate failed PSP arrangements in relation to the (insufficient) development of the proposed crucial factors.

Other countries already recognised that the urban water and sanitation sector is more complex and development does not only depend on the performance of utilities or PSP arrangements. Such countries (e.g. Zambia and Kenya) went back to the drawing board and started designing comprehensive programmes for reforms. However, it must be said that these countries were dealing with a highly complex situation where a great number of service providers operated mostly within the municipality systems. Trying to start restructuring each of the many service providers (50–100 and more) as entry point for reform was no option. This complexity was in some cases further increased when for instance the ministry responsible for local government became the line ministry for water supply and sanitation (Zambia) and where local government structures on national level opposed sector reforms (e.g. Kenya, Tanzania, Zambia, Mali, etc.). There was an advantage in going back to the drawing board (e.g. Kenya and Zambia) because it made reform champions look beyond the performance of the service providers, right at the start of reforms. Nevertheless, going back to the drawing board is not necessarily a guarantee for success either.

In the case of one national water company, decision makers have to be careful that the utility restructuring as entry point does not block further development in the sector once the utility is successful, like it seems to be the case in Uganda where the utility successfully blocked the development of a regulatory regime to be placed

[11] However, the interest of the private multi-nationals was limited to the service providers in the capitals (i.e. Dar es Salaam, Tanzania) or in other big agglomerations (e. g. Copper belt, Zambia).

[12] E.g. in Zambia, AHC-MMS Ltd. (refer to NWASCO reports 2004/05: 8). Another example is the concession contract with SAUR international in Mali, which was terminated in 2010 after the planned investments for asset development was not provided by SAUR international as stipulated in the contract. SAUR argued that the operation of the utility under the private partnership could not generate the necessary funds within Mali because of insufficient tariffs. Equally in Senegal, the contribution of the private operator (SAUR) to asset development was less than 7% (around 20 Million USD against movable assets) of the funds provided by the donors (World Bank, etc.). Hence there was a different expectation among the contract partners. Governments of low-income countries expected international operators to bring a significant amount of fresh money for investments into the sector while the international operators expected to raise such money through tariff increases. Both did not materialise and often the donors had to save the PSP arrangement by providing concessional loans.

outside the ministry.[13] Hence, considering the present development in Uganda, the positive picture Schiffler[14] paints has to be taken with care. There is little pro-poor orientation by the utility because most of the public outlets are on private plots where the plot owner sells utility water at exorbitant prices to the neighbours and the connected poor cannot benefit from a pro-poor block tariff structure.[15] Some municipality representatives criticised the insufficient transparency of tariff setting by the utility. This neglect of a simultaneous development of some of the crucial factors (regulation, sector information, etc.) and some relevant key principles (separation of key functions) is taking place when the dominant utility becomes a 'little republic', an expression used by Caiden and Wildavsky[16] for autonomous state institutions (parastatals) overstepping their intended limits.[17] It can eventually lead to a stagnation of even decline in access.

Comparing the reform process in the different countries, it became evident that reforms will take longer in a highly fractured sector such as in Kenya, Tanzania and Zambia to obtain substantial results than in countries like Burkina Faso where only one national utility is operating. Schiffler[18] notes after analysing 12 cases of utility improvements in several continents that it takes around 10 years to achieve a turnaround in the sector. In Kenya and Zambia, it took around 10 years to reformulate the sector policy, adopt a new Water Act and establish new sector institutions. Thereafter, it takes several more years to gradually improve the performance of utilities and increase investments for infrastructure development. In contrast to this, the utility in Burkina Faso reached already full cost recovery by the end of 1994 after 5 years of restructuring while in the other three target countries the sector

[13] In 2013 the establishment of a regulatory agency (department in the ministry) was prepared and in 2015 the establishment of several utilities to be formed by separating clusters of towns from the National Water Corporation was under discussion. Furthermore, legislation was prepared for a new Water Act and to establish an autonomous regulatory agency. According to some experts working with sector institutions, the National Water Corporation was fiercely opposed to such changes and the power struggle with the line ministry was described by one experts in September 2016 as follows: 'When the regulator (department in the ministry) faces a problem with the National Water Corporation it complains to the minister, when the National Water Corporation faces a problem with the ministry it complains to the president of Uganda'. This conflict was solved in the interest of the utility with the nomination of the director of finance from the National Water Corporation as permanent secretary of the line ministry who put a hold on the preparation for the new legislation to establish an independent regulator soon after his nomination in 2017 (own experience 2016 / 2017).

[14] Schiffler (2015: 143-159). The absence of a regulatory regime (mechanism) and the recent rapid expansion of National Water Corporation moving to a coverage from under 20 towns to over 200 presently will have a profound impact on its performance.

[15] Own experience during a visit in 2016 to the water utility, ministry and other national institutions and donor agencies.

[16] Caiden and Wildavsky (1974: 80, in Mosley 1987: 92).

[17] Even experts interviewed in Burkina Faso regarded the position of ONEA after restructuring as possibly too dominant in the sector although the BoD seems to set appropriate limits for the utility.

[18] Schiffler (2015: 171).

achieved only 100% O+M cost coverage after 10 years of reforms (Kenya and Zambia) or still have to progress towards this objective (Tanzania).

This deliberation indicates that the institutional set-up or mode of delivery should depend on the specific situation in the country and a generalisation is not conducive. In addition, the question of entry points of reforms is less important that the selection and enforcement of principles and key tools used in a coherent development in the proposed six crucial factors. Furthermore, there seems to be a limit how far a country can push the implementation of reforms depending on its development stage in the sector and the general context in the country. Thus, it has to be recognised that reform processes will never end at a certain moment in time especially when the environment is very dynamic and challenges are so substantial that the means are insufficient to respond adequately in medium terms. This means that the design of the reform must be carried out with a long term perspective where reform packages can be established, but such packages must interlink without producing significant turbulences as sustainable development is progressing. Endless stop and go movements and the introduction of 'innovative ideas' which have no success record documented by empirical evidence and are promoted by actors with little local knowledge will most likely have more negative than positive effects.

Different Champions and Structures for Reforms

Different countries had also different champions and institutional structures for the reforms. In Burkina Faso, it was mainly the BoDs of the utility with the input of the management of ONEA, which pushed for reforms in the sector during the restructuring of the utility. In two target countries (Kenya and Tanzania) the ministries responsible for water or specially created units within the ministry were the reform champions. In Zambia, the Water Sector Reform Support Unit (WS-RSU) was the reform champion, a temporary institution with members of several ministries, established outside the civil service structure and mainly financed by donors. In Kenya, it was the department of reforms later renamed the department of coordination within the Ministry of Water and Irrigation, which secured the role of the reform champion.

Most of these structures for reforms were dismantled soon after new legislation was adopted and the first new institutions established. This seemed too early in the process because advisors and decision makers who were instrumental for designing and commencing reforms were allocated to other institutions often outside the sector and no longer available for a transitional period where practical decision for implementation had to be taken. This is a crucial time for reforms because there are moments were the new orientation can easily be lost or undermined such as the case of lease fees payment by utilities to municipalities in Kenya demonstrates (refer to Sect. 4.3, regulation). It is also the time were the new institutions translate the sector orientation into their strategic plans which has an impact for many years to come and cannot easily be changed once adopted. The guidance for these institutions should be coming from the reform champions.

5.3 Overview by Country

Three of the four target countries are amongst the 15 poorest countries in the world. From 175 listed countries, Burkina Faso find itself ranked at 173, Zambia at 163 and Tanzania at 160. Kenya is ranked at 146, still among the third group of poorest countries in the world.[19] The demographic development brings a number of challenges to these countries. The urban population in Sub-Saharan Africa will grow from 37% in 2014 to (around) 55% in 2050.[20] The WB is estimating that the urban population will reach already 50% in 2030 (in a bit more than 10 years).[21] In 2015, the population in Sub Saharan Africa reached 1 billion,[22] which means that presently there are around 400 million urban dwellers. By 2060, (in just over 40 years) the population of Sub-Sahara Africa could be as much as 2.7 billion people.[23] At an estimated urbanisation of around 60%, there will be 1.62 billion urban dwellers, which means that the urban population will multiply by four in around halve or three quarter of a life time.

This breath-taking urban grow is already ongoing since decades. For instance, Tanzania had 34 million just 13 years ago and grew to 57 million presently, which is a 67% increase in just over one decade. It is projected that its population will reach 276 million by 2100.[24] The urban growth is combined with a high poverty rate,[25] which today is between 36% and 61% of the total population in the target countries. This rapid urbanisation combined with high poverty and inequality leads to a growing number of LIAs with a rising population density as newcomers in the urban areas need to find low-cost accommodation.

The population size separates the four target countries into two groups; Burkina Faso and Zambia with a population of just under 20 million and Kenya and Tanzania with around 50 million presently. The urban growth in the target countries is highest in Burkina Faso and Tanzania with over 5% and in Kenya and Zambia still over 4%. Also, the fertility rates have remained high. The average age in the target countries is between 17 and 19 years. The population density in three target countries has already reached more than half of the average of Europe. Only in Zambia it is less than a quarter (Table 5.1).

[19] http://www.spiegel.de/wirtschaft/ranking-die-aermsten-und-die-reichsten-laender-a-256276-3.html (last visited 02.2017).

[20] https://www.brookings.edu/blog/africa-in-focus/2015/12/30/foresight-africa-2016-urbanization-in-the-african-context/ (last visited 02.2017).

[21] http://www.worldbank.org/en/events/2015/06/01/urbanization-in-africa-trends-promises-and-challenges (last visited 02.2017).

[22] http://data.worldbank.org/region/sub-saharan-africa (last visited 02.2017).

[23] http://blogs.worldbank.org/africacan/7-facts-about-population-in-sub-saharan-africa (last visited 02.2017).

[24] http://worldpopulationreview.com/continents/africa-population/ (last visited 02.2017).

[25] The poverty rate is the ratio of the number of people who fall below the poverty line and the total population (Wikipedia, last visited 02.2017).

Table 5.1 General data of the four target countries – 1

2015, 2016, 2017	Burkina Faso	Kenya	Tanzania	Zambia
Population (mil.)	19	48	57	17
Pop. growth (%)	2.90	2.57	3.12	3.11
Urbanisation rate (%)	5.87	4.34	5.36	4.32
Pop. density (/km^2)	70	85	64	23
Land size 1000/km^2	274	569	886	743
Fertility rate	5.65	4.44	5.24	5.45
Av. age	17	19	17	17
Urban pop. (%)	29.5	25.8	31.0	38.5
Pop. below poverty (%)	46.7	43.3	36.0	60.5

Annual precipitation is lowest in Kenya and Burkina Faso. Both countries can be regarded as water scarce with several months of try periods. This has consequences on the technologies used for water supply and sanitation, particularly for sewer networks, which need a constant flow of water in order to remain operational. The GDP per capita is around half in Burkina Faso compared to the other three target countries. The Gini Index income coefficient is highest in Zambia and Kenya indicating a higher inequality level. The rapid urbanisation means that there is a growing necessity for funds for infrastructure development, which needs to be provided in a constant flow just to maintain in a first step the insufficient level of access. However, the low level of GDP indicates that there is a limit in what the governments can provide for the development of water and sanitation infrastructure.

The ranking by the index for corruption control indicates the substantial challenges Kenya is facing. It is placed on the same level as Nigeria and far behind Burkina Faso and Zambia. The latter are placed equal to some European countries. Challenges in the water and sanitation sector might be linked to these levels of corruption in the country. Nevertheless, the urban water and sanitation sector is often considered avant-garde in reforms and therefore, might be better placed to deal with corruption than a number of other sectors.

In the ranking of the human development index, Burkina Faso, Tanzania and Kenya is trailing well behind Zambia. Nevertheless, professionalization of service provision and partly of financing has attracted well trained personnel with high development capacity in the sector of all four target countries. Therefore, it can be assumed that it is less the professional knowledge of staff but more their attitude (corporate culture), which makes the difference in the four target countries when it comes to institutional performance (Table 5.2).

Despite the high level of corruption, Kenya is placed higher in the ranking for transparency and accountability than Zambia and Burkina Faso. Tanzania is trailing behind. Women are not particularly categorised as second class in the target countries but have to face noteworthy unequal opportunities in the society. Burkina Faso, in regards of gender inequality is trailing the other three target countries, which are about on the same level. In all four of the target countries, there are very few women in senior positions in the water and sanitation sector (expert interviews) despite of a growing number of women holding university degrees in all relevant fields for water and sanitation, including engineering (Table 5.3).

Table 5.2 General data of the four target countries – 2

2015, 2016, 2017	Burkina Faso	Kenya	Tanzania	Zambia
Precipitation mm	748	630	1071	1020
GDP, IMF int. $	1791	3360	3097	3899
Ranking among 189	172	150	155	142
Income Gini coeff.[a]	39,8	47,7	37,6	57,7
Corruption 0–100 TI	38	25	30	38
Corruption ranking	76	139	117	76
Human dev. Index	0.40	0.55	0.52	0.59
Ranking	183	167	173	139
Class	Low	Low	Low	Medium

http://www.worldometers.info/population/countries-in-africa-by-population/ (all last visited 2/2017)

https://www.cia.gov/library/publications/the-world-factbook/fields/2212.html

http://data.worldbank.org/indicator/AG.LND.PRCP.MM

https://en.wikipedia.org/wiki/List_of_countries_by_GDP_(PPP)_per_capita

http://www.indexmundi.com/g/r.aspx?v=69, http://hdr.undp.org/en/content/income-gini-coefficient

https://www.transparency.org/cpi2015/, https://en.wikipedia.org/wiki/List_of_African_countries_by_Human_Development_Index

[a]The Gini coefficient (sometimes expressed as a Gini ratio or a normalized Gini index) is a measure of statistical dispersion intended to represent the income or wealth distribution of a nation's residents, and is the most commonly used measure of inequality (Wikipedia, last visited 02.2017)

Table 5.3 General data of the four target countries – 3

2016	Burkina Faso	Kenya	Tanzania	Zambia
Total transparency and accountability	55[a] somewhat weak	67 moderate	42 somew. Weak	58 somew. Weak
Gender score, equal access	Burkina Faso	Kenya	Tanzania	Zambia
To employment	25	50	50	50
To national cabinet	50	100	75	75
To top judiciary	75	75	75	100
In legislature	25	75	100	50
Total	175	300	300	275

http://aii.globalintegrity.org/scores-map?stringld=transparency_accountability&year=2016 (last visited 2/2017).

[a]Points in the Global Integrity Scale: The Global Integrity scale on the Africa integrity indicators website is as follows: 81–100 (strong), 61–80 (moderate), 41–60 (somewhat weak), 21–40 (weak), 0–20 (very weak).

5.4 Development of Access to Drinking Water

In general, counting of sustainable access to safe water in towns on national level is based on utility services, regardless of service levels. Why informal service provision, neighbourhood sales and traditional sources are not included has been explained in length in Part I. It has also been shown that counting of access by the global monitoring is quite different and therefore, results of these two monitoring systems can differ significantly.

Progress in the sector can be measured on two levels. Maintaining the percentage of people with access to water in the same rhythm with the population growth over several years can be considered a noteworthy achievement, compared with the negative trend over many decades before reforms. However, keeping the percentage of access (people served) on a constant level in a situation with rapid urbanisation means that the number of underserved people is still increasing. Therefore, the second level of success, which should be reached by reforms, is to reduce the accumulated numbers of underserved people in order to move towards universal access. It is also interesting to see how far progress went towards universal access during reforms. Thus, the following questions should be answered for the four target countries:

- Did reforms lead to a stop or a reversal of the long lasting negative trend in access to water expressed in % of people reached?
- What level of coverage was reached (2012/13)?
- How did the number of underserved people develop (over eight years)?

Did Reforms Lead to a Stop or a Reversal of the Long Lasting Negative Trend in Access to Water Expressed in % of People Reached?

The JMP data in Fig. 5.1 show that in three of the four target countries the trend of access to water was negative at least since 1990 when JMP commenced recording.[26] Only for Burkina Faso, JMP indicates a positive trend since 1993. It is to note that the reform in the sector started already in 1989[27] and managed to show the first substantial results in 1995.[28] Hence the previous decline in the sector did not influence the straight line traced by JMP. The trigger for the earlier start of reforms in Burkina Faso was the excruciating shortage of water resources in Ouagadougou that became apparent in the 1980s and forced national decision makers to act earlier than other countries. Sector data from Burkina Faso covering a period before 1995 indicate that the population, which should have been served by the Office National de l'Eau et de l'Assainissement (ONEA), grew from 1980 to 1995 by 263%[29] while the

[26] JMP noted that in most of the developing countries there were no reliable data before 1990.

[27] Kenya and Zambia around 1998/99 with the establishment of commercial utilities and a regulator.

[28] Bertrand and Geli (1995: 167–187). Reforms in the other three target countries (Anglophone) started around 10 years later.

[29] Population grow, urbanisation and integration of new towns.

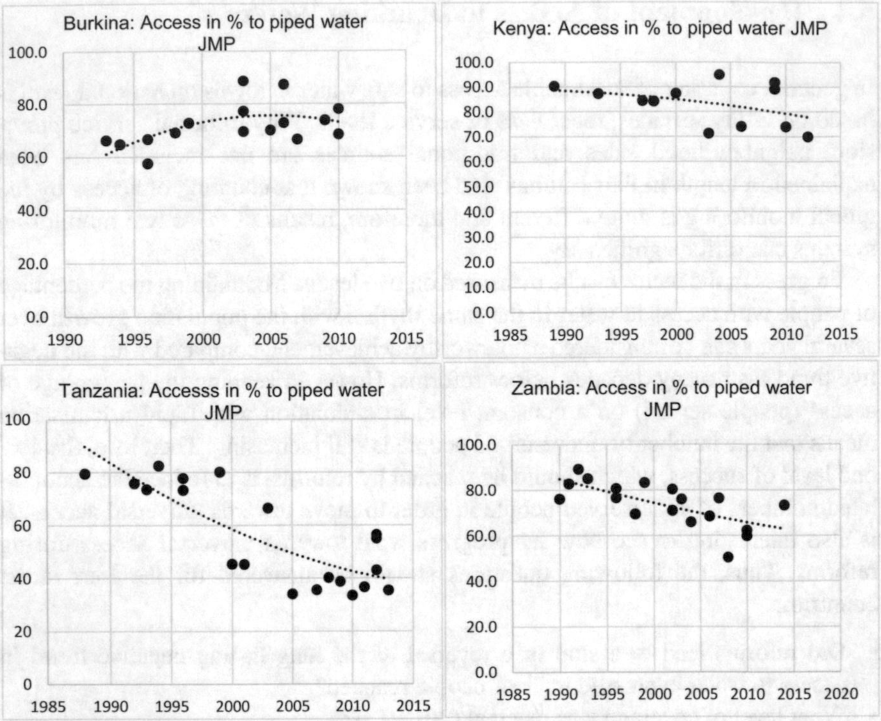

Fig. 5.1 Access to piped water in the target countries according to JMP. (Source: Data, Joint Monitoring Programme, UN; figure Werchota). It is to note that the presentation in this figure deviates from the method used by JMP (using straight regression lines)

population actually supplied with water grew by 145% only. According to the same source, the total number of the underserved grew during the same period from 371.000 to over 1.5 million. This indicates that there was also a negative trend in access to water before the reform in Burkina Faso as probably in all the countries in the region.[30]

The constant decline as shown by JMP in the three other target countries seems to have been a reality in the past but has to be questioned during reform periods. The use of a straight regression line from 1990s does not permit to show turnarounds obtained over a relatively short span of time. Furthermore, results of the surveys used by JMP sometimes vary quite substantially within the same year or from 1 year to another within countries which is most likely due to the selection of different areas for different surveys.[31]

---

[30] The 'contrat plan' 2007–2009, annex 1 (2007: 13–14).

[31] E.g. more than 20% in the year 2003 and 2004 and in 2009 in Kenya, more than 25% in Zambia in 2007 and 2008.

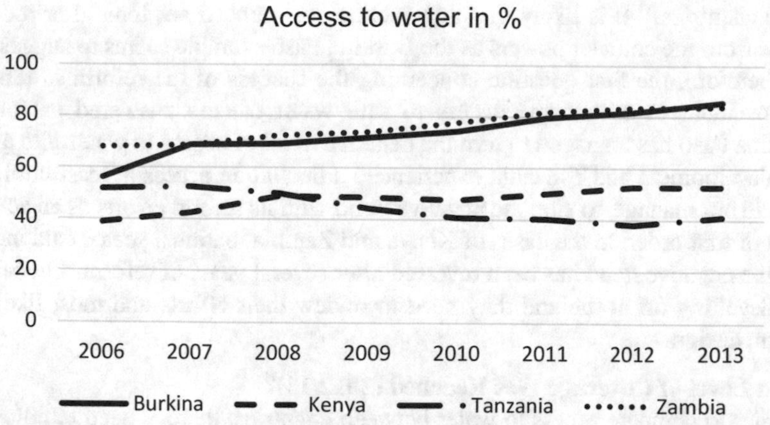

Fig. 5.2 Access to water in the target countries according to national monitoring

Utility performance data captured by new or upgraded national information systems have improved substantially during reforms in all target countries and make changes visible within a short time. Hence, national systems make changes in trends better apparent. Another advantage of the national systems is that water and sewerage is the sectors core business and data are collected countrywide at least annually and usually in the same (service) areas. This is not the case with surveys used by JMP where access to water and sanitation is a by-product[32] (e.g. Aids Indicator Survey). Therefore, data from the national systems seem to reflect better the situation on the ground than JMP data[33] especially when a relatively short period is under investigation (2006/7 to 2012/13).

Comparing the data from the national systems, Fig. 5.2, with the JMP data for the 8 years investigated, it can be noted that both monitoring systems indicate progress for Burkina Faso as best performer among the four target countries and for Tanzania, as worst performer, a decline in access to water. Concerning Kenya and Zambia, the trend lines of the two monitoring systems point in different directions. The national systems indicate a positive development, the JMP monitoring shows a decline. The latter seems to be due to the flaws of the global monitoring explained before. However, the improvement in access in Kenya and Zambia shown by the national systems seems to level off at the end of the period investigated.

The decline in access to piped systems in Sub-Saharan Africa since 1990 shown by JMP is another indication that the negative trend was a general phenomenon in

[32] Welfare Monitoring Survey, Integrated Household Budget Survey, Aids Indicator Survey, The Social Dimensions of Adjustment Survey, Demographic and Health Survey, Multiple Indicator Cluster Survey, Living Conditions Monitoring Survey, Population and housing census, etc.

[33] This might be the case for several Sub-Saharan countries.

these countries.[34] It is likely that this decline commenced not long after independence from the colonial powers as the Burkina Faso example seems to suggest.[35]

Therefore, the first question concerning the success of the reform in terms of improvement of sustainable access to safe water can be answered as follows: Burkina Faso has managed to turn the negative trend of access to water into a positive development and Tanzania experienced a decline in access at least until 2013 and did not manage, to stop the negative trend with its reform efforts as an achievement of first order. In the cases of Kenya and Zambia, national sector data indicate that the negative trend has been reversed after several years of reforms but because it is levelling off at the end they need to review their efforts and most likely the reform design.

What Level of Coverage Was Reached (2012/13)?

In order to compare access to water between countries, there is need to adjust figures because there is a difference in counting in each country. First the national monitoring systems in each country attribute different numbers of people to each type of utility outlets. The problem with this is that the attribution of people per household by utilities differs sometimes substantially from the household sizes given by the census, e.g. in Zambia. There is a limit the regulator should except when a number of people are attributed to a household connection (e.g. 5–10 in Zambia).[36] The utilities however, tend to use generally the maximum. For instance, the census in Zambia indicates a household size of 5.2 for the urban areas compared to ten consumers used per household connection by almost all utilities.

Attributing a number to an outlet is a way utilities can inflate access figures. In the case of Tanzania, some utilities and the regulator argue that many households sell water to neighbours and therefore, the number of people being serviced must be substantially higher than the average household size provided by the census. However, according to the rights to water, neighbourhood sales must be regarded as unacceptable (refer to Sects. 2.7 and 3.3).

Secondly, some utilities define their service areas according to their pipe network and not according to the administrative boundaries of towns, as the census does, or according to population density, as some regulators require (Kenya). Many utilities do not cover yet the entire area within the towns (municipality boundaries) with their pipe network, whereby mainly the LIAs are excluded. Limiting the analysis to the area covered by the network, but claiming that the coverage rate is for the town,

[34] According to JMP, Progress Report on Drinking Water and Sanitation (2014: 72, 73) update, in the period 1990 to 2012 in Sub-Saharan Africa access to sanitation stagnated on 41% and access to drinking piped water (without public outlets) declined from 42% to 34%.

[35] This should not mean that water and sanitation should be rated better during the colonial era. Access was also mainly restricted to the upper and middle class in many Sub-Saharan countries. However, it seems that the performance indicators of the public providers such as cost recovery, collection rate, etc. was acceptable.

[36] All utilities seem to choose ten people per connection according to personnel from the utilities interviewed.

means inflating access figures which a number of utilities in Tanzania practice.[37] The regulators have interest, not to allow such exclusion of the poor in the LIAs by the utility if they follow a policy to protect consumers and the human rights approach. However, changing this practice would mean that the coverage figures will suddenly drop which is difficult to explain to politicians and to the public. Such way of counting might still reflect the old approach applied by the ministries: utility services for the middle and upper classes and informal service provision for the people in the LIAs (refer to the section on water apartheid).

Another way to increase access figures is to use design capacities as some WSBs in Kenya do. Such figures are used with preference by the ministry[38] because of political convenience. For instance, the regulator reported a coverage of 55% for the year 2015/2016 while the WSBs report 68%. The annual report of the ministry provides an explanation about the difference: 'In some cases, the additional population by WSBs are calculated based on design capacities on water supply systems and not actual population having access to safe water'.[39] However, the ministry staff regularly use these inflated figures of the WSBs (without further explanation) in presentations and see no problem in contradicting the publishes figures from the regulator in its annual 'Impact' report.[40]

These examples indicate that it is necessary to adjust access data when comparing countries and sometimes utilities. To do this, the following approach is proposed; from the figures of the national monitoring, the number of people served in a household should not substantially exceed the number of the average household size provided by the census. Therefore, the assumption is made that in general, households in Africa have servants or visitors from the extended family living in their households over a longer period, but who are most likely not part of the stated household size for the census. Therefore, two additional persons are added to the rounded up household size provided by the census in order to calculate the corrected number of people per household connection, which actually should be used by the national monitoring systems. In addition, the number of people served in a town by the utility should at least be the urban population given by the census in order to avoid that utilities can inflate the percentage of people having access to water[41] by limiting their service area to their pipe network. Wherever the utilities indicate a higher number by including densely populated areas on the fringe of the towns (outside administrative borders) the utility figures shall be applied.

---

[37] Alone for the town of Dar es Salaam the figures of urban population according to the census and the people to be served reported by the utility differs around 1 million.

[38] Own experience in 2017 and 2018.

[39] Ministry of Water and Irrigation (2016: 70 1nd 71), 'The Annual Water Sector Review 2014/2015–2015/16'.

[40] Water Services Regulatory Board (2018: 19), 'IMPACT A Performance Report of Kenya's Water Services Sector 2015/16 and 2016/17, Issue NO. 10'.

[41] This is the case in Tanzania where some utilities are allowed to define their service areas according to the existing pipe network and by adding 200 m at its limits.

Such rectifications will allow comparing the coverage in percentage of the population to be served between the countries and most likely should also narrow the gap between JPM and national figures. It has to be recalled that the global monitoring of JMP does not make a difference between informal and formal service provision of water, which is inflating access figures for some countries. In addition, there is no sound data on informal service providers, which own and operate infrastructure. Therefore, when informal service provision is not considered by a country system there should be a difference between the corrected access data from the national systems and the JMP data. Although it is expected that the adjustments proposed will narrow the gap of data between the two monitoring systems, the obtained access figures must still be considered as an approximation of the situation. However, there are other independent sources which seem to confirm the results of the adjustments used in this book.[42]

Often experts demand that sector institutions in the countries should reconcile their results of the information systems with JMP outcomes, because of the assumption that the surveys used by the global monitoring are more reliable than the counting of national systems. This is questionable and does not sufficiently take into account the difficulties to reconcile data which are based on very different definitions of access and the fact that reliable information is missing e.g. on informal service provision, neighbourhood sales, etc. However, it should not stop countries to compare the results of the two monitoring systems and provide possible explanations of differences due to their context. In addition, results should be compared to other surveys often carried out by projects which allow identifying access to utility services and separation of neighbourhood sales.

Table 5.4 shows that the differences of access to water in 2012/13 between the two monitoring systems are in a range of −15% and +24% (fourth column in the table). The figures for Burkina Faso from the two monitoring systems are almost matching. There is a difference of only +3% although ONEA attributes two to three people (according to the sizes of towns) in addition to the average size of the households given by the census. This seems to support the assumption of this analysis that the number of people benefitting from utility services is to a certain extent surpassing the number of households served multiplied by the average household size provided by the census. It has to be noted that there are as good as no informal service providers in Burkina Faso (expert interviews).

After correction of the utility data, as proposed, the differences in the access figures from both systems are reduced to a range of −8/+6% (last column in Table 5.4) for the four target countries. The utility figure for Burkina Faso matches the JMP result while for Kenya it is lower than the JMP data. The number of informal service providers in Kenya could explain this. The fact that the utility figures for Tanzania and Zambia are still above the JMP figures after correction (+5% and +6%), although both countries seem to have informal service providers (Zambia,

[42] The result for the corrected access figure in Table 5.4 for Tanzania (37%) is almost identical with the coverage rate recently provided by the WB (2017:3) with 38% (household connections and public outlets, without neighbourhood sales).

Table 5.4 Access to water according to adjusted national data

Year 2012/13	JMP surveys – piped[a]	National systems uncorrected	Difference	Corrections					Difference to JMP
					Household size				
				Comments	Census	Utility	Corr. hh size	Corrected access to water[b]	
Burkina	83%	86%	+3	±0	6,0	8–9	8	83%	±0
Kenia	68%	53%	−15	>town boarders	4,4	4–6	6	60%	−8%
Tanzania	32%	41%	+9	+24% to utility data	4,8	10–15	7	37%	+5%
Zambia	60%	84%	+24	>town boarders	5,2	5–10	7	66%	+6%

[a]Piped water: Household and yard connections as well as water kiosks (public tap/standpipe)
[b]Refer to following Note:

Burkina: According to the expert interviews there are no informal service providers operating water or sewerage infrastructure. The corrected household size applied is eight

Kenya: According to the expert interviews and some studies (e.g. KfW for greater Nairobi 2014), there are a number of informal service providers, which could serve up to 10% of the market. The population to be served is defined by the regulator as more than 20 million and by the census (urban and peri-urban) around 15 million (2012/13), which indicates that the regulator obliges the utilities to serve as well urban fringe areas outside the administrative town boundaries which have urban character due to a certain population density. The corrected household size applied is six

Tanzania: According to the expert interviews, there are informal service providers, which could cover between 5% and 10% of the market. The population to be served, given by the regulator in its annual sector reports, is 24% lower than the census figure for the urban population. Therefore, the census data is used. According to the baseline 2010, around 2/3 of the urban population are living in the LIAs where 31% are served by the neighbourhood sales. This represents 20% of the urban population, which are considered as informally served. There are inconsistencies in the reporting of access figures as the multiplication of the number of household connections by ten produce a number of people served which is higher than reported by the regulator. The corrected household size applied is seven

Zambia: According to the expert interviews there are very few informal service providers operating water and sewerage infrastructure. Utilities use the upper end of household size the regulator is permitting to count (5–10). The number of people to be served given by the regulator is higher than the number given by the census, which indicates like for Kenya that some providers have to serve areas outside towns. The corrected household size applied is seven

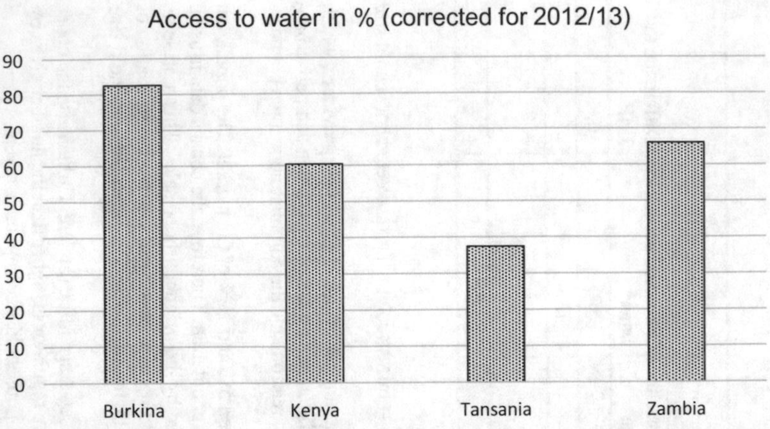

Fig. 5.3 Access to water in the target countries (2012/2013) according to national monitoring with corrected household sizes

however, very few according to the expert interviews), is most likely due to inaccuracies in both monitoring systems which might be more pronounced than for Burkina and Kenya.

Now the question of what level of access did the countries reach with the sector development at the end of the period-investigated can be answered. Burkina Faso has reached the highest level of people served in percentage at the end of the period investigated, progressing from just above 50% to slightly over 80% (Fig. 5.3). Zambia is above the 60% and Kenya reaches 60%, whereby both countries have made progress according to the reported data from the utilities as well as the corrected figures. Nevertheless, both countries have had a less impressive development than Burkina Faso during the reform process. Tanzania has obtained the lowest coverage level of the four target countries after a long and constant decline according to the data submitted by the utilities and JMP, and this decline seems to be still ongoing.[43]

The discussion on how access data from the different countries can be compared (and what corrections are necessary) shows that the use of quantitative data must be taken with care. It is helpful to use qualitative data (expert interviews) to indicate the limits of quantitative data provided by the sector institutions in order to obtain with adjustments a fairly realistic picture of the situation.

How Did the Number of Underserved People Develop?

It can be noted that the decline of access to water in Sub-Saharan Africa caused a substantial accumulation of the numbers of underserved people before the reforms in all four target countries. The aim of the sector reform is to reach the second level of success, which is to gradually reduce the accumulated number of underserved people.[44] This number can be obtained from the data reported by the utilities as well as from the numbers obtained by the corrected access figures (Tables 5.5 and 5.6).

---

[43] Concerning coverage level in Tanzania refer also to World Bank Group (2017).

[44] The first level of success was defined as an increase in access in percentage.

Table 5.5 Underserved people reported by the utilities

Indicators	Burkina Faso	Kenia	Tanzania	Zambia
Access to water 2006/7 in %	56%	39%	52%	68%
Access to water 2012/13 in %	86%	53%	41%	84%
People without access 2006/7	1.36 Mio.	(4.78 Mio.)[a] 7.60 Mio.	3.65 Mio.	1.62 Mio.
People without access 2013	0.63 Mio.	9.47 Mio.	6.08 Mio.	0.96 Mio.
Variation of the number of underserved	−54%	+25%	+67%	−41%
In % of population to be served 2013	14%	47%	59%	16%

[a]A number of big utilities have not reported in 2006/07. Therefore, the figure for the underserved as reported by the regulator from the same year (4.78) has been obtained by prolonging the line from the following years back to 2006/07 as an estimation

Table 5.6 Underserved people according to adjusted access data reported by the utilities

Indicators	Burkina Faso	Kenia	Tanzania	Zambia
Access to water 2006/7 in %	54%	47%	34%	53%
Access to water 2012/13 in %	83%	60%	37%	66%
People without access 2006/7	1.41 Mio.	(4.11 Mio.) 6.75 Mio.*	4.97 Mio.	2.35 Mio.
People without access 2012/13	0.76 Mio.	7.99 Mio.	6.46 Mio.	2.03 Mio.
Variation of the number of underserved	−46%	+18%	+30%	−14%
In % of population to be served 2013	17%	40%	63%	34%

Burkina Faso and Zambia have managed to reduce the numbers of underserved people, however, Burkina Faso to a much higher extent than Zambia. In Kenya, the numbers of underserved has still increased in the period investigated despite an increase in access to water in the percentage of the population. In Tanzania, the increase of underserved people compared to the population to be served is substantially higher than in Kenya. Both approaches show that the numbers of underserved people in Tanzania surpasses the people having access to utility services, and this in a situation where a downwards trend in access is still taking place.

Considering the results obtained for water in this chapter, it can be said that Burkina Faso is the best performer among the four target countries in term of the two level of success and the coverage rate achieved at the end of the investigated period. Zambia is second with positive developments in both indicators. Kenya is third placed with a positive change in the first level and a negative in the second while Tanzania is last placed trailing far behind the other three target countries with

Table 5.7 Ranking of target countries according to the development of access in percentage and the reduction in the number of underserved people

Country	Stop of negative trend	Increase access in %	Reducing the number of underserved	Reducing underserved by >50%	Score: total of +	Number of utilities
Burkina F.	+	+	+	+	4+	1
Zambia	+	+	+	–	3+	10
Kenya	+	+	–	–	2+	Around 100
Tanzania	–	–	–	–	–	>100

a rapid and constant decline in water coverage,[45] a substantial increase in the numbers of underserved people and the lowest level of coverage reached. It should also be noted that the reform process in Kenya has started about 2 years later than Zambia, and Kenya still faces the challenge to reduce the high number of small-scale utilities, which Zambia[46] has managed to overcome with its merging process undertaken just after the beginning of the reform implementation (in 2000).[47] The deliberation in this section allows a ranking of the target countries according to the qualitative data (Table 5.7).

5.5 Improvements in Utility Performance

In Sub-Saharan Africa, publicly owned and corporatized utilities became the preferred mode of delivery for urban water and sewerage service provision. The corporatisation of (formalised) service providers was one of the most significant steps in the sector development in many countries and often helped to reverse the negative impact decentralisation or civil service management had on urban water and sanitation service provision in the past (refer to Sects. 2.4 and 3.1). However, establishing commercial utilities is insufficient to solve sector challenges because without forming units, which can offer economies of scale, the benefit of commercialisation remains limited.[48] There is sufficient evidence that utilities with a professional management and a large degree of autonomy improve performance indicators. The questions, which should be answered in relation to utility performance, are:

[45] Except for a slight increase in the corrected access figures over the 8 years investigated.

[46] Zambia merged the existing municipal water providers into ten utilities at the beginning of 2000th.

[47] In Kenya, there is a tendency to cluster utilities since the new constitution came into effect in 2010 with the devolution of power from the national level to the counties. Hence the numbers of utilities has recently begun decreasing from over 100 to around 90.

[48] The number of utilities in Kenya and Tanzania remained very high (100 and more) while in Zambia it was reduced to 10 with the regionalization of (provincial) water service providers.

- Which utility indicators should be used to measure progress?
- How did utility performance develop during reforms in the target countries?
- What is the influence of utility performance on access to water and on the development of the sanitation chain?
- Is there a difference in performance according to size or even an optimal size for a utility?

To answer these questions, utility data provided by the information systems of the regulators in Kenya, Tanzania and Zambia and by the 'contrat plan' in Burkina Faso are used. The data available have an acceptable reliability. However, when the number of utilities is elevated (over 30 for instance), regulators have difficulties to carry out detailed verifications on the ground as their budget and number of employees are limited. This limitation regulators face is one of the reasons why a number of data from utilities remain unchecked and seems to be copied from 1 year to another without verification.

Which Utility Indicators Should Be Used to Measure Progress?
The selected performance indicators of the utilities can be clustered in two groups according to the level of reliability of the data provided. The first group includes four indicators: O + M cost coverage, collection, billing and staff efficiency. For these indicators, the regulator or external auditors can, for all large, medium sized and today for most of the smaller utilities, consult information systems based on software (accounting/bookkeeping, billing, collection, personnel and customer management). The utilities also use such systems to produce their annual financial statements, which, in general, are audited. Therefore, the data can be considered fairly reliable. The second group of indicators are composed of water production, water quality, non-revenue water and hours of supply. For these indicators it is more cumbersome and difficult to produce hard data. Regulators or external auditors depend on well-maintained technical equipment on the ground (e.g. bulk meters to be regularly calibrated for production and distribution) and utilities would need standards for measuring and calculating the indicators in order to become comparable between utilities.

For instance, the hours of supply can vary substantially in the different areas within a town[49] and the utilities in general do not register interruption of supplies for different areas. Often, the working hours of pumps are used as a proxy for hours of supply. This proxy can provide quite a distorted picture on water being delivered to the network because pump capacity can change with age, placement of the pumps is crucial and other technical reasons can influence such counting. Therefore, high performing utilities install bulk meter along the networks which measure the water quantity and sometimes register changes of flow. However, the average hours of supply as reported by the utilities does not expose the areas with frequent supply interruptions. Therefore, the hours of supply for a utility given for one or several

[49] 30–40% of the experts interviewed for this work reside in the LIAs and underlined that they are facing regular interruption of supply often stretching over days while medium and high income groups benefit from service improvements brought by reforms.

towns can be in general considered guesswork at best. In none of the four countries there are official standards (e.g. issued by the regulators) on how to register interruptions of services and calculate hours of supply for an entire town.

Concerning water quality, regulators try to introduce national standards for the numbers of samples to be taken for the different types of tests and for the different sizes of utilities. However, regulators have difficulties verifying compliance with such minimum requirements on the ground because recording by utilities is often unorganised and the quality of the utility or private local laboratories is difficult to verify regularly. Nevertheless, spot checks taken indicate that the water quality from utilities can be generally considered to be acceptable after reforms but fluctuations are not documented. The consequence is that for the indicators of water quality, non-revenue water and hours of supply, the utilities report almost constant values over time. Utility personnel seems to be aware of the limited risk that external specialists will show up and take time or have the capacity to verify the reported data for these indicators, particularly in a setting where the number of utilities operating in a country is high.

This is somehow different in Burkina Faso, where DANIDA (Danish Cooperation) for water quality and Gesellschaft für Technische Zusammenarbeit (GTZ/German Technical Cooperation)[50] for non-revenue water have provided extensive support to ONEA at the end of the 1990s and at the beginning of the 2000s. This helped introduce a system of reliable accounting for the indicators of production, water quality and NRW (expert interviews). Nevertheless, this is not the case for hours of supply. ONEA had stopped in the 2000s to provide one figure for each or for all of the towns for which it provides services, most likely, because it did not install a reliable system of registering supply interruptions in each of the different areas within a town.

Therefore, it can be said that the indicators of O + M costs, collection, billing and staff efficiency are the most reliable to statistically measure the influence of utility performance on access to water. Concerning the second group of indicators, data either need to be verified with extended investigations or qualitative data are used. However, from the point of the consumers three of the second group indicators seem to be very important and NRW is most important for measuring management quality (refer to Sect. 4.3, utility performance). Therefore, they cannot be ignored when evaluating utility performance.

Often, the progress in the number of household connections is used to evaluate performance of utilities. Using this indicator alone is questionable when considering the many underserved people who cannot afford to have or maintain a household connection. In the four target countries, the number of household connections increased substantially in Burkina Faso (increase of 152% during the 8 years of investigations) and Kenya (121%) but much less in Zambia and Tanzania (49% and 47% respectively). This development provides an indication that progress in household connection is not necessarily a good indicator to evaluate the association of

[50] GTZ is the German Technical Cooperation. The name was changed to Gesellschaft für Internationale Zusammenarbeit (GIZ) in 2011.

utility performance and sector progress. In Zambia, which has progressed better than Kenya in the reforms, the utility has only managed as little as Tanzania in respect to household connection. It seems that Zambia has more than compensated the limited progress in household connections with the promotion of shared facilities (water kiosks).

How Did Utility Performance Develop During Reforms in the Target Countries?

From the four key indicators, billing and O + M cost coverage not only depend on performance but also on the average tariff, which is generally set by politicians or regulators.[51] However, there is evidence that a performing and pro-active utility management can convince their BoDs or the regulator to adjust tariffs regularly. ONEA managed to have the tariff sharply increased in 1994[52] after the first performance improvements were achieved. Therefore, O + M cost coverage and billing can be considered as performance indicators for utilities although it has to be kept in mind that their value depends also on other influences.

In Kenya, for instance the regulator is criticising water utilities which do not submit tariff adjustment applications on time.[53] Sustainability of service provision under increasing demand is a main concern of reforms. The chart on O+M cost coverage in Fig. 5.4 indicates that only Burkina Faso managed to reach an acceptable level of self-financing for investments with a ratio of 150% O+M costs recovery. Zambia reaches over 120% while Kenya and Tanzania are only around 100%. It should be noted that before the reform O + M cost coverage was far below 100% in all four target countries. Equally, all four target countries managed to raise billing continuously over the investigated period and have also succeeded in increasing staff efficiency over the years. However, Zambia still has twice as many employees per 1000 connections than Burkina Faso. It should not be forgotten that in some cases the increase in staff efficiency is neutralised by an increase in expenditures for personnel due to salary hikes, especially for managerial staff or expenditure by the BoDs. It is the responsibility of the regulator to curb such counteracting development.

[51] E.g. According to legislation in the three Anglophone target countries, WSPs must operate under regulated tariffs with the objective of cost recovery. However, many small WSPs continue to operate with tariffs that can hardly cover their O + M costs. (refer to the Performance Review of Kenya's Water Services Sector 2012–2013: 10).

[52] The increase of tariffs was substantial and allowed for total cost recovery. Tariff increases for the poor were limited in comparison to the tariffs for the medium and high-income groups. Compared to 1990, the tariffs in 1994 were higher at water kiosks by 93%, at the first tariff block with a consumption up to 10 m^3 per month by 82%, but already at the second block (11–25 m^3) by 337% and at the third block 26-50 m3 by 400% (see Table 5.10).

[53] E.g. Impact report of WASREB (2014: 43, 54), Kenya: 'WSPs without justified tariffs need to urgently apply for tariff reviews to ensure revenues match the cost of providing the service.' and 'Rural WSPs should make regular tariff applications to Wasreb to enable them to have adequate resources to cover justified costs.'

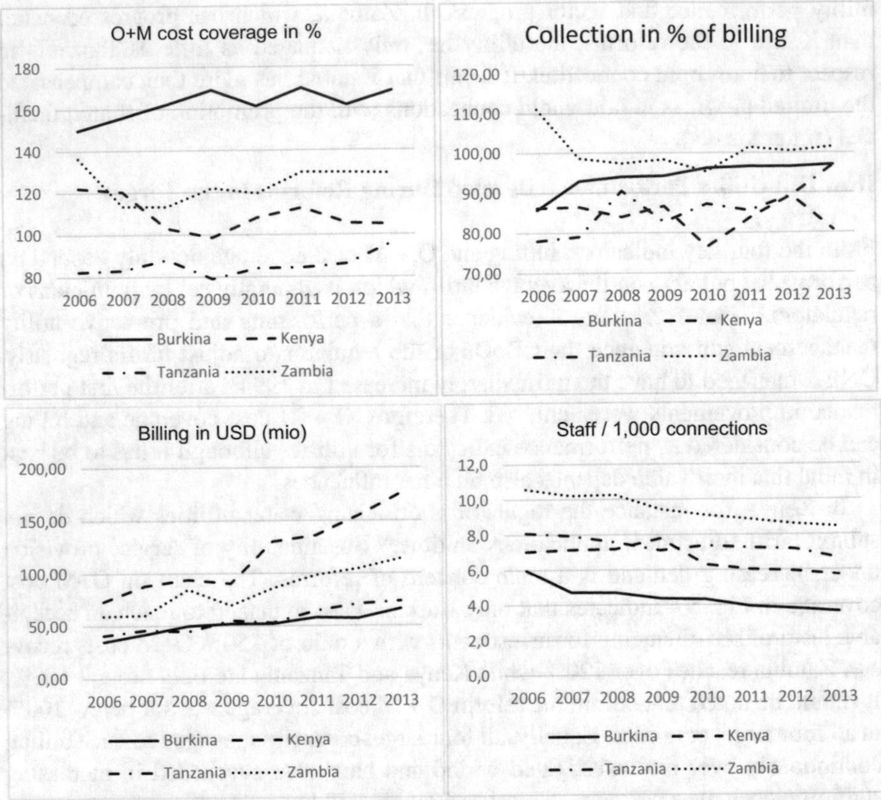

Note: The collection ratio can reach over 100% because outstanding payments collected in later years are often not attributed to the years when they appeared. Such phenomenon occurs especially at the start of reforms when utilities start concentrating on collection. In addition, in Kenya, only a part of the utilities (25) reported to the regulator in the first 3 years. Most of the small and some of the medium sized utilities reported thereafter. Because of a lower performance level of these utilities some indicators measured sector-wide deteriorated over the following years although the indicators of the initially reporting utilities improved further. The average exchange rate in the concerned years, which were applied in the calculations are as follows:

	2006	2007	2008	2009	2010	2011	2012	2013
BF XOF to USD	535	485	450	475	490	480	515	490
K KES to USD	73	68	68	77	81	89	85	87
T TZS to USD	1300	1250	1300	1350	1550	1510	1620	1700
Z ZMW to USD	4276	4345	5480	5214	7000	6381	6667	6400

Fig. 5.4 Key indicators for utility performance in the four target countries

Collection efficiency also depends on the ability of the utilities to overcome the resistance of the state institutions to pay their water bills.[54] Before sector reforms, in

[54] In order to be able to achieve a high collection rate on the water bills from government institutions the 'contrat plan' in Burkina Faso includes a specific paragraph on this issue. E.g. contrat plan 2007–2009 (2007: 9, 10) Article 19; 'Procédure de recouvrement par l'ONEA des factures de consommation de l'Etat et des établissements publics/apurement des créances irrécouvrables'. This must be seen as a reaction of challenges the utility faced in the past to collect water bills from government institutions.

many countries, the water bills from state institutions like police, military, universities, ministries, president offices, etc. constituted an important share of the turnover of the utilities[55] and usually the lion share of the outstanding payments for water bills. To overcome the non-payment and water wastage of government institutions is also one of the main concerns of reforms. A constant high level of collection efficiency is a good indicator that the utilities have managed this challenge. The chart on collection efficiency (Fig. 5.4) indicates that Zambia, Burkina Faso and Tanzania (the latter starting under 80% in 2006) have reached over 95% in 2013.[56]

Another indicator of interest in the sector development is litres per day per capita produced. It is also a rough indication of the level of average consumption (Fig. 5.5). Before sector reforms, high water wastage, despite a low coverage rate, was not an exception in many countries. This means that the limited number of people having access could afford to waste drinking water while the majority had to stand in front of dry taps. One of the main reasons for this was flat rates charged by utilities instead of block tariffs. However, the introduction of block tariffs needs the regular measuring and reading of consumption and consequently, the up-scaling of water metering. Underperforming and cash strapped utilities could not manage to ensure such investments and introduce the necessary complex organisational procedures to read regularly the water consumption of households.

Another reason for water wastage was the combination of flat rates and intermittent supply. When there was no water in the pipes, the consumers simply left the tap open for that time until water was delivered again. If the consumer were sleeping at night or out of the house when the water was running again, water was wasted and this was not reflected in the water bill because of the flat rates (expert interviews, Zambia). Very often consumers watered their gardens with drinking water and were surprised when hearing that using treated water for the garden transported through pipes to the customers must be considered a waste. There was a knowledge gap among consumers which had to be corrected during reforms by sensitisation and increased water tariffs. However, the indicators linked to the specific consumption can also be used as a pointer for an improvement of utility management as long as consumption is not limited by water rationing.

Furthermore, when the utility extends services to the LIAs on a large scale, thus, increasing the share of the poor among its clients, the average consumption will drop. The chart on average consumption (Fig. 5.5, in lt/d/c produced) shows a constant decline to a level between 50 and 100 litres for Burkina Faso, Kenya and Zambia, which is a sign that wastages through increased metering and the introduction of a rising block tariff were gradually controlled. Only Tanzania has not managed to do this and it can be noted that combined with the low level of access the utilities most likely still struggle to limit wastages as described above.

[55] E.g. for ONEA in Burkina Faso, it was around 20% of the turnover: Apercu sur l'état d'avancement de la restructuration de l'ONEA et perspectives (1995: 27).

[56] Tanzania has made good progress over the investigation period and Zambia has reached over 100% in some years due to collecting arrears and by not linking the years of collection to the years of billing in the accounting system.

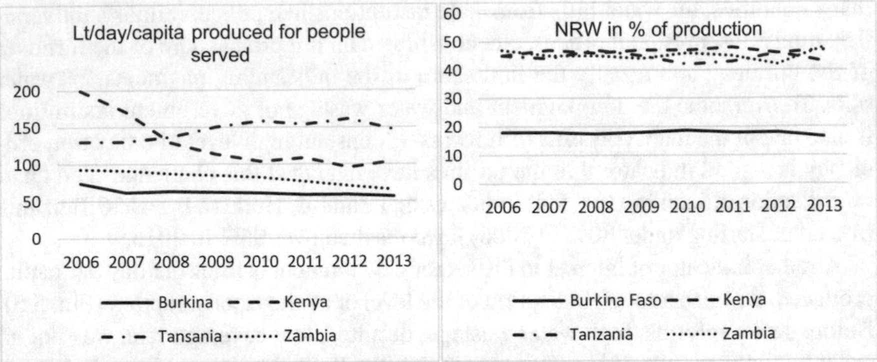

Fig. 5.5 Per capita production and NRW in the target countries

Table 5.8 Ranking of target countries according to utility performance indicators

Country	**O + M** cost recovery >100%	**O + M** cost recovery >150%	**Collection** of water bills >95%	Number of **staff** <4 = ++	Number of **staff** 4–6 = +	**Billing** of consumption: upward trend +	**NRW** <25%: +	Score
Burkina	+	+	+	++	n.a.	+	+	7+
Kenya	+	–	–	–	+	+	–	3+
Zambia	+	–	+	–	–	+	–	3+
Tanzania	–	–	+	–	–	+	–	2+

The utilities in the four target countries can be ranked according to the four key performance indicators (Table 5.8).

Despite some uncertainties of data quality for Non-Revenue Water (NRW) the chart in Fig. 5.5 provides a rough picture of the situation. Burkina Faso has managed to reach a remarkable level in curbing NRW with water losses under 20% calculated over all 56 towns serviced[57] while the other three target countries are still around 45% over the entire period of investigation. This achievement in Burkina Faso has been verified several times over the years with TAs and external technical auditors and can be considered as reliable but also as outstanding in the region. There are other medium sized water utilities in the three Anglophone countries which have succeeded to reduce NRW to around 25%.

What Is the Influence of Utility Performance on Access to Water and on the Development of the Sanitation Chain?

As indicated under Sect. 5.1 the degree of association between the proposed variables and access is established with Spearman's rho and by squared Euclidean distances. For Burkina Faso, the data show a very strong relationship (association) of access to water for the following utility performance indicators: O + M costs, col-

[57] According to the 'Etude sur la satisfaction des ménages aux services publics d'approvisionnement en eau potable fourni par les délégataires dans les quartiers non lotis de la commune de Ouagadougou', (2015: 8).

lection, billing, water production and number of staff. This is the same for Kenya and Zambia, whereby water production in Zambia is significant instead of high. Furthermore, Zambia is the only country among the three, which also shows a strong relationship between access to water and the water quality and hours of supply. Nevertheless, under the following question, the discussion explains a limitation of reliability of data for certain indicators, which includes water quality and hours of supply. In Tanzania where access to water has been gradually degrading, all variables but one does not indicate a significant relationship with access to water.

This analysis shows that an increase of utility performance measured by the proposed indicators can be linked to an increase in access to water in most of the cases (Burkina Faso, Kenya and Zambia) but this must not necessarily be the case as documented for Tanzania. The most likely reason why an increase in these utility performance indicators will not necessarily lead to increased access is that access depends on many other factors such as investments and especially the way funds for investments are used. Many utilities do not have an influence on this. As it is underlined in the next section also increased funding will not necessarily lead to increased access. Furthermore, as the Tanzania case shows, even a combination of increased utility performance and increased funding for asset development will not necessarily increase access, hence lead to a better sector performance. By this it can be seen that it is important to make a difference between utility and sector performance. Unfortunately, in many reports and studies this is not the case and therefore, they can generate misleading conclusions.

Concerning the link of water with sanitation, it can be noticed that in the three countries where the relationship between access to water and utility performance expressed in the four variables O + M costs, collection of bills, billing, water production and number of staff is high, there is an engagement of most of the utilities in the development of onsite sanitation and/or the sanitation chain. Burkina Faso is far ahead of Zambia and Kenya concerning the engagement in sanitation. Nevertheless, generally this engagement by the utility in sanitation development is rather documented descriptive (with activities) than with statistical data.[58] There is hardly any reliable data from utilities on the operation and development of sewerage networks. In addition, the importance of sewer networks for access is declining or stagnant on low level in the four target countries and most likely in all countries of Sub-Saharan Africa, leaving the vast and increasing majority of urban dwellers depending on onsite-sanitation. This is why it would be so important to involve the water utilities as professionals in the development of the chain for onsite sanitation and not only in piped water and sewer systems.

[58] In Burkina Faso, ONEA reports annually on the number of toilets promoted but not on the promotion of infrastructure for the decentralized sanitation chain. In Kenya and Zambia it is the Water Services Trust Fund and the Devolution Trust Fund (as pro-poor financing mechanism), which report on such activities and not the utilities.

Is There a Difference in Performance According to Size or Even an Optimal Size for a Utility?

In three target countries, there is more than one utility and in two of them (Kenya and Zambia) the regulator publishes data on performance for all of the utilities separately. This is an opportunity to verify if there are noteworthy differences among the utilities according to their size. The regulator in Zambia forms three different groups according to the number of household connections: >50,000; 15,000 – 50,000; and < 15,000. The 100 utilities in Kenya[59] were organised in the following three groups of utilities according to the number of household connections: >35.000; 35.000 to 12,000; and < 12,000.

The results of the association analysis of access to water and utility performance for the three groups (Figs. 5.6 and 5.7) confirm the results for the entire sector in the target countries. O + M cost recovery, collection, billing and staff efficiency are strongly associated to access to water at all utility groups. Nevertheless, this association seems to be more consistent with the small and medium utilities than with the large ones in Kenya and Zambia. In addition, large utilities in countries with several utilities seem to have more difficulties to maintain the same level of collection efficiency than small and medium utility groups. Another observation is that the collection efficiency of the small and medium utility groups is well above 100% for

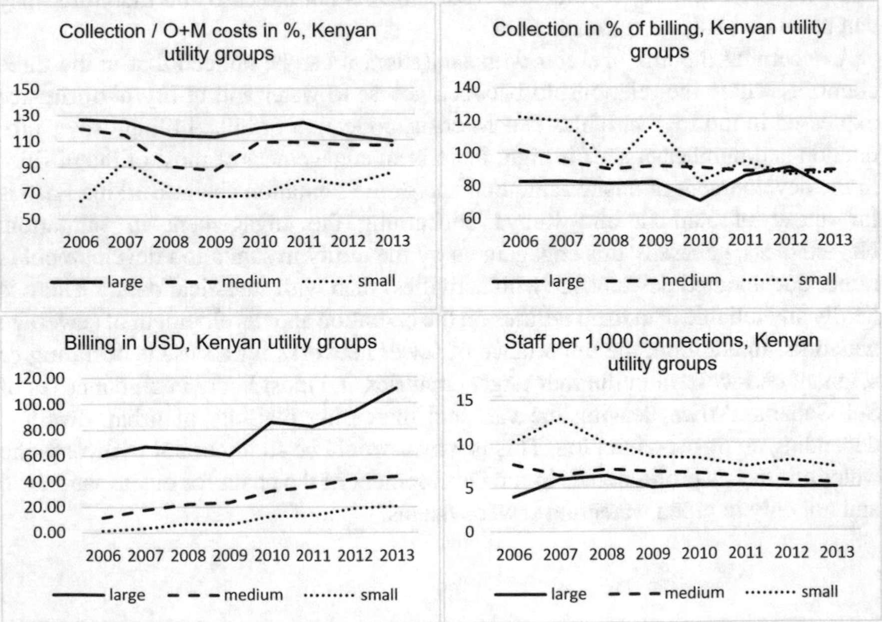

Fig. 5.6 Performance indicators of three utility groups in Kenya

[59] The number of utilities is changing in Kenya as merger processes are ongoing, now spearheaded by the counties.

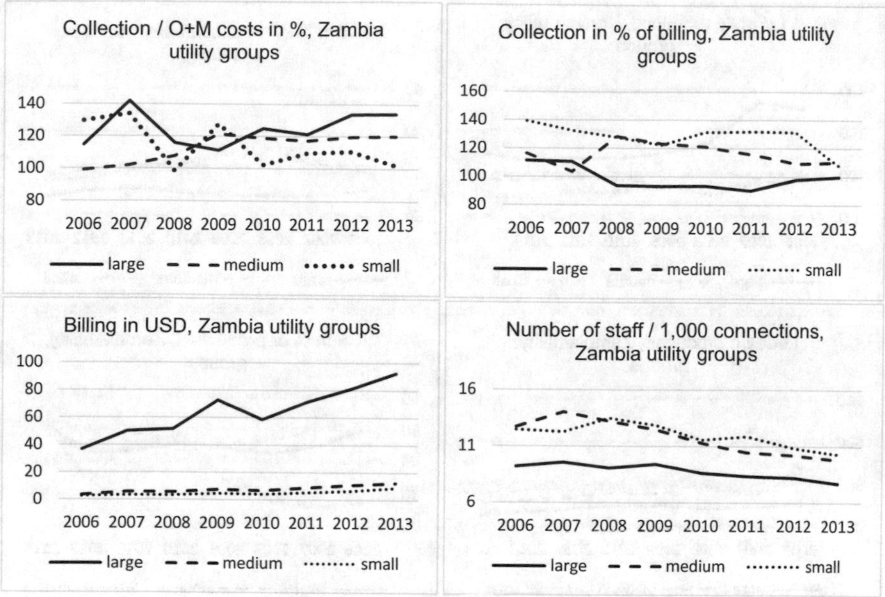

Fig. 5.7 Performance indicators of three utility groups in Zambia

the first years of reforms indicating that outstanding arrears must have reached a high level before reforms and these utilities managed to collect such arrears.

This suggests that medium sized utilities might progress better with reforms than the large ones at least in the first 8 years. It is understandable, that in big towns the water and sewer service deliveries are more complex and therefore, demands a higher level of organisation than in medium and small towns and most likely much more time to improve. The ranking of the utilities by the regulators in the two countries seems to confirm this statement. In Kenya, in the benchmarking system from the regulator medium sized utility were always best performer, except for the first year of investigation. Over the years, large utilities lost ground in the rankings with Nairobi reaching the third and Mombasa the 15th place in 2006, while descending in the rankings to tenth and 88th place, respectively, in 2013. The situation was very similar in Zambia where the best performers selected by the regulator were small or medium sized utilities in seven of the 8 years.[60]

Nevertheless, O+M cost coverage remained low with the small-scale schemes and is highest among the large group. This might confirm the challenges medium and small-scale utilities face to shield their company from undue interference and obtain the effects of economy of scale. However, professionalization has helped to improve management. It might also have to do with a lower number of big consumers in the small and medium towns. Equally, staff efficiency is higher among the large utilities than at the medium and small-scale groups in both countries. Without

[60] Ranking according to the sector performance reports of regulators in the period of 2006 to 2013.

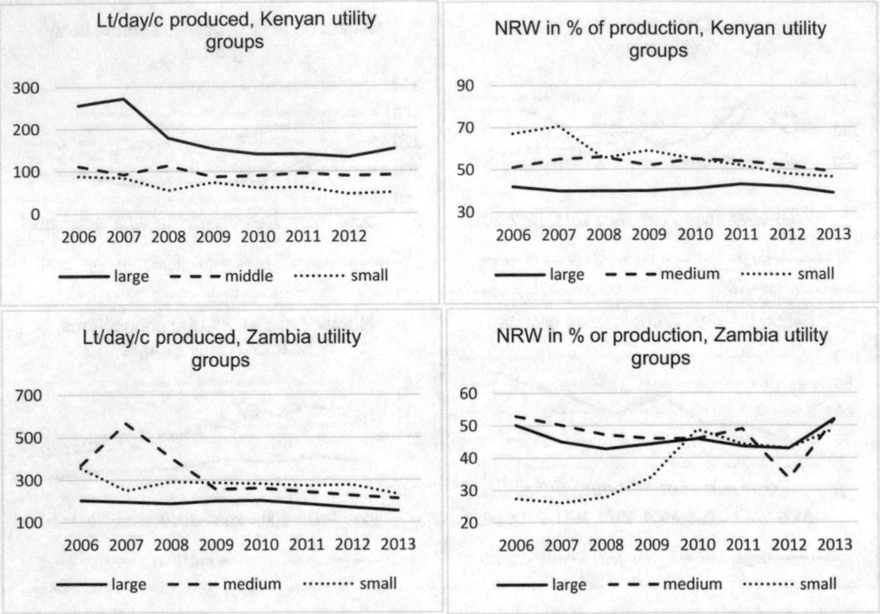

Fig. 5.8 Per capita production and NRW for three utility groups in Kenya and Zambia

doubt, large utilities such as ONEA in Burkina Faso have eventually an advantage if they can reach the same management quality than the medium sized utilities in the first years of reforms.

This does not answer the question of the possible optimal size a water and sanitation utility in Sub-Saharan countries should have. According to Oelmann[61] there can be no general agreement on the size of a water and sanitation utilities. Studies from different countries come to different conclusions because the (optimal) size depends on several factors such as growth of demand, population density, other regional factors, etc. and some studies also indicate that mergers of utilities did not necessarily offer efficiency gains or losses. However, Oelmann argues that there should be a minimum size. Also Rouse[62] underlines that utilities need to have a minimum size in order to offer attractive salaries and ensure scientific support functions. The empirical data on the small utilities from Kenya and Zambia support this recommendation for a minimum threshold of a utility size. Medium sized utilities have advantages as long as the management systems cannot ensure a higher quality in larger sized utilities. It should be remembered at this point that in the two Anglophone countries the utilities, which managed to reduce NRW to an acceptable level, are all medium sized (Fig. 5.8).

---

[61] Oelmann (2005: 28-30, 223).

[62] Rouse (2013: 4).

Concerning the indicators on litre per day per capita (lt/d/c) produced, there is an improvement in curbing wastage, which is generally achieved by increasing the metering rate and shifting from flat rates to block tariffs. In both countries, all utility groups managed to lower this indicator. NRW remained high among all groups of utilities in both countries. The example of the development of NRW reported by the small utilities in Zambia indicates the difficulties the providers face when trying to obtain realistic data on water losses. In the first years of reforms, the small-scale systems reported NRW less than 30%. At the end of the investigation period, the data on NRW was most likely more realistic with just under 50% and was in line with the level of the other two utility groups. In Kenya and Zambia, NRW also remained high after several years of reforms which indicates that all utility groups had difficulties controlling commercial losses (corruption) and install an adequate management system.

It can also be noted for the four target countries that with the establishment of an enabling framework and an improvement of performance in drinking water supply, utilities are more inclined to become involved in onsite sanitation by not only developing the sanitation chain but also by promoting access to safe sanitation installations at households, institutions (e.g. schools) and public places. Nevertheless, overall, the sector reforms in Kenya and Zambia have achieved limited results in sanitation compared to Burkina Faso which seems to be linked to the missing framework development for sanitation. However, there is one exception in Kenya with the project of Up-scaling Basic Sanitation for the Urban Poor (UBSUP)[63] which according to the project personnel has helped to build around 13.000 toilets in 23 towns and 10 decentralised treatment facilities from 2011 to 2018.

Finally, it is worth to note that the expert interviews stressed the fact that service interruptions in the LIAs are a huge problem for the poor especially in the three Anglophone countries despite the reforms. For the consumers in the middle and high income areas water rationing is less of a challenge as most of the houses are equipped with large water tanks, pumps, etc. Service interruptions even for days are hardly noticed. However, the poor in the LIAs usually cannot bridge long lasting service interruptions with water storage. Equally, the neighbourhood sales in Tanzania are a challenge for these poor. Only the utility in Burkina Faso seems to have made good progress in the service delivery concerning physical access (especially with water kiosks and subsidised household connections) and supply hours alike for all income groups (expert interviews).

[63] https://www.susana.org/en/knowledge-hub/resources-and-publications/susana-publications/details/2861 (last visited October 2018). Refer also to WSTF (2012)

5.6 Investments and Mobilisation of Funds

Access to urban water depends on the outlets of the water supply infrastructure where in the developing world people receive water either in the household, the tap in the yard or at water kiosks/public outlets. Concerning shared facilities, human rights require a cycle of maximum 30 minutes,[64] which should include waiting time at the kiosk.[65] For sewerage, it is the intake pipe, which leads from the house to the sewer network and drains the effluent to the treatment plant. In the case of onsite sanitation, it is the link to a collection and centralised or decentralised treatment system. Vacuum trucks, sucking the sludge from a toilet installation (e.g. septic tank) and depositing it at a treatment facility of the utility, can secure the collection in high and middle income areas.

It can be said that infrastructure and its development are preconditions for physical access and for acceptable services. Therefore, it seems to be obvious that there is a close link between financing of infrastructure and sector development in terms of access. Analysing the relevant data in the four target countries in respect to this association, surprisingly, the relationship between these indicators with access is only strong in Kenya and not significant in the other three countries.

An explanation for this phenomenon is that in countries with many people living in unplanned LIAs and under the poverty line, there are investments with different characteristics, as discussed in Sect. 3.4. They can be distinguished as first and last mile investments.[66] The underserved poor do not automatically benefit from first mile infrastructure (especially increase in raw water availability) and hence, first mile infrastructure development does not necessarily increase the number of people served. One example for first mile development is the enlargement of a treatment plant, which will increase water production, might help to improve water quality with better filters and might support a more continuous service delivery. Thus, first mile development improves foremost the service quality for the connected consumers and if financed with concessional loans (international cooperation) subsidises more the upper and middle classes and the few connected poor than the vast majority of the poor. The many underserved people are left out by these investments as long as they cannot have physical access to outlets of the utility system.

Nevertheless, first mile investments are a precondition to serve the poor, which unfortunately, is left unexplored in many cases. There are numerous locations where main water pipes cross LIAs and despite having the pipe so close to their homes, the poor do not have access. That there is a need to make the link between first and last

[64] 30 min from the home to the water kiosk, waiting time and the way back to the home.

[65] COHRE (2003: 8).

[66] It must also be mentioned that last mile needs first mile infrastructure. Therefore, the appropriate mix of fist and last mile infrastructure has a limit and most likely is around 10:1. Nevertheless, it will also depend on the ratio of the poor to the rest of the urban population and the percentage of underserved people.

mile infrastructure like the poverty funds in Kenya (WSTF) and in Zambia (DTF) have ensured[67] was recognised by all experts interviewed in all four target countries.

A Chronic and Widening Financing Gap

Despite the possibility of a missing statistical relationship between access and funding of infrastructure in the three target countries, it is obvious that infrastructure and the constant flow of funds for its development is essential, especially in a situation of rapid growth of demand and of a high number of underserved. The constant increase of funds during the early years of reforms in Kenya might explain the strong relationship. However, this constant increase in investments should not hide the fact that the level of infrastructure development was largely insufficient for decades and still is presently, and not only in Kenya (Fig. 5.9) but also most likely in most of the Sub-Saharan countries.[68]

The gap might even be widening, as the example of Kenya is indicating.[69] Nevertheless, the situation in Kenya would be worse today without the reforms, which have helped, at least, to gradually mobilise more investments just shortly after its commencement (Fig. 5.10). Reforms have also helped in some countries to make better use of the existing infrastructure by helping to increase access for the poor and by avoiding premature aging of assets and therefore spending for rehabilitation of infrastructure.

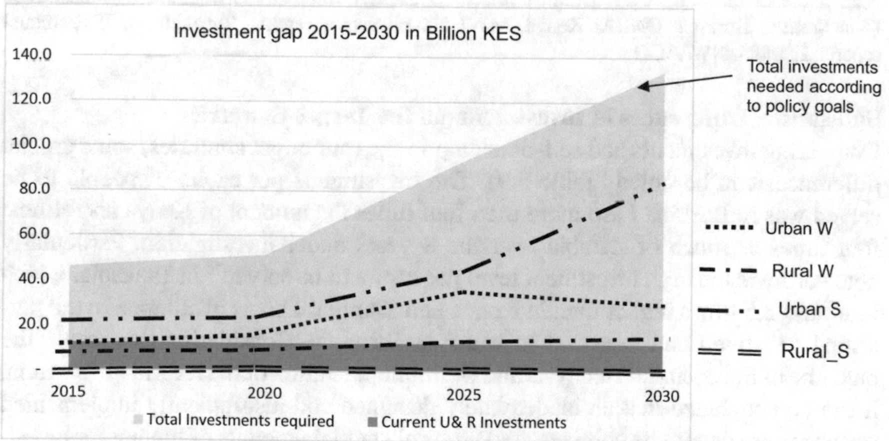

Fig. 5.9 Projection of the investment gap in Kenya. (Data Source: Sector Investment Plan 2014 and Annual Water Sector Review 2013–2014, page 49/50, figure Werchota)

[67] Own observation in Kenya and Zambia (2000–2017).

[68] OECD (2011: 4) Meeting the challenges of financing water and sanitation www.oecd.org/water, (last visited 05.2016). Refer also to Hutton and Varughese (2016).

[69] Strategic investment plan for The Water and Sanitation Sector in Kenya, Ministry of Environment, Water and Natural Resources, May 2014.

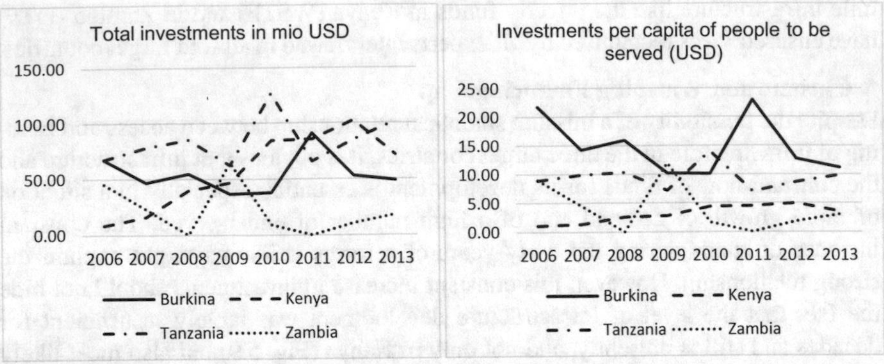

Fig. 5.10 Investments for asset development in the four target countries

Table 5.9 Average annual investments and self-financing in the four target countries over 8 years (period investigated)

Country	Investments in mil. USD	Self-financing ratio in %	Investment per capita to be served	Self-financing per capita served
Burkina	54.53	37	14.7	5.4
Kenya	56.88	9	3.5	0.3
Tanzania	84.02	0	9.5	0.0
Zambia	21.39	0	3.9	0.0

Data Source: Burkina, ONEA; Kenya, MWI annual sector reports, Tanzania, WSDP, annual reports; Zambia, NWASCO

Substantial Differences in Investments in the Target Countries

Comparing investments and self-financing in the four target countries, some crucial differences can be noted (Table 5.9). The investments per capita of people to be served was in Burkina Faso more than four times the amount of Kenya and almost four times as much of Zambia over the 8 years under investigation. Particularly noteworthy is the high investment level per capita to be served[70] in Tanzania, which is around 2.5 times higher than in Kenya and Zambia. This is in contrast to the persisting negative trend in access to water in Tanzania. Considering this result, the outcome of the document analysis and the implementation of the reform in Tanzania, it can be concluded that an inadequately designed and insufficiently implemented sector reform cannot be compensated with substantial amounts of money for investments.[71] It can also be concluded that the level of investments does not necessary depend on the wealth or poverty of the country (GDP per capita for instance) or on the assumed ability to pay for water and sanitation by the consumers.

Furthermore, Burkina Faso has a sector self-financing ratio (through water bills), which is more than four times higher than that of Kenya, and calculated for the popu-

[70] Population living in the service area of the utilities.

[71] 650 million USD over 7 years for urban W+S in Tanzania.

lation served it is 18 times higher. The sector self-financing ratios are as good as zero in Tanzania and Zambia. This persistent high level of self-financing infrastructure in one of the poorest countries in the world is an indication of what can be achieved in low-income countries with regard to consumer contributions for development of infrastructure. The political will for a higher self-financing of the sector was facilitated by the donor community in the first halve of the 1990s when the Ziga-Project was negotiated for Ouagadougou. It shows that the argument cannot be upheld that the consumer's willingness to pay has a tight limit and that self-financing cannot be stretched to a point where water bills can cover a significant part of the investments and the total repayment of loans.[72] It is also apparent (in the case of Burkina Faso) that it is not compulsory to have private sector participation in operation in order to raise sufficient funds for investment within the sector. It is equally obvious that in the three other target countries, there is resistance to use the full potential of self-financing[73] by holding the average tariff too low and not tapping sufficiently into the middle and high income earners (refer also to the expert interviews).

It seems that sector experts have difficulties to find convincing arguments when discussing tariff issues with politicians and the donor community do not use their leverage in most of the low-income countries to help the sector to move to universal access. Often the average tariff and tariff structure are not sufficiently distinguished in the discussions. In order to better serve the connected consumers and, at the same time, extend infrastructure to the LIAs the costs of service provision (around 150% O+M) needs to be covered, at least in the medium term. Once the average tariff is known then the decision on how to distribute the *'load'* between the consumer groups in an effort to reach social goals and reduce inequality can be taken. Hence, there is no competition between economic and social goals – one does not exclude the other. In Sub-Saharan countries, the argument that the tariffs are too high for the poor might have some relevance in some cases but generally not for the middle and upper income classes.

This was the strategy in Burkina Faso where the tariffs for the poor (postes d'eau autonome, kiosks and first consumption block at the household connection) increased only around 100% over 10 years while the other consumption blocks increased between 255% and 397% (Table 5.10, and it has to be noted that the FCA devaluated in 1994 by 50%). These increases, at no time, triggered unrest among the population in one of the poorest countries in the world although in some cases lead to very negative comments in the press and by NGOs. However, tariff increases in this case helped the poor to gain physical access to utility services. This underlines what many poor expressed in their interviews: physical access to utility water is the real concern and not the price demanded by the utility! Therefore, politicians and NGOs in general cannot claim to speak for the poor when they criticise regulated utility tariffs.

With an investment level per capita to be served in the urban setting similar to Burkina Faso (four times higher), Kenya would have been able to close the financing gap.[74] It indicates as well that Burkina Faso, with its investment level, is proba-

[72] E.g. Fujita et al. (2004: 84).

[73] All four target countries aim for full cost recovery in their sector policies / strategies.

[74] Refer to the Kenyan National Strategy for Water and Sanitation 2018.

Table 5.10 Development of water tariffs in Burkina Faso during the first 10 years of reforms

Water Tariffs	1990	1991	1992	1993	1994	1994[a]	1995	1998	99	2000
Postes d'eau[b]	46	46	54	64	76	87	87	89	91	93
Kiosks	90	90	107	127	151	174	174	178	182	186
1–10 m^3	90	113	123	134	147	164	164	168	172	176
11–25 m^3	95	120	158	209	276	320	320	338	356	375
26–50 m^3	200	250	312	390	486	840	800	860	924	993
50–100 m^3	255	320	376	441	517	840	840	860	924	993
+100m^3	280	350	402	461	529	840	840	860	924	993
Raw water	150	186	194	202	211	333	333	358	385	414

Source: ONEA

[a]Two tariff adjustments in 1994 because of the devaluation of the FCFA

[b]Public stand-post placed on a borehole and not linked to a pipe network like a kiosk

bly developing its water and sanitation infrastructure according to the present needs or, at least, has managed to narrow the financing gap significantly.[75] However, this comment might need to be more differentiated when considering the development in each of the many towns ONEA is serving because the lion share of the investments went to the capital Ouagadougou.

Financing Sources and the Mobilisation of Funds

In all four target countries, the main source of financing infrastructure is donor contributions. In Burkina Faso and Kenya, self-financing is the second largest source, although the Kenyan sector is far behind Burkina Faso (fifth column of Table 5.9). In Kenya, there are also contributions from the Gates Foundation and local banks. Both are quite limited in respect to the overall investments and almost as good as zero from the privates sector.[76] Government contributions through general taxes are also a small part of the sector funding if they are available at all.[77] Governments usually support sector reforms in order to reduce budget deficits by removing subsidy payments on national and local levels.

[75] Considering the remaining number of underserved people, it can be assumed that there is still a financing gap in Burkina Faso.

[76] Schiffler (2015: 6). Another example is Mali, where the expectation that the multinational company (Saur International) would invest as the contract indicated did not materialize. According to the experts interviewed, the multinational company explained that the insufficient tariffs did not allow them to generate the necessary funds for the agreed investments (own observation during a visit to the sector institutions in February 2016).

[77] The contributions of the national governments to asset development are approximately 6% in Kenya (sector report of the MWI and WASREB, 2016), under 5% in Zambia (NWASCO 2016), even less in Burkina Faso (limited to some payments to subsidize social household connections according to the experts interviewed) and around 10–20% in Tanzania (WSSR, 2014: 8, end of Phase I, the government dispersed 24% of the budget for the basket, which includes water resource management, rural and urban water and capacity building. However, there is no indication of how much of the percentage spent is for urban water and sanitation and how much of the government funds are used for asset development. Experts interviewed in Dar es Salaam, 2015 cautioned the figures by indicating that government has delayed disbursements and it is not expected that it will disburse all the committed funds. The estimates by the experts were around 10–15% of the entire budget for the basket).

Mobilisation of funds seems to be facilitated when the sector is well performing, like in Burkina Faso, or at least, when sector reforms have opened the possibility to increase performance of utilities, and other sector institutions like in Kenya.[78] In addition, the perspective to enhance self-financing of the sector through reforms might help to attract more donors. However, this alone does not secure a persistent high level of engagement by the donors. As already explained, a fund mobilisation mechanism with a number of tools (planning, financing models, monitoring) should be housed at professional institutions and not at the civil service structure like in ministries. Such a mechanism with the necessary instruments can also ensure a high level of aid effectiveness, which should be in the interest of all parties involved. Table 5.11 indicates that each of the target countries is on a different development level in regards to the use of tools for mobilisation of funds, the use of a professional fund raising mechanism and the monitoring of investments.

In Burkina Faso, the utility can accumulate funds for investment because it is protected by the Ministry of Finance which exempts the utility from paying tax for a surplus of income if it is earmarked for asset development. ONEA transfers such income to a special account, which is also monitored by the Ministry of Finance. In addition, ONEA as a fairly autonomous utility is managing investments as a professional which is not the case in Tanzania. Some decision makers in ministries do not like to delegate functions linked to investments to professional agents for some reasons and still can count on the support of donors. This observation is in line with the conclusion of Mosley[79] who discussed the motivation of donors when allocating aid and the areas of common interest for such allocations between donors and recipient countries. It is high time that donors evaluate receiving countries by hard facts before taking decision for funding and not by other criteria which are counterproductive for sustainable development.

The analysis of investments in the four target countries raises the following questions:

- Why was the level of investment per capita to be served in Burkina Faso substantially higher than in the other target countries?
- Why did Tanzania not manage to stop the negative trend of access despite a level of investment per capita to be served, which was substantially higher than in Kenya and Zambia?

The first glance on the results (Tables 5.9, 5.10, and 5.11) suggests the following answer for the two questions: The sector in Burkina Faso is more attractive and this already over a longer period. The self-financing level is high, the mechanism for

---

[78] Equally, in Mali in the last years where donor engagement and the number of involved donors increased substantially after the creation of a commercial utility and a professional structure for asset development. Nevertheless, the political turbulences and the loss of control of a part of the country by government might have also positively influenced decisions by donors to invest in the sector in Mali.

[79] Mosley (1987: 67).

Table 5.11 Existing structures and instruments for fund mobilisation (donors) in the four target countries

Country	Negotiation mechanism (all donors)	National investment plan	Sector financing model	Monitoring of investments	Professional investment mechanism[a]	Score
Burkina[b]	++	++	++	++	++	10+
Kenya[c]	–	–	–	Partly +	Regional +	2+
Tanzania[d]	++	+	–	Partly +	–	4+
Zambia[e]	–	–	–	Partly +	10 utilities +	1+

The score is ++ when the structure or instruments are available and +only when it is partly existing, as explained for each country as follows:

aInstitutions for investments separated from the state administration and operating on national level

bONEA acting countrywide as an autonomous service provider for water and sanitation, which includes onsite sanitation and sewer systems

cAsset development is organised regionally with eight Water Services Boards. There is no professional asset development agency established on national level and equally, key instruments on national level are missing. For last mile investments the Water Services Trust Fund was created, which is intervening countrywide but is receiving only a small fraction of the total investments

dThe Water Sector Development Programme (WSDP) is organised as a department of the ministry responsible for water and is lacking the necessary autonomy to act like an enterprise of the private sector. There are limited investments on last mile infrastructure and sanitation. With funds for urban water and sanitation of around 650 million USD over seven years only two million underserved were reached according to its reports, but according to the regulator only 300,000 additional people gained access through the regional utilities (75% of all utilities) during the same period

eThere is no financing mechanism on the national level except the Devolution Trust Fund (for last mile infrastructure) and the Ministry of Local Government and Housing responsible for water and sanitation has not been able to put the necessary structures and instruments in place

fund mobilisation and investments is more professional and transparent[80] and performance indicators of the utility are much better.[81] Contrary to this, the sector in Tanzania is still captured in the household connection for all paradigm and therefore, lacks an appropriate mix of first and last mile investments (insufficient poverty orientation[82]).

The advantage of Tanzania to have a sector-wide acting financing basket is not used sufficiently. The basket (WSDP) can also not develop a private sector like professional management as some of the utilities for operation can because it lacks sufficient separation from the state administration and politics. Some of the main donors have realised these challenges and have started to disengage from the WSDP after the first phase of investments (2006–2013), which reached around 1.2 billion USD for a seven-year programme for urban and rural water and sanitation. This impressive amount of money might be an indication that there are substantial funds

[80] L'intégrité de l'eau en action (2014), which documents good governance in investments in the case of the ZIGA project, implemented by ONEA.

[81] It can be assumed that utility performance has some influence on fund mobilisation and on the way the funds are used.

[82] The huge amount of spending for investments reported by WSDP compared to the limited number of additional people having received access in the period of investigation reported by EWURA.

Table 5.12 Contribution of poverty baskets to sector development and unit costs per beneficiary in Kenya and Zambia

Country[a]	Beneficiaries for water	%, additional people by poverty basket to total sector increase	Beneficiaries for sanitation	Water: USD/ beneficiary	San: USD/ beneficiary
Kenya, WSTF[b]	1,134,391	43.0%	89,773	18.9	21.5
Zambia, WDTF[c]	984,199	58.7%	–	15.9	–

[a]The Water Sector Status Report 10/2014, page 23, presenting a cumulated figure for the beneficiaries of 1,358,580 for the phase 1 of the WSDP while the regulator indicates an increase of addition people served of only 303.273 for the same period (2006–2013). Therefore, the available data do not permit an analysis for Tanzania
[b]Source: WSTF reporting sheet to ministry, 2013, which includes the first to the fifth call (2009–2013), average exchange rate to Euro is 1.35
[c]Source: WDTF analysis sheet, 10/2015 for 2006–2013, average exchange rate to Euro: 1.36

Table 5.13 Investments and people reached by poverty funds in Kenya and Zambia

Country	Investments by poverty basket in mio USD	Total sector investments in mio USD	Investments trough poverty basket in % of total	% of people reached by poverty basket
Kenya, WSTF[a]	21.40	390.62	5.5%	43.0%
Zambia, WDTF[a]	15.65	171.10	9.1%	58.7%

[a]for the period 2009–2013

available by the international cooperation for Sub-Saharan countries, which could be mobilised by the urban water and sanitation sector.

As explained before, Kenya and Zambia have established a specific financing mechanism for last mile investments (WSTF and DTF). This was one of the results of its poverty orientation in the reform. It seems that because of these poverty funds both countries managed to reverse the negative trend in access during the reform. According to the Tables 5.12 and 5.13, in Kenya, the investments of the poverty fund represented only 5.5% of the total investments in the sector, but this poverty targeted investments reached 43% of the total additional people gaining access in the sector during the same period. The figures for Zambia indicate 9.1% for investments and 58.7% for additional people reached.

Nevertheless, the absence of combining first and last mile infrastructure development on national level and the missing instruments for fund mobilisation on national level[83] combined with a low self-financing has posed a handicap on fund mobilisation for asset development (first and last mile) in Kenya and Zambia.

[83] Kenya has established eight regional Water Services Boards as assets developer and holder but have not established such an institution on the national level.

The Kenyan, Tanzanian and Zambian cases seem to suggest that professionalization of operation of infrastructure and regulation is not sufficient for successful sector development. It needs in addition, the mix of first mile and last mile investments and a professionalization of asset development within a framework of poverty orientation in order to make the sector significantly progressing in terms of access.

However, it should not be forgotten that putting low-cost infrastructure in place faces many challenges. Often politicians prefer to propose the highest service level (household connections) to its electorate despite the knowledge that the necessary funds are not available.[84] Utilities find it too cumbersome to operate low cost infrastructure and consider it often as commercially unattractive. Rationing of water by the utility will foremost start in LIAs and can stretch over months because the poor cannot intervene like the elite (expert interviews in the Anglophone countries). Rationing over long periods, for instance, makes kiosks operators and consumers reluctant to maintain or use the utility kiosks. There are still too many water kiosks and toilets, which never go into use or are abandoned shortly after being handed over to the utilities or the consumers. Inadequate planning, sensitisation, etc., and the resistance of many utilities to make the pro-poor infrastructure work are important obstacles to providing access for the poor. However, low-cost infrastructure from a formalised system such as utility water kiosks is highly appreciated by the poor, because they can escape dependency from informal service provision (expert interviews in all four target countries).

Therefore, sector institutions like regulators need to adopt a pro-poor approach and enforce it with the help of the ministries, financing mechanism and the international cooperation. The Kenyan example indicates that a well-designed pro-poor basket can reach many people and achieve a high level of sustainable use of low-cost infrastructure. The WSTF with utilities managed to obtain an average of 88% usage of its promoted low-cost infrastructure after around 5 years of going into operation (see following Table 5.14).

Table 5.14 Sustainable use of low-cost (last mile) infrastructure in Kenya, financed through the WSTF

	Infrastructure functioning		Infrastructure non-functioning		
Type of infrastructure	Number	%	Number	%	Functioning in %[a]
Water kiosk	264	84.6	35	11.2	88.3
Yard tap	164	73.2	37	16.5	80.0
Tank	46	95.8	1	2.1	97.9
Pump	2	100.0	0	0.0	100.0
PSF[b]	14	73.7	5	26.3	93.3
Total	264	84.6	35	11.2	88.3

Source WSTF, June 2013, survey on the use of low-cost infrastructure financed from 2009 onward. As explained by the WSTF, non-completed infrastructure was removed from the analysis and therefore, the sums of the percentage in the columns 3 and 5 do not reach 100% in all lines
[a]Corrected figures of total after removing non-completed infrastructure in the survey
[b]Public sanitation facilities

[84] Own experience in all four target countries (1989–2017).

5.7 Development of Access to Sanitation

The three questions answered for water are also relevant for sanitation:

- Did reforms lead to a stop or a reversal of the long lasting negative trend in access to sanitation?
- What level of coverage was reached in the sector development (2012/13)?
- How did the number of underserved people develop (over 8 years)?

Sector data of national systems is based on the reporting of utilities. However, utilities do not count access to sanitation facilities, except their contribution when promoting household sanitation. Regulators try to monitor access to sewer networks but do this with much less efforts than monitoring access to water. The only monitoring available for comparing countries with respect to access to toilets is JMP. However, until the SDGs, JMP was not making the link to a safe sanitation chain when counting access to sanitation (report 2017).[85] The safe storage of human waste in a pit was considered sufficient for safe sanitation in the rural and urban areas.

Nevertheless, improved toilets are an important contribution to safe sanitation as the design and recommendation for their use (hygiene education/behaviour) reduces the risks of humans being contaminated with faecal matter. This ability of the household to influence safety by the users of sanitation facilities is hardly possible for the water supply when the household is forced to frequent traditional water points because the urban user cannot easily influence the quality of the ground water and the state cannot expect from the users to monitor and ensure effective regular water treatment. Therefore, contrary to water, improved sanitation as defined by JMP can be used for the analysis, especially when looking at the situation at household sanitation facilities only.

However, as explained, the well-functioning of the sanitation chain in towns operated by professionals is in the public interest. This is the reasons why the state in some countries has started to use utilities to develop and operate decentralised sanitation[86] and promote/subsidise household sanitation facilities (toilets) which reply to minimum standards such as the possibility to be emptied. An outstanding example, even compared on global level, is Burkina Faso, where ONEA is promot-

[85] For instance, informal emptiers generally empty the pits of latrines and there are usually no regulations in place and enforced for how the sludge is transported and disposed. It can be said that usually informal emptiers do not treat the effluent collected. Furthermore, the figures for 2015 released by JMP in 2017 for safely managed sanitation only covers half of the global population and most likely more the industrialized world. Hence, it can be estimated that the situation of coverage in the world is worse than indicated in the 2017 report.

[86] Sanitation, which is not linked to a sewer system with pipe networks.

ing on average over 12,000 toilets annually for more than a decade.[87] ONEA started to be involved in onsite sanitation at the beginning of the 1990s, when reforms commenced. Similar programs, but on a much smaller scale have been started in the past in Zambia and Kenya which are implemented by the poverty funds for investments (DTF and WSTF).[88]

Figure 5.11 indicates that according to JMP, progress in access to sanitation facilities has been achieved in Burkina Faso, Kenya and Tanzania. The latter, in this case with remarkable progress, having started from a very low level in the 1990s. Zambia seems to have lost ground but on a high level, still with around 80% of access. With an access to sanitation installations in a range of 70–85%, the four target countries have by far surpassed access to water (35–65%), at least in three of the target countries. However, this offers no indication if progress in the development of the sanitation chain has been achieved. Only Burkina Faso scores high, with over 80% in access to water and sanitation alike. Therefore, it can be noted that in the urban setting access to sanitation is not lagging behind access to water, as it was claimed for the MDGs, when focusing on household sanitation only.

The picture is different when the sanitation chain is included in the analysis. In many countries in Sub-Saharan Africa, treatment of effluent by utilities from sewer systems is insufficient and for the sludge of onsite sanitation, generally there is little or no controlled collection and treatment at all. This causes pollution of the environment and has a negative effect on public health. According to Kloss,[89] in Kenya, which registers alongside Zambia the highest access rate to sewer systems among the target countries, 'Out of 14 million people in towns, the waste water of only 1.6 million reach the WWTPs [Waste Water Treatment Plants]' and that the treatment efficiency of the plants is around 20% only. Including the waste from the onsite systems, only 2–3% of the human waste is treated sufficiently.

There is also little attention on developing and monitoring the sanitation chain in Zambia and Tanzania. Although the regulator in Zambia is trying to include sewerage into regulations by reporting on sanitation services[90] there is no information in the annual sector reports on the quality of treatment of effluent by the utilities (equally in Kenya) in contrast to the monitoring of drinking water quality. In Tanzania, EWURA is reporting on the quality of effluent treatment but states that some utilities do not treat effluent or comply with standards such as the biggest in

[87] E.g. Programme National d'Approvisionnement en Eau Potable et d'Assainissement à l'horizon 2015, Rapport Bilan Annuel au 31 Décembre 2011, page 54, Ministère de l'Agriculture et de l'Hydraulique, Direction Générale des Ressources en Eau, Direction Générale de l'Assainissement and Office National de l'Eau et de l'Assainissement, ONEA.

[88] It is to note that in the meantime line ministries in the sector have changed their names such as in Burkina Faso in 2015 to Ministère de l'Eau et de l'Assainissement or in Kenya and Zambia. This might lead to a change in the orientation for utilities towards the involvement in onsite sanitation.

[89] Kloss (2009: 6).

[90] WSS Sector Report (2012/13: 12), The Status of Sanitation in Zambia.

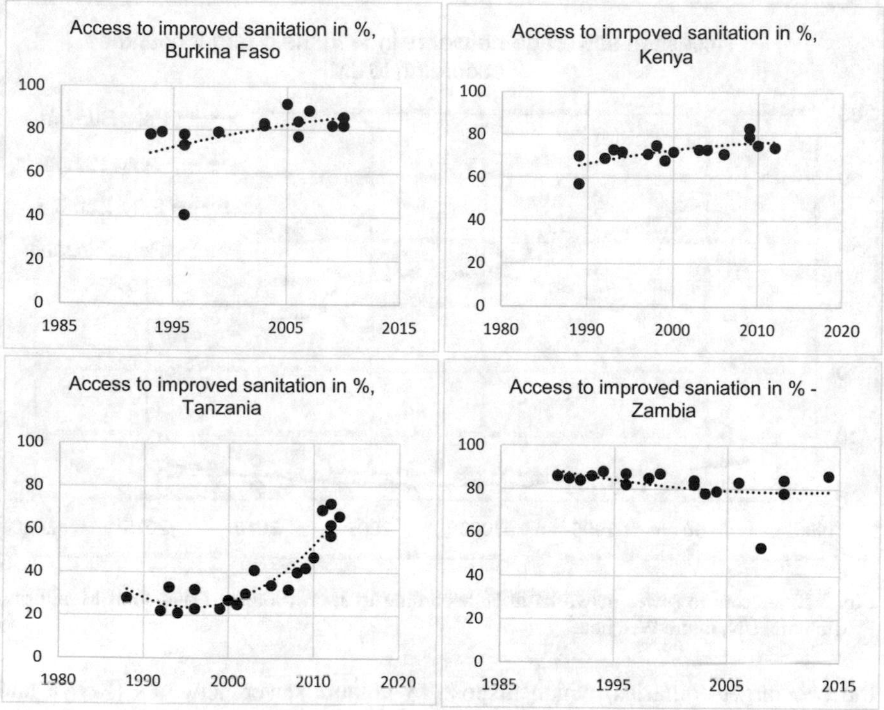

Fig. 5.11 Access to sanitation in % according to JMP. (Source: Data, Joint Monitoring Programme, UN; figure Werchota)

Dar es Salaam.[91] As this is persistent, it is one of several indications that attention for sanitation by sector institutions is still much lower than for water supply. In Burkina Faso where access to sewer systems is as low as in Tanzania (around 5%) ONEA does report on effluent treatment annually as it is required by the 'contrat plan'.[92]

Concerning the development of access to sewer systems in the last decades, there is little positive to report for the four target countries and most likely generally for all Sub-Saharan African countries. In many countries, water utilities are involved in operating and often developing piped sewer systems. Nevertheless, because of the high costs of net-bound sanitation and a lower attention for sanitation in general, the development of sewer systems is far behind water supply development. Hence, the need to concentrate on the chain for onsite sanitation is and will remain crucial in Sub-Saharan Africa, at least for the next few decades to come.

Also according to JMP (see Fig. 5.12) the percentage of people having access to sewer systems is either stagnant on a low level or declining. Referring to the data in

[91] Water Utility Performance Report for (2012/13: 115).

[92] E.g. 'contrat plan' (2010–2012: 5).

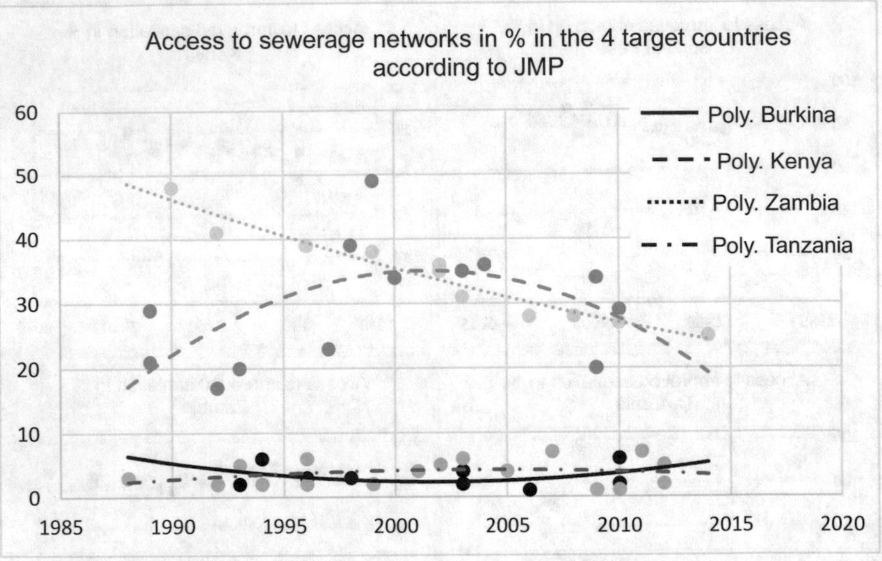

Fig. 5.12 Access to sewer networks in % according to JMP. (Source: Data, Joint Monitoring Programme, UN; figure Werchota)

the two target countries, which dispose of notable sewer networks (Kenya and Zambia), the national monitoring systems provide a sobering picture. WASREB, the water sector regulator in Kenya reported a decline in access to sewer systems from 19% in 2006/07 to 16% in 2013/14.[93] NWASCO in Zambia reported total sewer coverage of 58.7% in 2013, which included households with septic tanks not connected to a sewer network. The population with direct access to sewer networks in Zambia was 28% in 2013/14 compared to 34% in 2006/07.[94] It can be expected that this downward trend in access to sewer systems will continue in the coming decades. This decline in netbound sewer systems combined with the neglect of development of services for onsite sanitation in a situation of rapid urbanisation and densification of population explains the progress of Africa towards the 'new homeland of cholera'.

The relatively high percentage of access to sewer systems in Zambia compared to other countries is due to the fact that the mining companies in the copper belt provided such infrastructure and services before their privatisation under the Chiluba government. Nevertheless, this higher percentage of access to sewer networks does not shield the country (and Kenya alike) from cholera epidemics, most likely because of the neglected chain for onsite sanitation.[95]

[93] WASREB Impact report 8 (2015: 9).

[94] NWASCO, WSS Sector Report (2014: 45) and Urban and Peri-Urban Water Supply and Sanitation Sector Report (2006/07: 26).

[95] Spatial Analysis of Risk Factor of Cholera Outbreak for 2003–2004 in a Peri-urban Area of Lusaka, Zambia.

Summarising the development of access to sanitation, the following can be said: if the focus is only on the improved sanitation facilities at the household level, the percentage of access is higher than access to safe water in the four target countries. Bringing the sanitation chain into the equation, the situation can hardly be described as satisfying for both – access to sewer systems and collection and treatment of human waste from onsite systems. There are insufficient data on how (on-site) sanitation facilities are linked to a sanitation chain and therefore, access on sanitation reported by JMP must be taken with care.

Considering an estimated growth rate of 5% in the urban setting (especially in the LIAs), the population in towns double in less than 20 years. Comparing this growth to the development of access to sanitation according to JMP, it can be said that the number of households, which have no access to an improved household installation is still growing. The number of households, whose toilets are not linked to a controlled sanitation chain, is, without doubt, increasing very rapidly. Nevertheless, access to household sanitation facilities has not experienced a decline like access to piped drinking water. It has been indicated that it is not lagging behind access to water when considering only household installations.[96] This might suggests that after all, the desire of households to have sound sanitation at home has a higher priority than commonly thought or usually expressed in surveys by the head of the households.

However, it seems certain that in a majority of the countries in Sub-Sahara Africa where coverage to acceptable sanitation facilities in towns is progressing, it is not a result of reforms in the urban water and sanitation sector. One rare exception is Burkina Faso where the water utility is contributing substantially to access to sanitation facilities.[97] Supporting utilities on urban sanitation[98] might be less attractive for NGOs despite the fact that literature review reveals that NGOs generally criticise the lack of engagement by the municipalities when it comes to sanitation. On the other hand, some utilities point out that they would rather deal with local consultants or the local small-scale private sector than with local NGOs because many of them latter lack professionalism.[99] Equally, many NGOs may think that utilities are afraid to be too exposed when cooperating with them. Nevertheless, sector reforms in water and sanitation have started to engage the utilities in sanita-

---

[96] There are several contribution emerging which question the usual claim that access to sanitation is trailing access to water such as 'Does Global Progress on Sanitation Really Lag behind Water? An Analysis of Global Progress on Community- and Household-Level Access to Safe Water and Sanitation by Cumming O. et al. http://journals.plos.org/plosone/article?id=10.1371/journal.pone.0114699 (last visited 04.2016).

[97] With the promotion of around 16,306 toilets in 2012 and an estimated ten people using each facility (which includes shared facilities), ONEA ensured 163,000 people additional access to safe household sanitation; suivie PN-AEPA (2013: 10). This contribution represents 78% of the additional demand due to urban growth, which is estimated at around 210,000 in 2012/13.

[98] Especially when the sanitation chain for onsite sanitation is concerned.

[99] Arguments raised during discussions with middle and top management from utilities (own experience, 2015), especially in Burkina Faso where ONEA shifted contracts with NGOs to local private consulting firms.

tion in several countries, following thereby the sector principle of engaging professionals for service provision rather than leaving sanitation services with the state administration.

5.8 Development of the Crucial Factors in the Target Countries

The Sect. 4.3 elaborated in some detail the proposed crucial factors, their links and the key intervention areas. This reflection can be compared to the findings of the progress in access to water and sanitation, utility performance and investments/fund mobilisation in the four target countries (Sects. 5.4, 5.5, 5.6 and 5.7). However, the use of quantitative data alone cannot describe the development in all proposed crucial factors. Hence the present section will try to close this gap. It will help to understand why reforms generated such different outcomes in the target countries (see Chap. 6).

The Framework Development

The evaluation of the development in the framework was carried out by the means of a document analysis. Its objective was to see if there can be a link established between the development of the sector and the framework. It has been underlined that there was no literature found which could set an example on a systematic analysis of framework documents in the urban water and sanitation sector. Hence, it was necessary to establish a methodology for the document analysis which is described in detail in the above mentioned dissertation, volume 2.[100]

With the results of the document analysis, a ranking of the four target countries according to the extent the framework is enabling the sector development was possible. Comparing this ranking of the target countries to the ranking according to the sector performance as described in Sects. 4.4, 4.5, 4.6 and 4.7 it can be noticed that both provide a very similar picture. Burkina Faso as the best performing country has the most enabling framework, while Tanzania as the least performing country has the least enabling framework (Tables 5.15 and 5.16). The difference in the score for performance of Kenya and Zambia is very small and therefore, the different position of the two countries in the two tables below should not question the results of the document analysis.

The obvious differences between the target countries are for instance that Burkina Faso kept a very lean institutional structure despite decentralisation efforts in the country[101] and Zambia managed to merge the municipal depart-

[100] University of Vienna, International development, Werchota (2017)

[101] Before ONEA was established in 1985 as a public agency, municipalities were responsible for water and sanitation service provision. Thereafter, ONEA was transformed into a limited liability company in 1994. It might be the negative experience with water department of municipalities which avoided a break up of ONEA during the later decentralisation efforts of the country.

Table 5.15 Ranking of target countries according to the document analysis

Countries/score	Framework	Implementation	Additions	Total
Burkina Faso	21	23	4	48
Kenya	17	17	2	36
Zambia	11	18	4	33
Tanzania	11	10	0	21

Table 5.16 Ranking of target countries according to the sector performance

Country	Stop of negative trend	Increase access in %	Reducing the number of underserved	Reducing underserved by >50%	Score: total of +
Burkina F.	+	+	+	+	4+
Zambia	+	+	+	–	3+
Kenya	+	+	–	–	2+
Tanzania	–	–	–	–	–

ments for water and sanitation into (10) regional commercialised water utilities, registered under the company act despite substantial resistance from the line ministry, the MLGH. This clustering of utilities became a particular strength of the Zambian sector.

The other two target countries paid less attention to a lean sector architecture and today have to deal with around 100 and more providers, a complicated asset development structure and institutions which contribute as good as nothing to the sector in terms of quality and quantity (e.g. Water Appeals Board in Kenya). The separation of functions and regionalisation on extended scale in Kenya (13 sector institutions[102] and 100 utilities) is in stark contrast to the very lean structure in Burkina Faso (one sector institution and one utility). The institutional framework in Tanzania would also need a complete overhaul, especially a review of the numbers of utilities and the need for an asset holder in Dar es Salam. Also the involvement of the line ministry should be questioned in the interest of extending autonomy for sector institutions.

In Burkina Faso, the highly enabling framework and the lean institutional architecture seem to offer a strong support for implementation following largely all proposed key sector principles. In addition, there are sufficient detailed documents for implementation. A particular strong point in Burkina Faso is the sector coordination and harmonisation of stakeholders with the different committees as well as the regular meetings where the donors (FC only), government and ONEA discuss the results of the sector monitoring and link it to the sector financing. Some of these tools do not exist in the other target countries or are less effective.

[102] Ministry, WASREB, 8 WSBs, WSTF, WAB, KEWI.

Also in Kenya, the sector has a well-detailed policy and legal framework with a number of implementation documents, but this is compromised by an institutional framework, which is unnecessary complex, combined with a very weak oversight by the line ministry. The separation of service provision and asset holding and development followed the desire to promote PSP which finally did not materialise. However, in some countries (e.g. Senegal and Mali) the separation of asset holding/development and service provision helped to professionalise funding and asset development and to establish a badly needed fund mobilisation mechanism on national level. The Zambian sector has a policy and legal framework with many gaps but a more detailed implementation framework.

However, in all four target countries the documents for the framework were either newly established or reworked during the reforms although with different quality and coherence to the proposed sector principles. Good examples are the revised sector policy 2013 in Kenya, which has a much higher quality then the Sessional Paper of 1999. The same can be said from the new policy 2015 in Burkina Faso compared to the policy of 1998. Hence, sector reforms build capacities in the sector, which not only lead to a higher level of performance in service delivery or more funding for asset development but can also lead to framework documents with a higher quality and precision.

In Kenya and Zambia, the support through the poverty baskets (institutionalised pro-poor approach) has helped to progress access through last mile infrastructure development. However, the weak point in these two countries is the utilities, which in general still do not pay sufficient attention in their day-to-day operation to the poor (expert interviews in the LIAs). Hence, access to water and sanitation is not progressing as much as in Burkina Faso. In Tanzania, there is even less attention paid to the poor in the framework as well as by the sector institutions on a national level (regulator, WSDP and ministry) and the utility on local level. This might be one of the key elements which explain the negative results in access to water and sanitation.

In respect of sanitation, it seems that only in Burkina Faso reforms have increased the decision-making power of the line ministry for water and its sector institution (ONEA). This was possible by lifting policy decision for sanitation to the level above the line ministries (e.g. cabinet). In the other three target countries there was no sign that the framework for sanitation was significantly changed during reforms to introduce new sector principles such as professionalising services for onsite sanitation. It can be said that the dominance of the MoH in sanitation regarded as the 'mother of sanitation' (expert interview, Tanzania), hindered the development of a framework for sanitation in the water sector.

However, there is change ongoing which still needs to be driven further and gain quality. In Kenya for instance the executive order 02/2013[103] concerning the organisation of the Kenyan government gives the function of 'Public Health and Sanitation Policy Management' to the MoH and its implementation 'Sanitation Management' to the line ministry for water and sanitation. As there is no hierarchy of ministries

[103] Executive order 2/2013 (2013: 11, 13).

and a long-standing history of conflicts between these two ministries, it is difficult to imagine that one ministry would implement a policy for its key functions elaborated by another ministry. The solution would be to have the MoH elaborate a policy on public health and the Ministry of Water and Irrigation (now Water and Sanitation) elaborate a policy on sanitation infrastructure development and service provision. It is important to solve such conflicts because the result of the document analysis seems to confirm the link between the framework and sector performance which is also relevant for sanitation.[104]

The Development of a Regulatory Regime

In the three target countries where regulatory agencies have been established (Kenya, Zambia and Tanzania), almost all experts confirmed that regulation was crucial in bringing positive changes into the sector. It helped to improve utility performance, developed a better oversight of the sector and brought some improvements in corporate governance. Especially in Kenya, it turned out that the regulator became a stabilising factor in a very dynamic situation. The impact of regulation on the sector is for the stakeholder especially visible in three areas. Firstly, the annual report of the regulators on utility performance, which is based on elaborated information systems and benchmarking. Pollem[105] support this statement with his observation that regulation introduced with sector reforms in (four) Sub-Saharan countries has led to a notable improvement of transparency in the sector. Secondly, regulation helped to move the sector to at least 100% O+M cost coverage. Thirdly, regulation helped the utilities to make government institutions pay their water bills and therefore, succeed to maintain a constant high level of collection efficiency. Experts indicated that benchmarking has not existed and 100% O+M cost coverage has never been reached before reforms. For these and other areas (e.g. water quality) the regulators issued guidelines and established information systems.

Furthermore, regulation has helped to professionalize tariff setting in the three Anglophone countries and made this process more transparent. The previous influence on the tariff process from the central government seems to have subsided but unfortunately, often enough not from local government. Another most visible area of regulation was the improvement of customer services and especially complaint handling by the utilities. Furthermore, the regulators in Zambia and Kenya established consumer representation groups for the water sector (Water Watch Groups – WWGs, Water Action Group – WAGs) and provided them with power to handle unresolved consumer complaints before they are referred to the regulator as last level of decision making. In addition, the regulator receives reports from these consumer representation groups, which indicated other issues concerning consumers such as long lasting interruptions in water supply in the LIAs. Hence, with these consumer groups the two regulators established a prolonged arm of regulation to the ground.

[104] Burkina Faso is first and Tanzania last in both evaluations. Kenya is in second place according to the document analysis and third place according to sector performance. However, the score for Zambia and Kenya is very close in both evaluations.

[105] Pollem (2008: 459).

The effectiveness of such sub-structure of regulation can be seen with two examples from Zambia. In 2004, the regulator (NWASCO), before looking into it, rejected an application by Lusaka Water Company to increase tariffs arguing that there is a long list of pending complaints which has not been resolved. The list of these unresolved complaints was provided by the Lusaka WWG with the argument that it needs to be resolved before tariffs can be discussed. This temporary rejection by the regulator of a tariff adjustment proposal also improved the status of the WWG in regards to the utility. Before this incident the representative of the WWG was only received at one of the counters for consumers complains. After this incident the WWG was always received by the commercial director of the utility.

Another example of how useful the integration of consumer representation at the regulator can be, occurred in 2005 when the regulator instructed the Lusaka Water Company to halve the tariff in one of the LIAs, which only received water twice a week for as long as this drastic under-supply was carrying on.[106] The WWG for this town informed the regulator of the desperate situation the dwellers in the LIA are experiencing. However, despite the positive results of consumer representation groups both regulators (Kenya and Zambia) seem to lose the benefits of it because such structures need constant attention and a good management at the regulators. Nevertheless, it can be said from the experience in the four target countries that consumer representation structures placed on national level (e.g. Tanzania, Burkina Faso) seem to be less effective. The distance of such representatives to the consumers, especially in the LIAs, and the selection of their members (e.g. university professors, business people, instead of 'former' slum dwellers) left to ministers is generally of little help for the consumers and underserved people (Burkina Faso and Tanzania).

Concerning the autonomy of regulators, Pollem[107] studying four cases of water sector regulation in Africa, could not find a significant interference of central government on the regulators although he recognised the existence of such risks. He could also not find evidence that utilities have an undue influence on the regulator. He also found no evidence that the regulators have a financial interest in promoting high tariffs. The reason is the relatively low tariff level in the sector in the low-income countries. Equally, he found that with a high credibility of regulation, governments receive little pressure from the electorate to interfere in regulatory affairs.[108] These observations seem to be confirmed to a large extent with the present work in the three Anglophone target countries. In Zambia, the line minister appeared to be happy not to be involved in tariff setting and referred questions from journalists, during the launch of the annual sector performance report back to the regulator thereby indicating that it is no longer the responsibility of the line ministry.[109] In Kenya, the regulator had the autonomy to keep undue political interests at a distance

---

[106] Both examples are own experience.

[107] Pollem (2008: 453–455).

[108] Pollem (2008: 462).

[109] Press conference in 2003 in Lusaka, during the launch of the annual report of NWASCO for 2001/2002 on utility performance (own experience 2003).

while in Tanzania it seems that the regulator faces more pressure and interference by politics.[110]

Pollem[111] also attests that the regulator in Zambia (NWASCO) has a high effectiveness especially concerning the monitoring of the performance of utilities (CUs). What is interesting is that he compares the development of utilities with the (at the time still remaining) departments of water in local authorities[112] and finds that regulation did not have an impact on the performance level of these water departments. He qualifies their performance as 'catastrophic'. Hence, it becomes evident that autonomy of utilities is crucial even if it is not perfect and this generally cannot be secured with water departments being managed by the local authorities. Hence socially responsible commercialisation of service provision (including autonomous utilities and regulators) was a crucial move in the reforms towards improving performance of service provision and establishing a separation line between utilities and politics.

However, the development of regulation in the three target countries has also shown limits. In Kenya, many local/county authorities pressure management from utilities not to apply for tariff increases (expert interviews) and all the regulators could do so far is to expose such practices with the annual reporting to the public. In addition, regulators are concentrating almost solely on keeping O+M costs as low as possible. Hence, regulation in the three Anglophone target countries has not lead to a cost recovery significantly above the 100% of O+M. Regulation could also not influence much sanitation development and the strategy to replace the informal service providers.

In addition, regulation had limited influence on corporate governance and control of corruption in the utility as the persisting high NRW level is indicating. In many cases it is obvious that the autonomy of managers in the day-to-day operation is still reduced. Furthermore, the link of regulation to the provision of funds for investments remains weak in the three target countries and was limited in Kenya and Zambia to the exchange of the regulator with the poverty funds (WSTF, DTF). To this adds the tendency of stronger sector institution to start unnecessary competition (e.g. WSTF).[113] In Tanzania, a harmonisation with the financing basket (WSDF) seems to be totally absent because of the strong influence the ministry has on the sector institutions. Hence, there is need for better sector coordination by the line ministries and at the same time to roll out a private sector like management culture among professionals. However, most of the limits of regulation come from the insufficient support by the civil service system, the donors as well as an inadequate design of the framework.

---

110 The firing of the CEO of the (multi-sector) regulator in 2017 by the president over a tariff dispute indicates the limited autonomy of the regulator.

111 Pollem (2008: 473).

112 In the meantime, all local authorities have outsourced their water departments to the Commercial Utilities (CUs).

113 WSTF – FINISHING STRONG: WSTF Strategy towards Water & Sanitation Access for 100% of the poor by 2030, Version 1 (2017: 22), Competition with other Institutions e.g. WASREB', SWAT analysis

Nevertheless, regulation was useful also in unexpected areas. One example (Kenya) was the intervention of the regulator to secure the ring-fencing of sector income. Notwithstanding the huge and growing investment gap for water and sanitation, municipalities struck a deal with the line ministry to receive 'lease fees' from the water bills as compensation for handing over assets to the WSBs. There was no legal provision for such payments when transferring assets.[114] This income labelled 'leas fees' was used by the municipalities to subsidise other/non-water and sanitation related local services. Despite this drain on the self-financing capacity, donors first increased their funding during the reform until the regulator blew the whistle. Thereafter, the donors linked relevant conditions to the funding and hence, helped to end these counterproductive and illegal practices by the municipalities.[115]

Burkina Faso, which does not have a regulatory agency, has put in place a number of regulatory tools and mechanism[116] helping to secure a high autonomy for the utility and linking financing of the sector with good governance. The strong involvement of the Ministry of Finance gives Burkina Faso significant advantages in the development of the sector and provides a substitute for key functions of regulatory agencies such as tariff setting for cost recovery. This supports the effectiveness of the regulatory tools. The involvement of the utility in (onsite) sanitation development is especially successful. However, some experts interviewed underlined the weaknesses of ONEA in regards to the collection of information in the LIAs and the possible risk that the utility will become too dominant in the sector because of its superior knowledge and professionalism compared to the ministry.

Evidence points to the fact that regulation has a very high potential to improve utility performance, transparency and to reach O + M cost recovery where the sector has a record of low performance and inefficient governance.[117] So far, regulation was able to avoid a sliding back of the sector into a situation which existed before the reforms and helped the sector to develop certain resilience. Thus, regulation contributed significantly to make checks and balances work and safeguard benefits of the reforms, even in the case of Tanzania, although to a limited extent. This is one of the reasons why the negative trend in access was not reversed in Tanzania.

Utility Performance Development

The improvement of utility services in the four target countries has been described in Sect. 5.5. The high performance of ONEA in Burkina Faso has been highlighted and it is remarkable that with certain indicators, ONEA outperforms utilities in Eastern and Southern Europe. Its performance in NRW in many towns is better than

---

[114] It should be noted that the assets transferred by the municipalities where often previously created by the national government and the municipalities received such assets without conditions during decentralization.

[115] However, it was an issue, which over time raised concerns especially at the regulator and eventually, lease fees were gradually phased out.

[116] NANA (2015).

[117] The high level of performance of ONEA and the satisfying governance (political will) in Burkina Faso might be reasons why the introduction of regulatory instruments without a regulatory agency seems to be sufficiently effective in Burkina Faso.

most of the utilities in Great Britain. All target countries seem to have made significant advances in utility performance since the start of the reform although on very different levels. Concerning service provision to the poor in the LIAs, the utilities in Kenya and Zambia were supported by pro-poor funding mechanism in respect to the establishment of assets and capacity building among utility personnel on how to develop and operate sustainably last mile infrastructure. There are also some utilities in the other four target countries which reached the level of ONEA performance. The analysis also indicates that an increase in household connections, generally used in literature to document success of utilities to contribute to policy goals, is not necessarily a good indicator to measure progress in access to water in low-income countries (refer also to Chap. 2, Fig. 2.1). Only the combination of progress in household connections with other service levels (yard taps and water kiosks) can provide a realistic picture of how the sector is progressing towards universal access.

Also the outstanding performance of ONEA in sanitation has been described in Sect. 5.7. The recent renaming of some utilities in the target countries from water and sewerage to water and sanitation companies as well as the renaming of line ministries in this respect gives hope that finally the involved of professionals in the sanitation chain for all kind of facilities and a development of a framework for sanitation in the water sector will take place in due course in many more countries in Sub-Saharan Africa.

The Development of the Information Systems and Databases

The experts interviewed in the target countries indicated that because of missing or inadequate information in the sector before the reforms, decision makers did not have sufficient knowledge of how the sector functioned and which challenges have to be faced. For instance, decision makers were convinced that self-help groups of the communities or NGOs are the bet placed to provide services in the urban LIAs. Therefore, policy makers believed that small-scale community managed systems and NGO supported measures would help the poor to receive better services over time in LIA. As it turned out, this is generally not happening. Experts also explained that challenges were accumulating until a 'breaking point', which finally made 'firefighting' a daily event in the operations of utilities because of insufficient knowledge, orientation and a desperate financial situation (expert interviews, Kenya). However, due to new information systems at professional institutions, policy makers in the four target countries could be convinced that leaving the poor out of the reach of utility services produced the *'urban water and sanitation divide'* with a permanent discrimination of the poor. Enhanced information also improved planning.

Another element which led to an important change in the target countries in respect to information was the legal obligation of sector institutions to report to the public. The newly created professional institutions realised in a very short time that they would not be able to fulfil such obligations without installing and maintaining information systems. Regulators had to observe the sector and decide on interventions to enhance performance and to know the effects of their measures. Utilities had to submit standardised reports to the regulators, which demanded much more quality in data and information than the ministries previously were able to demand and analyse. This very different attitude by professionals compared to the civil servant staff from the ministries and the recognition of the crucial link between data and reporting developed the fertile ground for an improvement of the design and the sustainability of information systems.

Most of the experts interviewed recognised that the improved availability and quality of data as well as the annual reports produced by the sector institutions are some of the most visible achievements of the reforms. Thus, it seems that the sector during the reforms paid as much attention to this crucial factor as it did to the professionalization of service provision and regulation. A milestone worth mentioning are the reports of the regulators. It made the sector more transparent and allowed the regulators to make the sector development known to the decision makers and the public (according to the experts interviewed). It also became a trade mark for the regulators. For the first time politicians had a more reliable national source available to verify how far the sector develops towards their policy goals. However, there are signs that despite this improved information, politicians still do not grasp the urgency of the situation and react timely to upcoming bottlenecks.[118]

Nevertheless, the usefulness of information systems depends also on how data reflect reality. Experts in Tanzania reported that utilities felt obliged by the ministry to show progress annually towards the policy goal regardless of the real development on the ground. It seems that utility managers are afraid to face consequences if no progress is shown (expert interviews, Tanzania). The regulator unfortunately does not counteract on such misleading reports, which became possible because utilities are allowed to define the service areas on their own.[119]

> No there's lots to be done on data. Data is not realistic. They just may to please the donors, (to say) we have reached this and this – but in real sense, that's not the truth, I think they know what is happening, but they just inflate the data. The data is there, but they don't use it properly. I think we have been advancing on data, but the challenge is still there. (expert interview, Tanzania)

However, in all four target countries information on utility performance has improved, although with different quality levels. The same can be said for tariff adjustments. The systems for monitoring utility performance and tariff adjustments were separated because of the reasons already mentioned.

The tailor-made software in the three countries for utility performance elaborated for the regulators helped to obtain a more precise and comprehensive picture in regards to the challenges the sector is facing. The information systems WARIS (Wasreb Information System) in Kenya and the NIS (NWASCO Information System) in Zambia are the most comprehensive in the four target countries. WARIS was gradually improved over the last 10 years leading to three different versions.

[118] E.g. the water shortages in 2017/18 in Cape Town, despite experts blew the whistle since almost a decade..

[119] E.g. a limit of 200 m from the existing network (EWURA annual reports).

The present version of WARIS 3 developed in 2015 allows the sector to monitor and report on LIAs separately from the development of the entire towns. This can be considered a quality jump in monitoring, as it will be the first time that annual reports on individual LIAs are available. The WARIS and NIS were also designed to provide an incentive to the utilities to submit the data to the regulator. Because some of the utilities had no digitalised management systems, the two information systems from the regulators offered the utilities calculated results of their performance while entering the data into the computers.

Having two separate systems of utility reporting available, allowed the regulators to crosscheck data submitted by the utilities for different objectives. It was frequently observed that utilities submitted different data for the same areas of interest because for tariff adjustments they intended to obtain a higher tariff and for performance a higher ranking in the benchmarking. For example, utilities showed for the same year low O+ M cost coverage for tariff adjustments but a higher figure for the performance reporting. Comparing both reports, the regulator could correct such inappropriate reporting by utilities in many cases triggering inspection on the ground. The regular updating of baseline data would be crucial in a dynamic environment. Unfortunately, this is not done as required which makes baselines largely outdated after 10 years.

In Tanzania, a donor supported the establishment of a baseline for LIAs for some towns. This baseline was never used by the sector institutions for unexplained reasons. The baselines established in Zambia and in Kenya in the early and late 2000s were mostly used by the poverty baskets (DTF, WSTF), which obliged the utilities to make use of the data when applying for funding last mile investments. The most detailed baseline for the LIAs is MajiData in Kenya,[120] which includes around 2000 LIAs countrywide in more than 276 towns. The establishment of MajiData took almost 3 years and the cost for it was around 1.2 million USD. However, establishing baselines countrywide can be considered a huge task, does not directly produce impact for access and bear the risk of not being sustainably used. This might be the reasons why it is difficult to find donors to finance such important undertakings.

There is no countrywide baseline for LIAs in Burkina Faso. In the meantime, ONEA realised this gap and started in 2016 to prepare the establishment of such a database. It should be noted at this point that the costs of establishing a nationwide database might seem to be elevated, but this costs have to be compared with the savings which can be achieved by a better and more effective use of funds. To achieve such savings it implies that once the database is established it is used to the full extent over a longer period. In the case of the WSTF in Kenya, which invested for seven calls of proposals from utilities more than 30 million USD for infrastructure development in urban LIA, the cost of the data base represents only 4%. With the number of people to be reached being one of the key selection indicators for projects to be funded (in Kenya 102 over 10 years) it can be suggested that the costs of MajiData were more than compensated by the gains through to a better use of funds for last mile developments.

[120] www.majidata.go.ke (last visited 05.2017).

Although there has been commendable progress in data and information development, the area of investment has been left out in three of the four target countries. In Kenya and Zambia, the information on investments are only limited to last mile development and in Tanzania the information on investment for the first mile development is insufficient despite the implementation of a huge investment program. The experts interviewed in Kenya, Tanzania and Zambia confirmed that investment monitoring is an area where the reforms have not made real progress. Burkina Faso seems to be the only target country where information on investments is regarded as sufficient and reliable. In addition to the information systems described above, many utilities in the four target countries are now obliged to carry out regularly customer satisfaction surveys.

The Infrastructure/Assets Development

The only country of the four surveyed which seems to have managed to close the investment gap to a large extent is Burkina Faso, although the challenge to keep pace with the rising demand in the urban setting remains. This was achieved because there is a functioning/professional financing mechanism, the necessary tools for fund mobilisation are available such as national investment planning, financing model with tariff adjustment cycles, etc. and an effective approach of serving the poor. These are absent or insufficiently functioning in Kenya and Zambia, although in both countries a financing mechanism for the last mile development was established, which lost steam in Kenya and ceased to function in Zambia. The reasons are dwindling interest of donors in baskets for last mile infrastructure development which generally prefer to fund first mile investments through projects as well as the insufficient oversight by the line ministries and their preferences for the highest possible service level. This development might also indicate how much fund allocations by donors depend on individuals. When an engaged desk officer puts access for the poor in the forefront and has gained influence in the decisions in its development bank then the poor in the receiving country can benefit from pro-poor baskets. Once this desk officer is replaced by another who directs decision to business as usual for the bank (big money and projects, higher security for the funds, less cumbersome follow-up with TA and the partner institutions, etc.) then such baskets and with it the buckets of the poor are running dry.

Without the poverty funds in Kenya and Zambia, the sector in both countries would not have succeeded to reverse the negative trend in access in the last 15 years. These poverty funds have proven to be a formidable instrument to implement sector policy in regards to poverty and gender equality at utility level. In Tanzania, the financing basket (WSDP) has as good as no poverty orientation and is practically a department of the ministry responsible for water and therefore, has insufficient autonomy, which hampers it operation and functioning as professional.

ONEA in Burkina Faso found the appropriate mix of first and last mile infrastructure development and ensures that last mile development is as important as the development of the main component of the first mile systems. It seems to have been able to curb corruption not only to improve NRW but also in investments to give confidence to the donors and national government. ONEA is well coordinating its investments and thereby, helping the line ministry to mobilise funds. The coordination of infrastructure development (WSBs) in Kenya is not yet sufficiently secured by the line ministry. In Tanzania, there is no intention visible that the decision makers are ready to abandon the household connection paradigm soon which will mean that access will further decline regardless of how much funds are injected for infrastructure development. This finding is supported by the limited impact the massive support for asset development has had in the investigated period. It is becoming apparent in the three Anglophone countries that the investment gap will further grow until a radical rethinking on investments and funding is taking place.

Burkina Faso also seems to be the only country where the national utility has acquired sufficient knowledge how to serve the poor in the LIAs and an up-scaling of this knowledge has taken place country wide. ONEA generally starts with water kiosks in non-served areas and later adds connections (yard taps and household) as much as it is feasible. Once the area is largely covered with individual connections (into the dwelling) water kiosks are gradually phased out. As the municipalities include continuously non-served adjacent areas into the towns, ONEA is obliged to constantly increasing the total number of water kiosks despite closing down kiosks in areas where it achieved full coverage with individual connections. In the other three target countries there are only a handful utilities which are successful in serving unplanned LIAs with water in such a systematic way. However, all of the four target countries are well behind in infrastructure development regarding the chain for onsite sanitation.

The Development of Sector Financing and International Cooperation

The development of financing the sector and the association between investments and sector development observed in the four target countries has been described in Sect. 5.6. Hence, it leaves the deliberation under this sub-heading to the issues of international cooperation except for one general comment: Burkina Faso is the only of the four target countries where the utility managed to obtain a solid financial status. The active involvement of the Ministry of Finance was crucial for it and is unfortunately absent in the other three target countries (confirmed by experts interviewed):

> So, that strategic guidance comes from the Ministry of Finance in terms of dealing with donors. It doesn't work in the water sector because Ministry of Finance is not really engaged in the water sector. Each donor must figure out their own way of dealing with this government … because the partner (government) doesn't drive the car. (Expert Interview, Zambia).

Referring to the international cooperation, there is very limited evidence in the four target countries that aid in the sector has the negative effect described by Moyo.[121] The poor have no disadvantage from the creation of infrastructure which follows generally strict procedures overseen by donor representatives. It limits the risks of corruption in the sector regarding constructions projects. This is less the case for projects entirely financed by government structures, which however, are not so many.[122] In addition, many experts interviewed confirmed that reforms have

---

[121] Moyo (2009: 52).

[122] E.g. the National Pipeline Corporation in Kenya, which carries out government funded projects and has never succeeded to attract donor funding in the last decade.

helped to control corruption better within the partner system, with the help of donor financing. Control of funds for last mile is certainly more difficult than for first mile investments. However, the needed amount of funds for last mile can be estimated at around 10–15% of the total investments required, which reduces the risk of diversion of big amount of funds, and there are good examples, e.g. in Burkina Faso, of how large funds for first mile investments can be controlled.[123] There are also little signs that donor interventions harm the framework in the sector except when insufficient coordination allow stakeholders not to aligning to new sector policies, such as promoting informal service provision for instance. Contrary, many experts confirmed that international cooperation was detrimental to create a new enabling framework for modern reforms.

However, despite these few positive notes, it is obvious that aid effectiveness can be substantially improved in regards to people reached and the enhancement of sustainable development (especially improved corporate governance). The expert interviews carried out in the four target countries provide a mixed picture in this regard about the donors. Some experts indicated that the donors do not set sufficient conditions, insist sufficiently on compliance with conditions or use the benefits of close cooperation between TA and FC in order to help enforcing good donor and partner governance.

> I think it [TA] is very helpful, ensuring these institutions are becoming sustainable, must go together [TA and FA], making sure there is a transmittal of knowledge, because that is what matters in the long run. As opposed to just a donor comes in and draws in some money. The X[124][donor] came to put many billions in the system. They refused to acknowledge there was a governance issue, not distancing yourself. Now, I like Y [TA] approach, because, Y takes in the governance challenge, it needs to be addressed. But for example, an organization, like X is, you know, refused to get involved in serious governance issues. But then, after they will sit at dinner and complain about how corrupt everybody is. They are not doing something and I see it a lot among the donors, for the past two years. Before 2014, it was a bit more effective. (Kenya)

Nevertheless, conditionality seems to have worked in Burkina Faso[125] where a high level of success has been achieved. Also in Kenya there are examples that conditionality can work.[126] Therefore, it can be concluded that donors can convince partners to take a certain development path and accept conditions, which however need to be closely monitored and acted upon. Donors can also link the release of funds to delivery if their representatives are sufficiently engaged, willing to look into the matters and keep an appropriate distance to the stakeholders involved. As

[123] A best practice example is the investments in the Ziga water dam for Ouagadougou. Water Integrity Network, Water Gouvernance Facility UNDP/SIWI and ONEA, 2014, L'intégrité de L'Eau en Action Endiguer la Corruption dans la Construction de Grands Barrages. L'expériences du Barrage de Ziga, Burkina Faso.

[124] The name of the donor agencies was replaced by X and Y.

[125] Own experience in the 1990s.

[126] As it was expressed by a permanent secretary in Kenya in 2012 after a donor disagreed with a decision (BoD and CEO) of a supported parastatal in the water sector: 'I rather lose a CEO [from a parastatal] than a donor'. (own experience in 2012).

Mosley[127] puts it 'However, for political pressure to be effective, it must not only be applied; it must also be yielded to'.

A good example for this is Burkina Faso where donors finance an annual review with external technical and financial competences. They do not rely on utility or ministerial data only. Donors participate in conferences where the compliance with conditions by the government and the utility is verified. At the same time, the donors have to account for their contributions to the ministry and utility. Such mechanisms do not exist on national level in Kenya and Zambia and are not as effective in Tanzania.[128] Therefore, Burkina Faso has been able to reach a higher level of development in the sector and mobilise more funds per capita to be served for investments.

However, the support for first mile asset development is judged to be generally helpful because it can directly or later help to include the poor with outlets for water and inlets for sewerage. Without doubt, the support of donors during the reforms has strengthen public institutions by acquiring more competences and thereby, has made it possible that the sector can better resist the inappropriate use of power in the sector by the political systems in the receiving country. A good example for this is Kenya where such donor support was available at certain time in the development process.[129] The same strengthening of public institutions took place at the utility (ONEA) in Burkina Faso, which helped to improve and maintain good governance in the sector as the experts interviewed underlined. Sector reforms supported by donors have also increased the efficiencies and the effectiveness of the public utilities with the introduction of regulatory tools.

Reforms in all four target countries have led to an increase in donor funding for investments. This increased confidence is mainly due to the improvement of sector indicators and the resilience of sector institutions. However, funding in the three Anglophone countries seems to confirm that the preference of donors is still the first mile investment and that this seems to please national decision makers who know little about serving the poor (household connections as prestige, 'first round effect' leading maybe later to a 'trickle down effect' Mosley).[130] This difference between what donor banks want to offer and receiving countries want to have compared to what the poor really need has to be overcome.[131] The expert interviews also suggest that many countries are lagging behind in reaching the same quality of cooperation as achieved in Burkina Faso. The most noteworthy challenges remaining are the improvement of formulating conditions, the mix of TA and FC and the anchoring of the donor actions into national institutions. The latter includes monitoring and the

[127] Mosley (1987: 39).

[128] E.g. the report of WSDP claims that in the urban setting the fund has reached 2 million people while the reports from the regulator provides a progress in access for the same period of under 100,000.

[129] Refer to the cases cited regarding declaration on water for free.

[130] Mosley (1987: 169, 170, 171).

[131] 'What the poor want…is small and simple; what the aid agencies offer them is very often large and complex.' and '…Third World will always tend to favour the capital-intensive option.' and 'The donor spends his budget and the recipient politician (and civil servant) earns the support of his home 'constituency', Mosley 1987.

use of the appropriate national institutions such as national financing baskets for the sector, especially in the case where several utilities operate in one country and last mile development is concerned.[132]

Referring to the recommendation of the high-level meetings (OECD) the following remarks are possible when looking at the development in the four target countries. Next to the already mentioned problems linked to conditionalities and the hesitance to support financing baskets, many experts consider the ownership of partner governments as weak. They even question the interest of the partner governments to exercise strong ownership explaining that this would also generate pressure for more transparency in the receiving countries. It can be observed that in many countries neither the line ministries nor the donors seem eager to submit themselves to a closer scrutiny by establishing regular donor-partner conferences for investments. In addition, donors seem increasingly to shy away from supporting partner institutions responsible for asset development on national level maybe because of risking to be dragged into closer scrutiny and investigations into corruption. It seems obvious that the discussion of what development banks must improve cannot be over yet.

However, donors should not be made scapegoats for the shortcoming of national governments (Mosley).[133] Donors can easily be accused of failure when funding is interrupted because of noncompliance with conditions or insufficient governance and thereby the works remain incomplete. Nevertheless, to waste only halve instead of all the funds must remain an option. Hence, donors would need to be very transparent and convincing when the press exposes uncompleted investment projects. On the other hand, partner countries do not use the full potential of mobilising donor funds. Expert interviews revealed that the necessary decisions and instruments to help motivating donors to engage more in the sector are often missing.

The deliberation in this section should be sufficiently convincing that the six proposed factors are crucial for the sector development. Nevertheless, the proposal does not claim that there are no other factors which have certain influences on the development of the sector such as training programs financed by the international cooperation for instance. However, it is interesting to see how utilities which have no support from outside significantly can improve performance just because the framework (e.g. autonomy) and regulation helps to improve good governance and provides the right condition to advance the financial situation of utilities. This is a condition for utilities to offer decent salaries and adopt a policy to recruit the best possible staff from the labour market and offer interesting careers. This seems more important for utility performance and the sector development than an enlarged program of training often unfortunately provided to unmotivated staff delegated from the state administration or nominated by protégés of politicians. Generally, the available local capacity to improve utility performance is often underestimated by the international cooperation.

[132] The DTF in Zambia has not received substantial funds in the last few years despite a successful track record, the WSTF in Kenya is running out of funds since 2017 for the urban sector despite having provided a substantial contribution to access (water and sanitation) in the sector and the WSDP in Tanzania has never received funds with conditionalities to target the LIAs on large scale.

[133] Mosley (1987: 43).

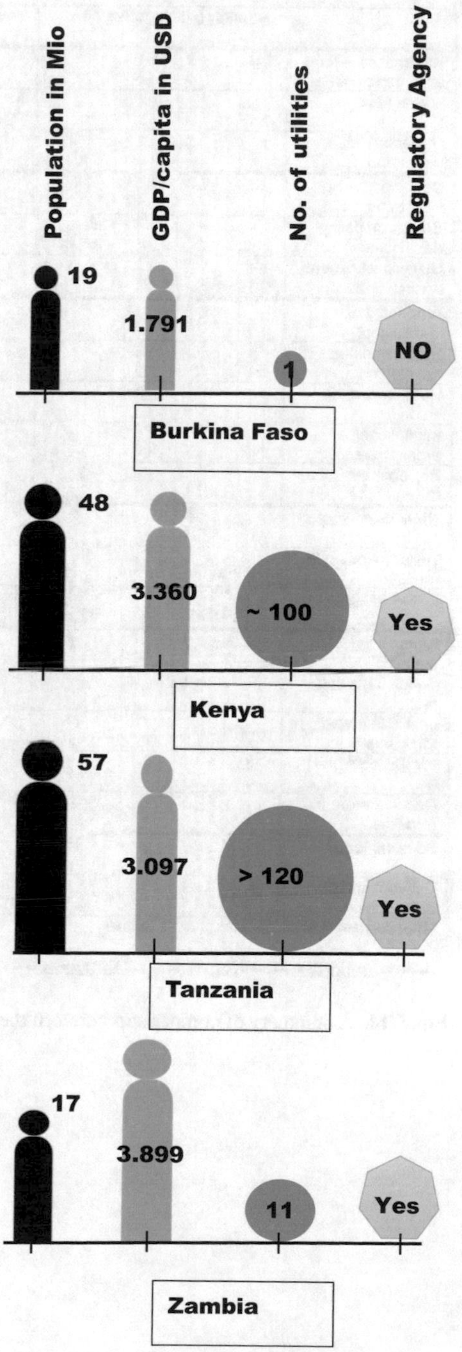

Fig. 5.13 Overview of general information about the target countries

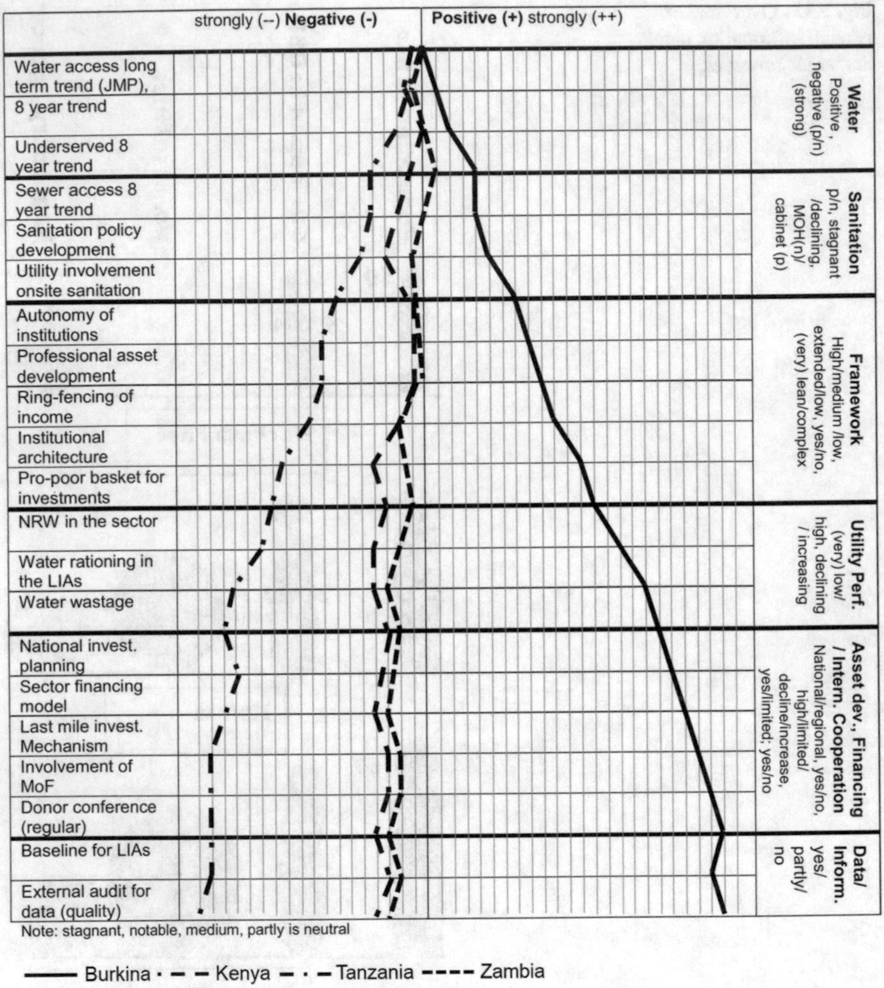

Fig. 5.14 A summary of comparison between the four target countries for the six crucial factors

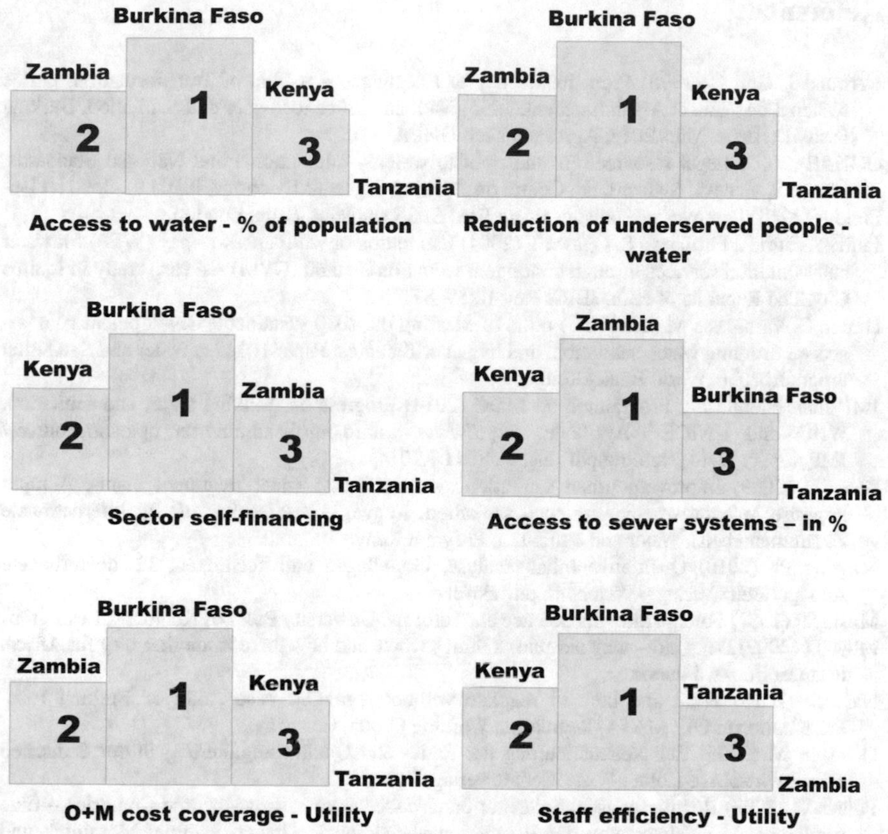

Fig. 5.15 Ranking of target countries according to sector and utility performance indicators

References

Bertrand J, Geli H (1995) From Insolvency to Excellence – A Tale of Transformation, Office National de l'Eau et Assainissement, 1995, National Office for Water and Sanitation, Burkina Faso, ID Bleue, Mars 2015, AgroParisTech-ONEA

COHRE (2003) Legal resources for the right to water – International and National Standards, Source 8, Geneva, Switzerland, Centre on Housing Rights & Evictions, ISBN 92-95004-11-6

Field A (2013) Discovering statistics using IBM SPSS statistics. Sage, London

Fujita Y, Fujii A, Furukawa S, Ogawa T (2004) Estimation of willingness-to-pay (WTP) for water and sanitation services through contingent valuation method (CVM) – a case study in Iquitos City, The Republic of Peru. JBICI Rev 10:59–87

Hutton G, Varughese M (2016) The costs of Meeting the 2030 sustainable development goal targets on drinking water, sanitation, and hygiene. Technical Paper 103171, Water and Sanitation Program, 2016, World Bank Group

JMP-Joint Monitoring Programme on MDG (2014) Progress on drinking water and sanitation. WHO and UNICEF, Available: http://www.wssinfo.org/fileadmin/user_upload/resources/JMP_report_2014_webEng.pdf (last visited 04.2014)

Kloss R (2009) Improving urban sanitation systems: waste water treatment plants, A rapid response to improve environmental sanitation. Kenya, GIZ (Gesellschaft für Internationale Zusammenarbeit), Water and Sanitation Program Kenya

Mayring Ph (2010) Qualitative Inhaltsanalyse, Grundlagen und Techniken, 12. überarbeitete Auflage, Beltz Verlag – Weinheim und Basel

Mosley P (1987) Foreign aid – its defence and reforms. University Press of Kentucky, Lexington

Moyo D (2009) Dead aid – why aid makes things worse and how there is another way for Africa. Penguin Books, London

Nana D (2015) What and how to regulate without regulator: case study of Burkina Faso. Presentation at GIZ MATA, Regulation Working Group, Germany

Oelmann M (2005) Zur Neuausrichtung der Preis- und Qualitätsregulierung in der deutschen Wasserwirtschaft. Kölner Wissenschaftsverlag, Köln 2005

Pollem O (2008) Regulierungsbehörden für den Wassersektor in Low-Income Countries – Eine vergleichende Untersuchung der Regulierungsbehörden in Ghana, Sambia, Mosambik und Mali. Dissertation, Carl von Ossietzky Universität Oldenburg, Fakultät II, 2008

Rouse M (2013) Insitutional governance of regulation of water services – the essential elements. IWA Publishing, London

Schiffler M (2015) Water, politics and money a reality check on privatisation. Springer, Cham/Heidelberg/New York/Dordrecht/London

Werchota R (2017) 'The crucial factors for a development of the urban water and sanitation sector towards sustainable access for all in Sub-Saharan Africa' Band 1 und 2, University of Vienna, Austria

World Bank Group (2017) Urban water and sanitation in Tanzania, remaining challenges to providing safe, reliable, and affordable services for all. World Bank Brief, WASH Poverty Diagnostic, Series

WSTF – Water Services Trust Fund (2012) Up-scaling basic sanitation for the urban poor (UBSUP) – UBSUP Preparatory Study, document 8, water services trust fund, Kenya and GIZ – Deutsche Gesellschaft für Internationale Zusammenarbeit, Germany

Chapter 6
Reasons for the Different Reform Outcomes in the Four Countries

Abstract The previous chapter described the development in the different crucial factors of the target countries. In this chapter, the reasons for the different outcomes in the target countries are identified. Burkina Faso has the most enabling framework and is the only one of the four where a framework development for sanitation took place. For several reasons explained, utility performance with its strong poverty orientation is outstanding and regulatory instruments are in place and effectively working without the existence of a regulator. Sector data are verified regularly by external specialists and asset development is highly professionalized. Tools/mechanism have been put in place to ensure effective mobilization of funds, the alignment of donors and that all parties involved in financing and asset development are made accountable. In the three Anglophone countries the framework is less enabling, utility performance is mixed, sector data are not verified by external specialists and asset development is either not sufficiently or not at all professionalized. In addition, the basic tools for fund mobilization and regular donor rounds are as good as missing. Other specific reasons why sector reforms had limited impact in the three Anglophone countries are outlined per country.
This chapter summarises the sector development during reforms for each country. It indicates common elements across countries as well as differences which help to better understand why the reform outcomes are so different and what causes success and failures during reforms.

Burkina Faso as Best Performer and the Remaining Challenges
It has been shown that Burkina Faso has managed to reverse the negative trend in access to urban water in both, the percentage of people having access to controlled water delivery and the reduction of the number of the underserved people. The sector also reached the highest level of coverage in water among the four target countries and provided a significant contribution to increase access to safe sanitation for urban households. There are several reasons for this outstanding success.

Burkina Faso has started the reforms around 10 years earlier than the Anglophone target countries and has developed the most enabling framework in the sector. The relevant sector documents address most of the key issues in all of the crucial factors

R. Werchota, *Empty Buckets and Overflowing Pits*, Springer Water,
https://doi.org/10.1007/978-3-030-31383-8_6

and thereby, respond well to the multi-dimensional nature of urban water and sanitation. The challenges the sector is facing are addressed by the framework to a large extent with an effective sector orientation. The sector (in this case largely the utility) consistently implements 11 of the 12 proposed sector principles which include the four concerning professionalization and autonomy of sector institutions, the two which are relevant for financial performance[1] as well as formalisation of services and poverty orientation. The separation of regulation from other functions is not reflected in the institutional framework but is catered for in a number of regulatory tools, which are in place without having established a regulatory agency.

The often observed gap between framework documents and implementation is closed with a number of concepts helping the management of ONEA to translate the intentions of the framework documents into impact. This is especially evident with ONEA's contributions in sanitation development. The utility received clear instructions from government regarding its responsibilities in sanitation development and management. It is the only of the four target countries where framework development in sanitation has taken place in the water sector during the reforms.

This clear vision on the different levels helps the sector institutions (ministry and ONEA) to guide other stakeholders to alignment to government policies and strategies. Furthermore, the almost nonexistence of informal service provision, which gradually phased out with the 'aggressive' extension of utility services to the LIAs, does no longer constitute an obstacle to the present and future sector development and ended the water discrimination of the poor. The balanced focus on social and economic goals limits the risk that main objectives in the sector are side-lined. The utility seems to pay as much attentions to the promotion and operation of the (lowest) service level for the poor than to connections for the middle and upper income classes. Burkina Faso has the leanest institutional framework of the four target countries and at the same time succeeds to contain the risk of conflict of interests.

A number of tools and mechanism in place ensure that there are enough incentives and pressure on the utility to improve performance. The 'contrat plan' and its systematic evaluation by external specialists as well as a regular supervision of ONEA's performance by the sector stakeholders are such examples. Also from outside the sector, there is a constant pressure and support coming. The Ministry of Finance provides ONEA with sufficient backing to develop its financial means and remain accountable in financial matters.[2] It has also obliged ONEA to commission tariff studies elaborated by professionals, link investments/repayment of loans to water and sanitation tariffs and accumulate/ring-fence substantial amounts of funds as co-financing by the utility for big projects. The involvement of the Ministry of Finance and the public in tariff decisions has successfully kept politicians from holding tariff unnecessarily low. This substitutes functions which are in other coun-

---

[1] Professionalization service provision, asset development and sanitation as well as autonomy of management. Cost recovery and ring-fencing.

[2] However, the involvement of the Ministry of Finance can also bear the risks that a performing utility becomes a cash cow for the state treasury.

tries secured by a regulator. Despite a solid poverty orientation, the decision makers in Burkina Faso obviously understood that only a financially strong utility can fulfil social goals to a large extent.

At the same time, the balanced composition of the members of the Board combined with the strong involvement of the Ministry of Finance in the water sector seems to have avoided that ONEA grew into a 'little republic' and block necessary developments in the other crucial factor as this is sometimes the case in other countries with very strong utilities (e.g. Uganda[3]). With this and the control of corruption within ONEA, the sector is in a position to make a strong case, especially at the dialogue mechanism with the donors which is also used as a platform to verify if all parties (ministry, utility, donors) honour their engagements. This is not only mutual responsibility in practice but also strong leadership by the receiving country. Thereby, the Ministry of Finance in support of the line ministry plays a major role in fostering sector credibility and fund mobilisation and hence, significantly influences the sector development. This helps to harmonise the stakeholders of the international cooperation and secures a continuous flow of funds for investments. The utility plays the role of a (national) basket for asset development. Thus, as indicated before, Burkina Faso seems to have managed to close the financing gap during the time investigated to a large extent.

The utility performance is outstanding in the region. In the ranking of the utility performance, Burkina Faso occupies the first place among the four target countries. The excellent level of all performance indicators, being maintained on a high level since more than a decade, is a result of the autonomy of management, a sound financial position and a very stable body of staff, especially among the middle and upper management levels. Staff efficiency is twice the value of Zambia. The low level of NRW maintained since a very long time shows that ONEA has managed to control corruption in the company as well as the use of the outlets by intermediates (kiosk operators, delegated management in LIAs) and consumers. It demonstrates the existence of a management quality well above the level observed in the countries and which seems to be only matched by very few single utilities in Africa.

The substantial increase in household connections with the help of promoting social connections was accompanied by a continuous increase in water kiosks (low-cost technologies). This appropriate mix of first and last mile infrastructure development and the well-functioning of the system in the LIAs (expert interviews) document that ONEA acquired the necessary knowledge of how to serve the poor and at the same time maintain control of its infrastructure. Providing water kiosks with the same attention than other service levels has ensured that such low-cost technology is sustainably used wherever there is demand for it. ONEA's contributions to sanitation development, especially access to safe sanitation facilities in the households (for the poor) is outstanding even compared on global level. This involvement of ONEA professionalised sanitation functions in the country going

[3]Government took care of the repayment of loans for the water utility and only recently, the utility started to pay back some of the contracted loans (own experience in 2016 and 2017).

beyond the management of the few sewer systems. However, expert interviews revealed some challenges in customer care linking it to insufficient consumer representation in the sector and data availability on LIAs.

Concerning information on utility performance and investments, Burkina Faso is well advanced with its system of data collection and more advanced than the other three target countries in data verification. Next to the professional monitoring of NRW and water quality, it is the only of the four target countries which can provide solid data on investments in terms of hardware created and its impact. ONEA regularly updates a national investment plan, a long-term sector financing model and ensures an acceptable monitoring for investments after projects have come to an end. None of the other three target countries has something similar for the entire sector and as effective working. These functions are well separated from the line ministry which is not the case in the three Anglophone countries.

ONEA as professional service provider, acting as a sector basket, being able to significantly co-finance investment projects and paying back all loans has helped government to consolidate its budget and the sector to become more resilient. However, the recent droughts (2016) with the need to start rationing of water again in Ouagadougou despite an (huge) investment in storage and abstraction of raw water undertaken at the end of the 1990s provides a good indication of how climate change will increase challenges in the sector in future and what limitation semi-arid countries will face to support a high number of people in such a difficult environment. In the opening speech of the Ziga dam in 1999 it was claimed that the capital will now have sufficient water forever.[4] Less than 20 years later, the situation has put in question such perceptions. Many other utilities in Africa are in an even more difficult situation because the necessary investments of tapping raw water are already overdue for decades. This and the rapid grow of demand due to urbanisation indicate that the battle of reaching and thereafter sustaining universal access to water and sanitation is by far not over – also not in Burkina Faso where achievements of reforms are outstanding.

Common Areas of Progress and Limitations in the Three Anglophone Countries

There are common developments which facilitate or hold back sector advance but also a number of striking differences in the tree Anglophone countries compared to Burkina Faso. Reforms helped to improve the framework documents although to different extent in the three Anglophone target countries as shown by the document analysis. The implementation of the proposed sector principles is not as consistent as in Burkina Faso. Six of the 11 principles are half-heartedly followed or only partly implemented such as the autonomy of utilities, professionalising funding and asset development, formalisation of service provision (in practice), full cost recovery, ring-fencing of income and involvement of utilities in sanitation. Some of the proposed crucial principles where not or not sufficiently included in the framework documents and implementation concepts.

[4] Speech of the line minister – own experience at that opening ceremony.

In Kenya and Zambia, but much less in Tanzania, the regulators provided a significant contribution to close the gap between framework documents and implementation with concepts and guidelines. For various reasons, there was no framework development possible for sanitation. Consequently, professionalization of onsite sanitation development which concerns the majority of the urban population has not taken place despite of an upsurge in cholera outbreaks in the recent years.

Like in Burkina Faso, commercialisation/corporatisation and the introduction of a regulatory regime has helped in the three Anglophone target countries to achieve progress in service provision. However, regulation faces substantial limits. In all of the three countries the utilities do not receive sufficient support for tariff increases although tariff reviews have been professionalised with the lead taken by the established regulatory agencies. The sector succeeded in the four Anglophone countries to move or come close to 100% O + M coverage but remained stuck in this level. The regulators do not receive the support from ministries on national level to enforce higher tariffs. Regulators have also difficulties to ensure ring-fencing of income above 100% O+M costs for investments at utilities which achieve a higher level of cost recovery. This and an underdeveloped investment planning and fund mobilisation mechanism have hampered the three countries to narrow the investment gap substantially, although investments increased due to reforms. Adding to it, the insufficient attention for the poor either on national (Tanzania) or local level (Kenya and Zambia), is limiting progress of coverage.

Most of the performance indicators from the utilities improved although there are big differences among the service providers. Isolated cases reach a level which is comparable with ONEA in Burkina Faso but many, especially smaller, utilities trailing far behind. The sector average in performance of utilities is therefore much lower than in Burkina Faso. However, in the three Anglophone countries, subsidies for O + M have either phased out or in the case of Tanzania have been substantially reduced. Although the improvement of O + M coverage are qualified by experts interviewed as a milestone achievement of the reforms, because it helped to improve the maintenance of assets and avoids premature degradation of installations, it is much too low to ensure a sound financial situation of the utilities.

The opportunity to raise tariffs for the middle and high income classes according to their willingness to pay has not been used so far. This insufficient financial and managerial strength as well as the knowledge gap among utilities on how to serve difficult LIAs has held back progress in access to water. The mix of first and last mile is still not appropriate for the contexts in the countries although progress has been made in Kenya and Zambia where investments in last mile infrastructure is more driven by promotion of programs on national level than by strategies from utilities. In all three Anglophone countries, according to expert interviews, it seems that the poor are the first concerned with water rationing over long periods. This is another indication that utilities still don't pay enough attention to serve the poor and therefore, remain far from reaching universal coverage.

Almost all experts interviewed in the three target countries underlined that reforms have helped to improve the water quality. Nevertheless, in contrast to Burkina Faso most of the experts (except the dwellers in the LIAs) recognise an improvement in water treatment but do not trust the pipe network and believe that it is at the distribution where the water quality is deteriorating to a questionable level. Most likely, utilities do not inform the public sufficiently about their progress in guaranteeing good water quality at the outlets. Some recent surveys[5] in several towns in Kenya at water kiosks and yard taps confirm that utilities in general deliver satisfying water quality to their clients as a result of the reforms.

The pressure regulators exercise on utilities to improve performance is undermined in many cases by insufficient corporate governance (BoDs) and the undue interventions of governments either on local (Zambia), county (Kenya) or national level (Tanzania). Nevertheless, reforms have helped in all countries to fight off undue interference in some areas of utility management by making consumers, especially state institutions, pay their water bills although a persistent high level of NRW indicates that substantial commercial losses (e.g. illegal consumption) in the three target countries persist. Even though reforms helped to increase water metering and reduce lump sum billing to curb water losses, the utilities in general have difficulties to combat successfully corruption which leads to water losses. Hence the management quality in the three target countries is well below the level of Burkina Faso.

Concerning effluent treatment, all three Anglophone countries continue to face challenges, which the reforms have not eased. The regulator in Tanzania includes indicators on effluent treatment in the annual report and the regulator in Zambia includes households with septic tanks in the counting of coverage, but there is as good as no enforcement in place in the water or by other sectors (e.g. environmental agencies) which would build enough pressure to ensure progress in wastewater treatment.

Another milestone mentioned by the experts interviewed are the annual reports of the regulators on utility performance based on tailor made information systems for utility monitoring. This has improved transparency tremendously and provided the decision makers a better overview of the sector. However, the quality of the sector data is not as consistent as in Burkina Faso. There is hardly any regular external verification of data and the regulators face limits in carrying out inspections because of the high numbers of utilities, especially in Kenya and Tanzania. In all three Anglophone countries databases for the LIAs have been established, but either they are not used by the sector institutions (Tanzania) or are not updated regularly (Kenya and Zambia). In addition, the information systems on investments in urban water and sanitation are either insufficient (Tanzania) or as good as absent on national level (Kenya and Zambia). Investment monitoring is an area where the reforms have not made substantial progress in the Anglophone target countries and this is one of the results of the incomplete professionalization of asset development.

[5] GFA (2016).

A Model of financing the sector is missing in the three Anglophone countries, which would oblige the line ministry or the regulator to link funding for development with tariff adjustments. Furthermore, a well-designed and systematically used fund-raising mechanism driven by professional institutions on national level is either underdeveloped (Tanzania), fractured (Kenya) or as good as not existent (Zambia). Hence, the combination of insufficient self-financing of assets and the inability to demonstrate how loans will be served makes the sector unattractive for investors. It seems that despite this crucial handicap, the ministries responsible for urban water and sanitation are unlikely to outsource sufficiently asset development to professionals, which would have the capacity to oversee or handle autonomously big amount of funds for investments and convincingly account for.

Another difference to Burkina Faso is that funds are not used effectively enough to promote good governance in the sector. The push by the (OECD) Paris agenda for mutual responsibility and ownership by the receiving countries has not led to an adequate engagement of the donors in the three Anglophone countries. Instead, donors seem to gradually pull out or keep staying away from national financing structures (e.g. baskets). Financing single projects through utilities are still preferable to donors than integrating their actions with cash contributions or aligned projects integrated in a national basket, which would be particularly necessary where a higher number of utilities operate in the market, like in Kenya and Tanzania and last mile investments are undertaken.

In all target countries, the committees/working groups created with the Sector Wide Approach for Planning (SWAP) are largely ineffective for funding because they are dialogue ('talk shows' as one expert in Tanzania expressed it) and not fundraising platforms. Only in Burkina Faso the SWAP structure was complemented with donor rounds (FC only) chaired by a national ministry. In the case of Kenya and Zambia such a badly needed mechanism on national level for fund mobilisation and far-reaching accountability is missing. Furthermore, in the three Anglophone target countries investments are not sufficiently monitored by national systems and either FC is not combined with TA or both do not cooperate sufficiently. A desk officer from donors sitting at the embassy of the receiving country is often acting like a civil servant, making sure that the contracts for FC and, if available, TA are fulfilled by the letter. At the same time such desk officers try to avoid anything which could interrupt the disbursements of funds or worst, to bring the project to a halt, because it could damage the professional career.

Furthermore, the leadership of national partners in donor coordination and fund raising is insufficient in the three Anglophone countries. Donor coordination seems to take place rather in the donor rounds (self-coordination). There, the partner countries are maybe guests sitting on the fence while the meetings are chaired by one of the donors. Expert interviews suggest that in such an inadequate set-up for financing, FC mainly concentrate on securing the repayment of their funds and are free to ignore possible negative effects of their intervention for sustainable development in the partner countries. The very limited engagement of the Ministry of Finance in the sector and the large absence of the donors in the policy dialogue are not helping in sector development of the low-income countries.

Specific Development and Challenges in Kenya

Kenya managed to reverse the negative trend in access to water expressed in percentage of population to be served but could not stop the increase in the number of people having no access. One of the handicaps in the sector is the institutional framework, which is unnecessarily complex. This makes coordination more cumbersome and increases costs for consumers. Unfortunately, it also creates pointless competition between institutions (e.g. regulator and WSTF, WSBs and WSTF) where more cooperation would be necessary. There are too many unviable small scale utilities and a blown-up structure for asset development which has its significant limits. The intention to up-scale PSP and thereby attract funds from the private sector by the separation of public assets from service provision did not materialise in one lasting case. Although funds for asset development increased during reforms, the level of investments remained way below expectation leading to a further increase of the investment gap.

With the separation of public assets from operation, eight regional Water Services Boards (WSBs) as asset holders and developer were unnecessarily created. This regional structure was not completed with a professional supervision by an umbrella institution on national level. Such an incomplete professionalization of funding and infrastructure development combined with a very weak oversight by the ministry let to different regional standards, procedures and monitoring systems in the country. Many utilities describe the WSBs as arrogant, complaining that they are not consulted when first mile investments are planned and assets are developed in their service areas despite the fact that the utilities have to repay the loans. There seems to be even cases where the WSBs received the transfer of funds for the repayment of loans by the utility but are not able to document that this money has been forwarded to the treasury for payment to the donor. These examples show impressively that a stronger involvement of the ministries of finance and a closer coordination with the line ministries would help the sector in its development. It also shows that a laissez-fair attitude by the line ministries and the donors facilitates the sector institutions to become ‘a little republic’ especially when the power of money (for investments) is abused.

To all these problems, the provision of the constitution to transfer service provision to the counties, including asset development, was added. After 10 years of reforms, it undid with one stroke all the cumbersome and very costly efforts to create and make this (mega) construct functioning (e.g. transfer of assets, personnel, etc.). The complicated process of transferring assets had to be restarted again while the WSBs were stuck with the previously transferred personnel and became heavily overstaffed because of reduced functions. However, having established a strong lobby during its existence, the WSBs started to fight for their survival. Until today, 8 years after the constitutional changes, no solution has been found to make the institutional framework leaner and at the same time finalise the professionalization of asset development although the Water Act 2016 has a provision to abolish the WSBs. This development indicates how an inappropriate institutional framework, once in place, can hinder development and haunts the sector for decades and how a blind reliance by partner countries on donor preferences, in this case PSP, can limit sector development for a very long time.

The weaknesses of leaving functions to the state administration, which should be better carried out by professionals, can be seen with the example of the so-called Sector Investment Plan (SIP 2014) commissioned and overseen by the line ministry. This document only estimates the financing gap in the sector by calculating the demand in cubic meters per person multiplied by an estimated amount of investment needed. The SIP covers the different sub-sectors (water, sewerage, onsite sanitation, urban, rural) and refers to the policy goals on access. However, the SIP is by no way an investment plan, but still labelled and defended as such. Until today, the claim that the ministry has an investment plan provided the decision makers with the argument not to establish a proper asset development plan for the sector which would indicate areas of investments, priorities, choice of technology/service levels, costs, financing, time frames, impacts, etc. Thus, investment efforts remained piece work often enough based on ad hoc decisions and preferences of donors, without being integrated in a comprehensive development plan and linked to a sector financing concept. One can wonder how donors with the order to support receiving countries in their development can neglect their mandate and keep on injecting public money in such an unfavourable set-up for sustainable development.

Next to regular and improved reporting, the regulator has introduced a benchmarking system, which covers the main challenges the sector is facing such as corporate governance. Benchmarking used in the reporting to the public exposes best and worst performing utilities. Regulation has also led to national standards in the sector, which are gradually accepted by the utilities. The high number of informal service providers in the country and the support of some stakeholders to maintain their existence do not make it easier for the ministry to streamline its policy. Nevertheless, the regulator in Kenya has covered new ground in regulation by obliging utilities to report separately on the development in the LIAs.[6] WASREB was also detrimental to end the practice by the municipalities to demand a share of the utility income from water bills to cover costs in other sectors.

Most likely, the most important success of WASREB is that it could safeguard the key principles and the achievements of the reforms during turbulences the sector had to face when obliged to align to the Constitution 2010. Key elements of the reform were carried forward from the Water Act 2002 to the Water Act 2016. Without the regulator the sector would most likely have moved back to square one and lost 15 years of reform efforts. In the ongoing dispute of the two government levels the regulator is not only functioning as a stabilising force but has also become a professional institution recognised by committees of the parliament, the line ministry, the responsible of water and sanitation in the 47 counties reporting to the governors, the courts, etc. Without this stabilising effect and professional knowledge the potential of further progress would have been lost. It is interesting to see that although national regulation was fiercely rejected by the counties during discussions on the Water Bill and even from a number of counties after the enactment of the new legislation of 2016, today the line ministry as well as the responsible for water and sanitation in the counties seek advice from the regulator.

[6] With the introduction of the information system WARIS 3 in 2015.

Concerning utility performance, the sector has not managed to break through a less than an average performance level. The up-scaling of the high performance level achieved by some utilities like Nyeri will take much more time and requires additional efforts by the regulator, the county governments, the line ministry and the donors. The sector has not yet found the right minimum size for utilities in order to ensure that all of them become viable. Remarkable is the established water and sanitation baseline on LIAs, the MajiData. It is very comprehensive and has been made accessible to the public. However, there is a challenge in its updating, which would be necessary because of the constant increase in the numbers of LIAs and the rapid growth of their population.

The sector framework in Kenya has an outstanding poverty and human rights orientation[7] but this is not yet sufficiently streamlined among utilities. The poverty basket (WSTF) is also exceptional among the Anglophone countries, which was until now supported by a number of donors. The contributions of the WSTF to improve access to water and sanitation with the development of the last mile infrastructure was until recently very significant but lost steam because the fund started to compete with the WSBs for first mile investments. Unfortunately, since some time, the aim of the WSTF is to become more prominent in the sector and thereby compromises its pro-poor orientation. Donors become a part of this regrettable development because they do not easily find a functional national institution to promote their 'innovative' financing approaches such as blended financing and misuse the poverty basket.

Main Reasons for Tanzania Falling Further Behind

Tanzania is the only of the four target countries, which could not improve the two indicators on sector progress – reverse the negative trend in access in terms of percentage of people to be served and stop the increase in the number of underserved people. In addition, coverage with water is the lowest in the four target countries and less than halve of Burkina Faso despite the implementation of a huge investment program. Some of the main reasons for this is the insufficient development of the framework where documents for implementations are missing and the sector orientation is inadequate. Tanzania is a good example that neglecting only one of the proposed crucial factors for development can result in significant limits in progress. Even lots of money for investments could not substitute the inadequate development of the framework and the resulting governance weaknesses. Any efforts will have little effects on the sector performance as long as such bottlenecks remain.[8]

Missing checks and balances allow the line ministry and local governments to exercise an influence on the utilities, which seems to be stronger than in the other

[7] Refer also to the brochure of the line ministry, MWI (2007) *'Water Sector Reform in Kenya and the Human Rights to Water'*

[8] It is the same with technology and management. The introduction of new technology (e.g. prepaid meters for shared facilities) or management concepts (e.g. delegation of service provision by utilities) cannot substitute measures to improve the framework or compensate weaknesses in organisation and management. See also Heymans, Eales and Franceys (2014: viii).

three target countries. This limits the autonomy of utilities and regulator and hence, the sector performance. It is the only country of the four where experts from the utilities interviewed stated that the line ministry with its civil service personnel and the politicians have a right to intervene even when it concerns day-to day operations! It is also the only country of the four where subsidies provided by the ministry are still common for a number of utilities. Therefore, corporate culture among managers in many utilities has not much changed during reforms.

Another reason for the poor sector performance in Tanzania is the high complexity of the market which is as pronounced as in Kenya. It has the highest number of utilities among the four target countries. Tanzania is another excellent example for the difficulties the sector is facing when an institutional framework has been introduced and cannot be modified even after the reason for its existence is no longer relevant. It is worth remembering the brief history of private sector participation with the involvement of a multinational in Dar es Salaam which left an asset holder and developer behind. Like in Kenya, the institution had established a lobby to stay alive after the arrangement with the private company was abruptly ended. With many BoD in the sector there is ample room for a minister to reward supporters of the ruling party with a seat at the sector institutions.

However, utility performance improved to a certain extent, but this improvement and a huge investment programme did not lead to a sufficient increase of access to water and sanitation to make the table turn. Next to the neglect of last mile investments into the LIAs, the increase of household connections was modest, although decision makers are still caught in the technology paradigm – household connection for everyone – and the line ministry did not outsource asset development to professionals. Thus, the last mile development with low-cost technology and professionalization of asset development are not part of a national concept despite the low coverage rate.

The establishment of a national structure for fund mobilisation and financing of investments and the provision of substantial funds was supposed to boost the development in terms of access. However, the planning and implementation of infrastructure designed as a program housed in the ministry and carried out by civil servants fell far short of what a professional funding basket would have done. In addition, the high poverty and inequality levels would have required a stronger poverty and gender orientation. Also in this country like in Zambia utility and sector performance was not the driver for the sector credibility. The potential to create a sound financial position among utilities in order to reach an acceptable level of self-financing was not used.

The most disappointing element in the reform was the low aid effectiveness in terms of access. Considering this, there was little justification the sector could offer for such a massive donor contribution to be disbursed in such a short period. The sector was simply not ready for it and aid was not used to overcome the sector bottlenecks. From the perspective of the poor this money was largely wasted. It confirms what Mosley said regarding donor decision on allocations of aid and also on aid effectiveness. Aid as promotion of sustainable development is often not in the centre of decisions.

Furthermore, there are weaknesses in the information systems and in regulation. Many utilities define their service areas by the existing network and thereby, exclude LIAs in counting, which are part of the urban population according to the national census. The fact that the regulator accepts this and also takes no measures to stop the utilities counting people served by neighbourhood sales as having access indicates that there is little understanding of what a pro-poor and human rights approach should include. Such counting produces misleading results of access to water and sanitation. To this is added the missing verification of sector data by external experts. This allows the utilities to produce improvements on paper because of pressure by the ministry and the municipalities to show progress (expert interviews). Also for investment, data collection and reporting had substantial quality issues which have never been overcome during reforms. In addition, investments were not used to strengthen the enforcement of regulation.

It is most likely that informal service provision and with it the *'urban water and sanitation divide'* is still growing despite the reform and substantial funds made available by the donors for investments. It can be concluded that improving utility performance and the injection of a huge amount of funds for asset development was not sufficiently supported by the development in other crucial factors and that the orientation towards overcoming sector challenges by adopting the relevant sector principles was insufficient, especially poverty orientation, autonomy of sector institutions and self-financing of the sector (sustainable development). According to experts interviewed, there is an insufficient involvement of the donors in the policy dialogue. There is still much to be done in the implementation of the recommendation by the Paris Agenda especially in mutual responsibility for sustainable development.

Specific Development and Challenges in Zambia

Zambia is the only Anglophone target country, which managed to reverse the negative trend in access expressed in percentage of the population to be served as well as in the number of underserved people (like in Burkina Faso, but less pronounced). Zambia benefitted from a well-designed sector framework with sufficient detailed implementation documents. However, it lacked next to the inconsistent application of the retained sector principles (e.g. total cost recovery, autonomy) the crucial principles of ring-fencing of utility income for investments and the involvement of utilities in onsite sanitation.

The contribution of the pro-poor fund (DTF) to access achieved with last mile infrastructure for water and sanitation in the first 10 years of the reforms and its close link to regulation[9] were some of the reasons of Zambia's success. The many

[9] The legislation gave NWASCO the functions of regulation and asset development through the WS-DTF (Water Sector Devolution Trust Fund). The latter was intended to support the CUs (corporatized utilities) during the clustering process. However, NWASCO decided to make the DTF a 'poverty basket' in order not to create competition with the utilities on asset development and separate it from regulation in order to attract donors, which refused to provide funds without such a separation of functions. The separation was carried out on the level of the BoD which guaranteed that decision making for the DTF was strictly separated from regulation (different committees) and in the institutional framework where two separate entities within NWASCO were created and strictly separated.

utilities extensions into the LIAs with low cost technology (e.g. water kiosks) compensated the limited progress in household connections. However, the engagement of donors at the DTF has in the meantime come to an end, leaving little hope that the positive development in sector progress due to the increase in last mile investments will be maintained in future.

Until recently, the function of oversight in water and sanitation which included fund mobilisation for the sector was exercised by the MLGH while the regulator reported to the Ministry of Energy and Water Development.[10] This led to a constant conflict in the water sector between the MLGH and the regulator NWASCO. Consequently, the resistance of municipalities to regulation was fuelled by the line ministry for water and sanitation. It partly paralysed sector development during some years because the regulator was ignored by the water departments of the municipalities. Only repeated outbreaks of cholera in the respective towns made the MLGH rethink its opposition to regulation. To stop the outbreaks, it finally asked NWASCO for help to transfer municipality departments into the established corporatized commercial utilities.

However, the initial resistance by the line ministry to regulation did not stop NWASCO to introduce successfully a number of tools such as benchmarking and reporting to the public and press ahead with the establishment of utilities. For the latter some donors provided support in the form of feasibility studies and in some cases additional funds for investments as incentives. Today, this process has been completed and there are only utilities which operate urban water and sewer systems in the country. The reduced number of utilities compared to Kenya and Tanzania is one of the big advantages in Zambia and is due to the work of the regulator. Regulation has helped in the start-up phase but also ensured that utility performance increased over time and sustainability was achieved. Since the reforms there was no further occurrence of total breakdowns of service systems like it was the case before the reforms. Zambia is also a good example of the difficulties utilities have to face to extent their services with shared facilities into to LIAs (Chingola) in order to increase access for the poor. It was the use of local knowledge which brought the breakthrough to overcome the aggressive resistance from informal service providers and the few consumers who had connections and used them for reselling water (neighbourhood sales).

The function for asset development remained with the utilities. However, the necessary support from national level in the form of standard setting, fund mobilisation, etc. was absent like in Kenya. The MLGH, covering many other sectors and having little knowledge about water and sanitation service provision, simply neglected its oversight functions. Hence, Zambia like Kenya has no structure in place to secure professional countrywide investment planning and financing modelling which limits fund mobilisation. With this, the sector cannot make a strong case for convincing lenders to engage and the utility performance is not yet on a level to

[10] It has already mentioned that Zambia established the Ministry of Water Development, Sanitation and Environmental Protection to which the functions for water and sewer was transferred from the MLGH in 2016.

compensate this weakness. Zambia was the only target country where until recently no information on investments in the sector was collected on national level; not at the MLGH or at the regulator.

However, the regulator was one of the first in Africa to develop a wide range of regulatory instruments, which many of them are becoming standard among regulators in Sub-Saharan Africa.[11] With the introduction of regulation, the influence on tariff setting by the central government significantly reduced, although the above-mentioned limitation due to local government interventions still remained. There are a number of tariff adjustments, which were carried out in the run-up of general elections, which would have been unthinkable before the reforms. However, like in Tanzania and Kenya, the regulator concentrates more on keeping production costs low in the name of consumer protection and less on enabling the utilities to obtain a sound financial position in order to be in a position to expend their services to the underserved people. Every sector institution, including utilities, seems to wait for donors to present themselves.

In Zambia like in Tanzania, dwellers from the LIAs interviewed indicated that the situation in water supply, after some improvements, has deteriorated again in the past years because of massive water rationing. Investments for tapping sufficient raw water for the urban sector have been neglected since decades and climate changes seem to aggravate the situation. This most likely will not change soon because of the limits explained regarding fund mobilisation and the insufficient attention of utilities to maintaining shared facilities for the poor. Like in Kenya, the inability to put in place an effective fund mobilisation and asset development mechanism is one of the biggest bottlenecks for the sector development.

References

GFA (2016) Development of the water and sanitation sector – evaluation of water kiosks and water quality in the urban low income areas in Kenya, Deutsche Gesellschaft für Internationale Zusammenarbeit (GIZ) GmbH, first and second study. GFA, Hamburg

Heymans C, Eales K, Franceys R (2014) The limits and possibilities of prepaid water in urban Africa: lessons from the Field. WSP/World Bank Group, 90159 v1, Washington

Pollem O (2008) Regulierungsbehörden für den Wassersektor in low-income countries – Eine vergleichende Untersuchung der Regulierungsbehörden in Ghana, Sambia, Mosambik und Mali. Dissertation, Carl von Ossietzky Universität Oldenburg, Fakultät II, 2008

[11] Refer also to Pollem (2008).

Part III
At Least Full Buckets and Clear Pits on the Way Forward

This part of the work commences with a discussion on how to end the urban water and sanitation divide and winds down with conclusions and recommendations on how the sector can develop towards universal access in low-income countries, especially in the context of Sub-Saharan Africa. Although it can be expected that in a more distant future all urban dwellers will have access to piped water and sewerage systems in this world, it would be irresponsible not to use intermediate solutions which fit into a sustainable development concept in the meantime and keep neglecting the justified importance of water and sanitation for people and the society. It can also be assumed that the decline in the sector is not a deliberate act of a few or many but owned to missing knowledge and an inadequate balance of power and interests. Therefore, the following chapters are important since they indicate how sector reforms can be designed and implemented in order to move more swiftly to universal coverage in urban water and sanitation in our world.

Chapter 7
Ending the *'Urban W+S Divide'* by Serving the Poor

Abstract Once the upper and middle classes are covered with W+S services by utilities through household connections for water and a chain for sanitation (either piped or onsite), progress in access tends to come to a hold in the low-income countries. The reason for this is that serving the poor is a particular challenge for utilities and regulators. Hence, reaching universal access to safe W+S needs rethinking and more attention for urban low-income areas. Stakeholders in the sector need to build more knowledge on how to reach the poor sustainably. A pro-poor orientation has to be streamlined in all six crucial factors. Furthermore, all (three) service levels have to respond to minimum requirements of human rights in order to end the discrimination of the urban poor. Utilities have to have different strategies (as proposed) for the different types of low-income areas. Main challenges are to protect infrastructure from vandalism, control water loses and at the same time offer a mix of service levels which can secure universal access in all transitional phases. For sanitation, the differences between the approaches proposed by public health officers and urban W + S specialists are indicated and a strategy for a goal-oriented sanitation strategy is briefly outlined.

In most of the low-income countries the middle and upper classes are already connected to utility services. However, they are plagued by insufficiencies in the services provided especially water rationing, insufficient water pressures, undocumented water quality, etc. Such deficiencies lead in general to certain inconveniences in the daily use of water but the households of the middle and upper classes can do something about it. Additional water tanks, the installation of pumps on the plot and complementing drinking water supply with bottled water from the super markets can help to secure continuity and security of water for increased comfort which includes toilet flushing, showering, house cleaning, dish washing, etc. Still, there are other urban dwellers who have (limited) access to utility services but can hardly help themselves in this way like has been shown with the example of Agnes. Such consumers at the lower end of the middle class or very near the poverty line depend on the landlords. These tenants don't have the means for additional installations or have little incentives to pay for it, as the costs are substantial in comparison to the

R. Werchota, *Empty Buckets and Overflowing Pits*, Springer Water,
https://doi.org/10.1007/978-3-030-31383-8_7

monthly house rent and many of them have to shift to other living places when work demands it because of high transport costs (matatus). Agnes has been living in various multi-story buildings over the last years and still faces difficulties to find a place where she can access utility water with ease.

Referring to Sect. 2.5, it has been argued that the utility should have a responsibility beyond the point of delivery for water quality in the case of water kiosks. The same argument can be used for the case of partial access to regulated water supply as described above. The regulator could encourage utilities to improve services for the concerned consumers in the LIAs by suspending connections until owners of the buildings fulfil certain requirements for their tenants such as providing a connection for each household equipped with a meter (to avoid over-charging), installation of sufficient water tanks and if needed provisions for pumping water up to the top floor. Thereby, utility service provision can be used to enforce minimum standards for the benefit of consumers which in the case of sanitation might also be set by other sectors. A good example for this is the promotion of safe sanitation facilities in the households by utilities (e.g. Burkina Faso) which follows the minimum requirements established by public health authorities for the safe use and pit emptying.

However, while improving utility services for connected consumers remains a concern, the most pressing challenge is to extent utility services to the underserved people. Therefore, this chapter concentrates on the people who do not have access to utility outlets at all in Sub-Saharan Africa and argues that the state needs to concentrates on two areas of development in the sector: Bringing sufficient raw water to the towns and offer the poor who depend entirely on informal providers or traditional water points access to formalised services. Bringing utility services to people who depend on informal service provision and traditional water points must be regarded at least as important as improving services for the already connected. Once the legal obligation for the formalisation of service provision is in place, the state has to control utilities, which hold a monopoly, that they fulfil their mandate to serve everyone in town. In addition, the state has the obligation to ensure that informality is phased out.

Regrettably, there are still a number of experts arguing that a water utility in Sub-Saharan Africa can hardly be in a position to satisfy the rapid growth of demand in the LIAs and at the same time reach cost coverage and control its infrastructure. Others argue that the problem for the poor is the tariff at the utility kiosks or utilities services need to be complemented with SIPs. All of this seems unfounded because some utilities are very advanced in providing access for everyone such as ONEA in Burkina Faso (86% average coverage in the country) or Nyeri in Kenya (91%). Talking to the underserved people one will understand that for the poor access to formalised water and sanitation services is more important than utility tariffs and escaping informal service provision is their top priority.

However, many utilities in other countries still refuse, hesitate or simply do not know how to serve the poor. There are too many unsuccessful attempts and fears to lose money and control over assets, which seems to be justified in some cases. Particular difficult is service provision in the (hard-core) slums and generally in the unplanned LIAs where utilities not only risk losing control over their infrastructure but also have to deal with a situation where their personnel are threatened or sometimes even physically attacked by informal service providers (e.g. Chingola, Zambia). Nevertheless, the difficulties utilities encounter cannot be an excuse to exclude the poor from formalised services because access to water and sanitation is a human right and the number of unserved people is growing.

A Stronger Focus on W+S for the Urban Poor

Until today and after years of reforms, many unplanned areas are still either not connected to the utility network or the population is insufficiently served. This indicates that moving to universal access in towns is by no mean a quick fix. In the view of the exploding demand in the urban areas in Sub-Saharan countries it seems that it will need much more than two or three decades to reach everyone. Universal access can only be achieved if there is a stronger focus on serving the poor than in the past and effort are constantly undertaken in first and last mile development. A limited investment project undertaken here and there from time to time is insufficient to respond to the rampant urbanisation taking place in Sub-Saharan Africa. Unfortunately, many utilities even neglect the infrastructure for the poor in the LIAs which has already been provided with the support of development partners. The poor in the LIAs are usually the first deprived when water rationing becomes the order of the day. Awkwardly, it just needs a change of one decision maker at the utilities such as a new managing director or a member of a board for assets in LIAs to fall into disarray because either commercial consideration pushes the obligation to serve the poor into the background or uniformed decision makers at ministries, regulators, etc. allow many to return to the household only paradigm.

With long lasting supply cuts people have to abandon water kiosks forcing them to go back to informal serviced providers, neighbourhood sales and traditional but contaminated water sources (refer to Chap. 2, Fig. 2.1). It is the responsibility of regulators and ministries to counteract such behaviour and take swift action when utilities neglect their mandate to serve everyone in town. However, when the blame falls on the utility for the failure to serve LIAs, insufficiencies created by undue political interference, an inadequate framework, their bypassing by NGOs and donors, and etc. should not be forgotten. International cooperation needs to understand the urgency in the urban setting and the importance of access to water and sanitation. The preference for the rural areas (Moyo 2009)[1] has to end by finding the appropriate urban/rural mix and providing more priority for water and sanitation development in general.

Pro-poor Measures in Each Crucial Factor: A Comprehensive Approach

An approach needs to be comprehensive in order to be effective. In the urban water and sanitation sector it means that pro-poor measures have to cover all relevant intervention areas in the proposed six crucial factors. Hence, the pro-poor focus has to be reflected in the framework and streamlined with pro-poor implementation concepts at each of the sector institutions. Regulation has to guide and oblige

[1] E.g. the use of unimproved sanitation facilities in the urban areas grew by 165% but only 11% in the rural areas over the last 25 years)

utilities to serve the urban poor. Utilities have to prepare their organisation for extending services into the LIAs and financing baskets have to concentrate on last as much as first mile development. Regulators and financing baskets have to offer incentives and if needed, expose underperforming utilities which do not reach the poor. Information on LIAs regarding water and sanitation has to be available. Reporting on underserved LIAs by the utilities must become a standard requirement in Sub-Saharan countries and funding must consider the affordability of infrastructure development in the low-income countries. Such a streamlining of a pro-poor vision and its translation into actions in all crucial factors for development will make the sector move forward towards universal access.

Minimum Requirements for all Service Levels

In the developing world, progress in the sector was mainly held back because of the 'household connection only' paradigm. As argued before, not everyone can buy, maintain or be offered an individual utility connection because of several reasons outlined but at the same time experts interviewed showed that underserved dwellers in the LIAs strongly desire access to utility services. Consequently, the move from informal service provision to the utility services has to have highest priority for the development and this can only be achieved by accommodating the poor with a basic (utility) service level they can afford and the system can maintain. In addition, the decision makers need to be realistic in their approach. The long accumulated and still growing investment gap compared to the limited funds available for asset development requires a gradual process of development where people can move along a service ladder. However, the bottom line is that all service levels offered to consumers have to respond to minimum requirements for water quality, unrestricted access and proximities of outlets (allowing a 30 min cycle) as well as a certain continuity of services during the day, at least.[2] Hence, it is suggested that countries in Sub-Saharan Africa will have to follow a similar development route in urban water like some of the transitional and the industrialised countries have already went through, with centralised systems offering different service levels managed and overseen by professionals.

A Strategy to Reach the Poor with Water in the LIAs

Utilities need to prepare themselves before moving their services to LIAs. Recognising that conditions in the LIAs can be very different in respect to the layout, security, composition of the population (e.g. tribes), etc., utilities have to have strategies for the different types of areas. Such strategies need to be written down in order to be known and streamlined. They also document that the institution has the will and a vision to fulfil the national and global policy goals. Strategies also need to include an indication of the possible technologies to be applied and the way utilities want to maintain and operate them.

One particular reason why systems in the LIAs are often failing to reach everyone is that utility employees want to top up their salaries with illegal activities regardless of the cost for the poor and the counterproductive effects on policy goals. A limited number of household connections in a slum, like in Kawangware where

[2] According to the expert interviews in the four target countries, at least every day.

Betty lives, will not only be a lucrative business for resellers but also for some company staffs when kickbacks can be demanded. Utility employees can be very useful for such illegal businesses, ensuring that not too many individual connections (resellers) are established in an area and utility kiosks are regularly deprived of water which eventually makes the operator lose its business. Such corrupt activities are often combined with the manipulation of the water meters from resellers which helps to increase the kickback payments to utility staff (refer to the case in Kericho, Kenya). The utility management has to be aware that it will have to deal with internal and external resistance or even aggression when trying to change such a situation. It has to take courage and strong preventive measures to root out this kind of corruption because it is not an exception that management of the utility (or technical advisors) receive (death) threats, possibly from their own involved personnel, when restructuring efforts are undertaken (e.g. Bobo Dioulasso, Burkina Faso, 1991).[3]

To overcome such external and internal resistance, utilities need to use local knowledge (Chingola, Zambia 2004). An approach which fits all will fail. Utilities should make use of local NGOs with proven track record in solving such problematic situations. In addition, dwellers in the LIAs do not have much trust in state institutions (expert interviews) which means that the utility cannot call upon the police or army. It has been interesting to see in a number of cases how effectively dwellers in LIAs supported the utility to overcome resistance of informal providers and malpractices within the company.[4] Hence, utility strategies have to address these risks and propose effective measures and control systems right from the start when planning to move into LIAs with services.

Once the strategies provide clear directions, the next step in moving services to the LIAs is to make the necessary means available for a pro-poor approach. This includes the nomination of a qualified responsible for the LIAs or, if the utility is large enough, the establishment of a department for service provision to the poor. As soon as a pro-poor structure is in place and the procedures and mechanism of reporting back to the supervisors/management are defined the rest of the staff like areal managers etc. has to be brought on board with internal information and training campaigns. Before moving into a new area or undertaking an upgrading of an existing system the utility needs to collect sufficient information in order to classify the area and select the appropriate strategic approach.

Where vandalism and illegal connections cannot be easily controlled or basic information is missing the utility is advised to move first with water kiosks close to the border of the area. Only when the kiosks have been operating for some time without great difficulties negotiations with existing and potential consumer groups can start in order to establish a system of local oversite. Many dwellers in the LIAs will have interest in having the shared facilities moved closer to their dwellings and therefore in exchange will cooperate with the utility, especially under the threat that service provision can be put on hold at any time if water losses increase or assets are damaged. The relationship between the utility and the kiosks operators as well as the operators and the consumer groups is thereby crucial. There must be a clear

[3] Own experience.

[4] E.g. Chingola in Zambia in 2004 (own experience).

contractual arrangement between utilities and kiosk operators which regulates responsibilities and obligation of each party. In addition, water kiosks have to be frequently visited by utility personnel and swift actions have to be taken if operators are failing or vandalism is an issue. There must be a channel through which the consumers can complain at the utility about the quality of the service offered by the kiosk operator. Also the kiosk operators need to have the possibility to complain to a decision maker at the utility.

Furthermore, the water kiosks must be spaced in such a way that the kiosks operator can make a modest living from it. Servicing around 1000 people with one kiosk and three taps will ensure a certain commercial viability. Wherever a well-tested pre-paid system can be installed and is secured (payment by mobile phone for instance) the utility should take this opportunity if the users accept such technology. Also the placement of the kiosks must ensure that the facility can be controlled and the safety of the users are not compromised, for instance through heavy traffic around the kiosk. Water kiosks must not be high jacked by groups, political movements (refer to the case of Chipata, Zambia) or individuals. Therefore, selection of kiosk operators must avoid that cartels are established or that a family obtains many kiosk contracts for their members or relatives. The attribution of kiosks should also be gender balanced. Often utilities have difficulties to hold someone responsible if a group (community, women, etc.) is receiving the contract to manage a kiosk. However, all these elements will have different relevance in different contexts. Important is that the poor is not punished if a kiosk operator fails by simply closing the kiosk down when problems arise. A swift replacement of the operator by the utility is recommended while the pending issues with the sacked operator are clarified.

The introduction of the next service levels (yard taps and household connection) should go hand in hand with increasing security for the utility assets. A close monitoring of water losses is therefore the key to success. This means that zoning of the area and the placement and regular reading of bulk meters are indispensable. In the case of increased water losses in the area a swift control of the consumer data base and a verification of internal procedures (corruption in billing, meter reading, disconnection, etc.) are crucial. Hence, individual connections and yard taps can only be installed where the utility can easily identify the households and thereby is in a position to deliver a bill and disconnect consumers in case of non-payment or water theft.

Maintaining water kiosks while connecting plots will avoid that the poor who cannot have a connection will have to depend on neighbours to fetch water. If the beneficiaries of water kiosks choose to buy water from the neighbour because it is closer to their house and they have a friendly relationship at least they remain with the choice to go back to the public outlet at any time where water is more affordable and access is unrestricted. With a gradual increase in individual connections the water kiosks could phase out progressively. Closing all water kiosks before a coverage of less than e.g. 95% is reached means that too many of the poor will be pushed back to mercy of informal service providers, neighbourhood sales or traditional water sources. Maintaining some pre-paid kiosks in these cases will help the remaining poor without an individual connection and reduce the costs of operation for the utility.

It is not a coincidence that successful utilities (e.g. ONEA, Burkina Faso) increase individual connections and shared facilities simultaneously. This is necessary because of the rapid urbanisation and the increase in the number of LIAs to be served. Therefore, when utilities in Sub-Saharan Africa discard prematurely shared facilities, access will decline in the urban setting. The utilities have to find the appropriate mix of first and last mile investments, the right strategy for the different types of LIAs, the most viable mix of service levels and accept that the development and operation of low cost technologies need more follow up and attention than connections in the residential areas of the middle and upper income classes. Donors need to understand that introducing last mile infrastructure as an accompanying measure for first mile development is insufficient and prone to fail and that the promotion of low-cost technologies on scale needs the help of a national institution which makes use of local knowledge.

Many utilities and all donors, it seems, lack such knowledge and often think that the promotion of social connection in all kind of LIAs can lead to universal access. The promotion of social connections can also be seen as another way to hold on to the household connection paradigm and the desire of some staff in the company to make money aside on the expenses of the utility and the poor. It has also been observed that urban dwellers receiving a social connection make sure to be disconnected soon after its establishment in order to reconnect illegally often with the help of utility staff because the pipes have now moved closer to the dwelling ('Why not steal water if you can', expert interview in Kenya). Hence, utilities are well advised to progress in steps in serving LIAs as described above and not jump blindfolded into a program of social connections just because funding for it is available.

The poor know very well the value of sustainable access to clean water and therefore, there is no need to offer utility services for free. However, as has been explained in length, there is need to offer subsidised tariffs to the poor with the means of rising block tariffs and social connections whenever sufficient raw water and water production capacity is available and security for utility assets can be expected. Otherwise, water kiosks are the appropriate solution until the bottleneck of water availability for towns is lifted. The delegation of services by utilities to subcontractors (delegated management), often hailed as a best practice by some stakeholders, must be taken with care. It seems that there are more failures than (proven) success stories especially when monitoring is carried out over medium and long term. There is a risk that such delegation brings substantial disadvantages for the poor and the utilities alike.

Wherever utilities are not able to move into LIAs within an acceptable timeframe and support for a community managed system is available such a temporary solution should be accepted as long as a later integration into a utility system is part of the plan. Therefore, transferring assets of small-scale systems to the users of the communities in towns must be avoided because it will finally delay sector development and foster discrimination of the poor (refer to Sect. 3.4). Any transfer of rural water experience (one of the core businesses of NGOs) to the urban setting must be seriously questioned.[5] The argument that the community and the civil society institutions know better what the urban poor need has been proven wrong in many cases in the urban setting. Keener et al.,[6] while studying water stand posts and the involvement of informal serviced providers, found:

> Management by Community Organisation:…where there is not enough social cohesion, strong local power structures, and no oversight from supporting institution, the model can also lead to corruption and mismanagement.

It is not surprising that Keener et al. also found cases where 'community involvement was captured by local elites as soon as the mediating NGO left.' NGOs should work together with utilities to extend formalised and controlled services to the poor instead of venturing into areas where professionalism for service provision and asset development is required and massive investment funds are needed.

Harmonizing the Aspirations of Public Health and Urban W+S Development Expertise

The perception of what has to be done to increase sustainable access to safe sanitation is very different among stakeholders depending on the sector they are involved. It is useful to know these differences and the issues in sanitation in order to propose a strategy for the water sector regarding its contribution towards universal access. The Kenyan Environmental Sanitation and Hygiene Strategic Framework 2016–2020 from the Ministry of Health[7] and the National Water and Sanitation (2019–2030) from the Ministry of Water and Sanitation Strategy are good examples of the conflicting situation in Sub-Saharan Africa which is also a result of the global discourse on sanitation. Furthermore, the long standing rivalries between ministries (e.g. MoH and MWS) in sanitation[8] is a fact which needs to be addressed and defused because after all, both sector pursue the same national goal for universal access to safe facilities and services. The root of the conflict lies in the different understanding of what each sectors has to contribute and how they should be organised to achieve the common goal.

The situation on sanitation described in the national sanitation strategy of the MoH in Kenya is largely shared by the sanitation strategy of the line ministry for water and sanitation. Hence there is some common ground. However, the approaches how to move forward to reach universal access to safe sanitation cannot be much different in the two sectors. The MoH, like the global discourse, makes little difference between the realities in the urban and rural settings and their consequences for sanitation infrastructure and services. It is not difficult to understand that a well-

[5] Refer to Sect. 4.1.

[6] Keener et al. (2010: 20).

[7] Ministry of Health (MoH) (2015: 19, 23, 24, 25 m, 26)

[8] This becomes evident when comparing the National Environmental Sanitation and Hygiene Policy from the MoH and the National Water Services Strategy 207–2015 and the draft 2019–2030 from the Ministry of Water and Sanitation (before Ministry of Water and Irrigation), where sanitation development in the water sector is included. The word sanitation has been avoided in the title of the strategy paper in order to appease the MoH.

functioning sanitation chain is detrimental in towns but not necessarily crucial in the rural areas. Concerning the organization in sanitation, the MoH proposes that the costs of operation in sanitation should be financed by the government (municipalities) and professionalization of sanitation services in towns is not addressed at all. Therefore, the strategy does not mention the involvement of utilities in the different types of sanitation development. Key principles guiding water sector reforms such as professionalization of service provision and asset development, separation of key functions, autonomy of sector institutions, pro-poor orientation, etc. are as good as absent or very differently addressed than in the water sector. Although there is a strong note on the deprivation of the poor, subsidies for sanitation are questioned and only considered for the poorest of the poor.[9]

> … In Kenya, basic sanitation services are not accessible to the majority of the population. The result is that the poor are deprived of decent and dignified lifestyles leading to a deterioration of health, wellbeing and human environment.' and it continues 'The policy discouraged sanitation subsidies [for household facilities]' and '…including new types of cash transfer and social subsidies to enable households in the lower wealth quintiles to purchase through the market.

Furthermore, the perception of what the institutional framework for sanitation should look like cannot be more different either between the two sectors. The separation of the function for environmental health (MoH) and service provision (Ministry of Water and Sanitation) is seen as a problem instead of a necessity because of their fundamentally different nature. The proposed solution by the MoH is the creation of a (mega) structure for sanitation as lead institution – 'multi-sectoral and interagency agency' – which should cover all sanitation functions. In addition, the existence of two separate regulators for the protection of the environment and for water and sanitation service provision is considered as a blockage for sanitation development.

Contrary to this, the water sector prefers the involvement of professionals and considers this as a precondition for a turnaround in the sector. In the water sector stakeholders underline the multiple advantages of linking water and sanitation service development under one framework. This recognition has led in many of the low-income countries to the decision that utilities are responsible for urban water and sewerage service provision, at first for piped systems and lately also for decentralised (onsite) systems forming a sanitation chain. Equally, that (national) regulation of water supply should include sanitation service provision. Also the approach for financing the sector is very different. Cost coverage for operation should be secured by the water bills for sustainability and subsidies for the poor should be generated within the sector by cross-subsidisation between big and small consumers and if needed between water and sanitation services.

With this understanding of the differences in the two sectors it becomes obvious that the conflict can only be solved if the separation line is adequately traced between the responsibilities of public health officers and the developers for infrastructure

[9] MoH, National Environmental Sanitation and Hygiene Policy, Kenya (2015: 1, 2, 61).

and its operation for service provision. Hence, there is a need firstly, to find the adequate separation line and an effective cooperation mechanism between the two (and other) sectors and not to place all possible functions somehow linked to sanitation under one umbrella and authority within a sector. Secondly, there is need to take on board the available academic knowledge and the lessons learned from the past on service delivery organised in an industrial process. It is helpful to recognise that the organisation of urban sanitation services and its development has many parallels to the water sector and lessons learned from reforms can inform decisions on sanitation.

A Strategy to Reach the Poor with Sanitation Services in the LIA
Because of the significant and increasing number of unplanned settlements where the density of the population is growing and a controlled evacuation of effluent is absent, sanitation issues become increasingly front stage importance. For instance, the Kenyan Environmental Sanitation and Hygiene Strategic Framework 2016–2020[10] states that in the urban areas only 12% have access to piped sewer systems, 5% of the effluent is effectively treated and 80% in the LIAs have to use shallow pit latrines. This example indicates that there is urgency to improve access to safe sanitation facilities and the development of a formalised chain starting from the household where the users can defecate with dignity and ending at the safe release of treated effluent, regardless of piped or non-piped systems. However, not many utilities have experience in creating and managing on-sanitation systems (the chain) which is very different from piped water supply and sewer management but it seems that wherever they are involved, sanitation activities are more effective and sustainable than when civil servants (municipalities) take over such functions.

Toilet facilities include both capture and containment of the human waste and are located generally on private property. Therefore, in towns, utilities have to connect privately-owned interfaces and storage with services further along the sanitation chain, managing the associated decentralised infrastructure situated on public and utility owned ground. These off-piped services are labour intensive and more complex in operation than piped sewer systems. Thus, many decision makers in the urban water and sanitation sector tend to discard such responsibilities, like it was the case in the past for water, and propose to leave decentralised sanitation services for the poor to the civil society organisation or the small scale private sector with an oversight by municipalities. This involvement of stakeholders which are extremely difficult to control and do not guarantee sustainability of their actions does not help the poor. It leads to the same inefficiencies and discrimination in sanitation as observed in water supply. In fact, water service provision before reforms suffered the same or similar insufficiencies then sanitation still does today.

Looking at the household level again, the middle and high income classes have the possibility to construct adequate sanitation facilities in their homes as well as septic tanks on their premises and pay for a tanker service to collect and evacuate the sludge regularly. Tanker services can be controlled on local level because of their limited number and their visibility. However, controlling tankers does not

[10] Ministry of Health (2015: 11)

solve the problem in many LIAs and especially in unplanned areas where their access is restricted. But pits in densely populated areas still need to be emptied which generally is left to individuals without control how the sludge is collected and disposed. The risks for public health are thereby obvious. Hence, the big difference to the rural setting is that in towns storage of human waste in the ground (septic tanks or traditional pit latrines) does not necessarily mean safely managed sanitation.

The case of Betty in Kawangware shows that in many urban LIAs households cannot find a place to construct a toilet in or near their rooms or the toilets building cannot be shifted when the pit is full. Hence sharing of facilities and the provision of emptying the pit without a destruction of the toilet or some of its elements (e.g. slab) are necessities. Experts interviewed indicated that sharing toilets is not a problem for the poor as long as the number of households is limited and the toilet is placed on the plot of the users. Referring to the expert interviews and especially to the women in a LIA in Ouagadougou, Burkina Faso, heading a household where people have to defecate in the open, there is little doubts that the poor accept to pay for sanitation facilities and services when provided as subsidised low-cost technology. The challenge remaining in organising sanitation is to bring together the use of appropriate technologies and management systems combined with incentives to adhere to public health regulation. This is more likely to happen where users and landlords have common interests and where the framework delegates function to professionals who know how to make use of local knowledge. Concepts to respond to these challenges have been tested on scale (e.g. UBSUP in Kenya) but the experience of rolling it out nationwide seems still to be insufficient.

Thus, what has to be done to move towards access for everyone to safe sanitation in the urban setting? Sanitation development for the poor needs first and foremost the acknowledgement of the importance of sanitation infrastructure development as passive health care and the need for an appropriate framework with an orientation which reflects the contexts. The deliberation so far suggests that as for water, policy making as well as infrastructure development and service provision for sanitation should fall under the line ministry responsible for water and sanitation. It should be this line ministry and not the MoH or any other multi sectoral structure to develop an enabling framework for sanitation service provision. It is helpful when either the office of the Prime Minister (Burkina Faso) or the President of the State (Kenya) decides on the functions in sanitation for each concerned sector. It should also be recognised that sanitation service provision is based on an industrial process and is better delegated to professionals (utilities and investments baskets), including its regulation. Urban sanitation for the poor, like water supply, should not be left to the civil service structure, to NGOs or *'sanipreneurs'* making a business case for collecting cartridges from households, etc.[11] Constructing and operating decentralised treatment facilities are a matter for professionals.

An overhaul of the framework documents for sanitation has to be complemented with detailed implementation concepts to be elaborated by the institutions respon-

[11] Refer to Sanergy in Nairobi, Kenya or to Cape Town, South Africa.

sible for sanitation according to the framework. Regulation should make sure that utilities pay as much attention to the development of off- and onsite sanitation as they do for water supply and like for water, several service levels within a regulated system should be offered to the urban dwellers. It should not be left to the utilities to decide how they are engaged in sanitation and how they report on progress. Like proposed for water, national financing baskets should integrate the promotion of onsite sanitation and its chain and donors should provide support to these national institutions. Utilities should develop and operate (with the involvement of the small scale private sector) decentralised collection, transport and treatment facilities. If a regulatory regime is in place for water, there should be no hesitation to investigate how functions for onsite sanitation can be included in the regulatory regime.

As mentioned before, there is experience which can be used by utilities for developing onsite sanitation such as the Up-Scaling of Basic Sanitation for the Urban Poor (UBSUP) project in Kenya[12] where the WSTF provides funding for the utilities to establish (decentralised) infrastructure for the chain of onsite sanitation and helps the poor to access adequate and subsidised toilets in the LIAs. The development of onsite sanitation is also spearheaded by the utility in Burkina Faso. However, this should not stop the development of piped sewer systems where population density is high, sufficient funds are available, the system is able to cover O+M costs and sufficient effluent flow permits such systems to function.

In addition, the Burkina Faso case documents that utilities can also be involved in the promotion of school sanitation and facilities at public places, especially when donors are willing to provide funds for such support to professionals. Also the WSTF in Kenya promoted public toilets owned by utilities and operated by individuals at sanitation hotspots like market places for instance. This concept is similar to water kiosks where the utility owns the infrastructure and engages a private operator.[13] For the slum in Kawangware, Kenya such public toilets owned and operated by professionals would help dwellers like Betty tremendously until a slum upgrading program can ensure that all households have access to their own or to shared toilets placed on their living space.

References

Keener S, Luengo M, Banerjee S (2010) Provision of water to the poor in Africa, experience with water standposts and the informal water sector, WSP 5387, Public Research Working Paper, Africa Region Sustainable Development Division, The World Bank

Moyo D (2009) Dead aid – why aid makes things worse and how there is another way for Africa. Penguin Books, London

[12] https://www.waterfund.go.ke/stories/safisan-ubsup, last visited August 2018.

[13] Sixty four public toilets were financed by the WSTF, handed over to the utilities and are still successful under operation, some of them as old as 10 years.

Chapter 8
Conclusions: The Quintessence of W+S Sector Reforms in Sub-Saharan Africa

Abstract Reforms have changed the sector in all four target countries and brought many benefits. However, there is an unstable equilibrium with a permanent risk that achievements are undone with a stroke. The sector is still receiving insufficient attention and there are too many unqualified players being able to act uncontrolled. The buy-in by politicians especially for a new sector orientation is insufficient what prevents institutions from using sector potentials and compromises reform implementation. An ill-designed institutional framework will haunt the sector for decades. Informality and uninformed self-proclaimed representatives prevent the poor from gaining adequate access. Limited professionalism and weak leadership restricts sector credibility. Regulation fosters sector resilience but faces limits. Considering recent developments, it seems that sanitation development will remain neglected. There is also much room to improve the partnership between receiving countries and international support (donors, NGOs, etc.). The contributions fall short because the complexity of the sector and its specific features are not understood or ignored by the bankers and development banks are not concerned enough about their role in promoting sustainable development. The good news is that the Burkina Faso case indicates that even the poorest countries can be best performers.

The mixed results of reforms and the fact that soon hundreds of million people in Sub-Saharan Africa will share the difficult living condition of Betty and Agnes in the LIAs show that there is an urgency to act and close the knowledge gap in urban water and sanitation development. The literature review and the analysis of sector reforms guided by the proposed model of development allowed a profound insight into the sector and permit the following conclusions.

Reforms Changed the Sector but There's Unfinished Work and an Unstable Equilibrium

Although there is no doubt that the sector is in difficulties, many ignore or even deny that there is an urban water and sanitation crisis in the low-income countries and that it is gaining intensity. Irrefutable signs are overlooked. However, a number of countries realised the decline and carried out sweeping reforms in the recent past,

R. Werchota, *Empty Buckets and Overflowing Pits*, Springer Water,
https://doi.org/10.1007/978-3-030-31383-8_8

some with remarkable results. It brought changes in regards to outdated paradigm which led to new sector orientations and an anchorage of new bearings in low-income countries. Despite this and in some cases even good progress in development, the reform processes cannot be considered sufficiently accomplished yet because often the sector framework remained incomplete and continuous corruption and unsatisfactory governance are indicating that checks and balances are either missing or not sufficiently working.

Also the investment gap has been continuously growing in many countries despite reforms and negligence in pursuing certain sector principles has slowed progress in access. There seems to be a fatigue after 15–20 years of reforms or according to experts interviewed, 'reforms got stuck'. Because of increasing challenges due to the effects of climate change, etc.[1] complacency is no option. There is need to carry reforms to the next level and enlarge them into areas so far neglected (e.g. sanitation, financing). The risk that hard-won results are undone in a stroke is omnipresent when for instance new legislation is adopted, decentralisation efforts are undertaken, new political movements come into power, etc. It needs a professionalised sector with increasing resilience to counteract the many confounders.

W+S Importance Neglected, Attracting Too Many Unqualified Players
International cooperation has provided substantial support to strengthen the sector in partner countries with declarations on global level and concessional funding. However, there are a number of burning issues remaining. Urban water and sanitation development in the low-income countries does not receive the attention it deserves despite its crucial importance for society and the rising number of people being deprived of adequate access. Health specialists are too dominant in the discourse of water and sanitation which results in misleading messages about the importance and the status of urban water and sanitation in the developing world. The development of infrastructure for water and sanitation as passive health care seems to be less important than active health care although it is becoming increasingly apparent that curing repeatedly infections stemming from insufficient access to safe water and sanitation with medication damages people's immune system over time (e.g. antibiotic resistance). This development can be described as an aberration in the light of the responsibility of the state to ensure a healthy workforce and sound living conditions in towns. To disguise the negative development in the sector, success is produced on paper. This weakness, already recognised by the General Assembly of the UN in the past, has unfortunately not been sufficiently corrected in the transition from MDGs to the SDGs.

There is a history of repeated unrealistic goal-setting by the UN for more than 40 years. The tradition of missing goals and the messages of improvements in a

[1]E.g. Kenya, The Guardian June 2017. 'Thirsty city: after months of water rationing Nairobi may run dry'

https://www.theguardian.com/global-development-professionals-network/2017/jul/24/thirsty-city-after-months-of-water-rationing-nairobi-may-run-dry?utm_source=Global+Waters+%2B+Water+Currents&utm_campaign=f1203502e4-Water+Currents+2017_08_22&utm_medium=email&utm_term=0_fae9f9ae2b-f1203502e4-25799197 (last visited 05.2017).

degrading situation blur the urgency and send the wrong signals to the politicians in the partner and donors countries alike. It encourages decision makers to reduce their attention for the sector even further and donors to walk away from it. This is aggravating the already precarious situation. Furthermore, there are too many stakeholders who use the vacuum left by the utilities and try to make a business case out of the deplorable situation of the poor. This leads to a two class system in towns which discriminate the poor and creates an overcrowded field of actors pulling in different directions which makes the management of development in the sector today much more difficult in low-income countries than it was for the industrialized countries in the past.

Many stakeholders do not embed their actions into a comprehensive view of the sector and in national policies. Besides this, there are insufficiently tested (often unworkable) ideas sold to the low-income countries and short cuts taken in development. Numerous 'innovative' ideas and solutions praised as solving the urban water and sanitation crises in the low-income countries remain at best niche products. The problem with this is that sometimes partner institutions rely on such promises which finally do not materialise. Thereby, the sector loses time and money, and many people needlessly their lives.

The Responsibility of the State Is Undermined by Undue Interference
Lessons learned from history backed up by academic knowledge shows that the state has to intervene in the sector and that it is not in the interest of development to leave urban water provision and sanitation to individual households or uncontrolled to business people. Sector reforms have responded adequately to this with the obligation to formalise service provision guarantying minimum standards according to human rights for water quality, access, price, sustainability, effluent treatment, etc. and by recognising that centralised systems are the best option to deliver such services for everyone in towns. However, the advantages of centralised systems under regulation are often compromised by governance issues and corruption because politicians use water and sanitation institutions too often for patronage and furthering their political career. Thereby, basic sector principles are ignored. Undue interference of politicians' limits the autonomy of professionals trying to reach a higher performance level and prevents utilities to obtain the necessary financial strength. Political considerations often enough determine water tariffs rather than goals of development. BoDs are generally political appointees, instead of professionals, who have no consequences to fear if the institutions they should guide and oversee underperform permanently. This means that even deep-rooted and well-designed reforms are undermined in the process of implementation. It is not helpful that the state delegates' functions like regulation and service provision to professionals and then hamper their efforts with weak leadership and personal interests when sector institutions need support.

Limited Political Buy-in Combined with Unclear Vision
Many of the decision makers do not take sufficiently into account the importance of sector principles for the orientation of development. During a situation analysis, reform champions often get lost in the multitude of challenges and thereby risk set-

ting incorrect priorities or opting for unworkable approaches proposed by incompetent or biased stakeholders. Consequently, the set of sector principles is often incomplete or contains elements with secondary relevance. Successful reforms would require an early consensus on the most important sector principles and a buy-in on the highest possible political level. Unfortunately, some reforms are designed by current fashions (PSP, blended financing, etc.) and not with a long term view and a sustainable development concept in mind. The global discourse has its own share in this because it prominently suggests answers with little relevance for urban water and sanitation and at the same time rejects badly needed low-cost solutions such as shared facilities. In addition, national decision makers during implementation of reforms make concessions which waters down sector principles. Also the streamlining of sector principles right down to the implementation agencies is generally insufficient.

Ill Designed Reforms Can Haunt the Sector for a Long Time
Although it seems less important how and in which area sector reforms commence, it remains very important to get the framework right either at the start of the reform or later when utility restructuring and more investments have taken place. Thereby, the number of utilities in a country has an influence on the suitability of the entry point of reforms. When reforms start with restructuring of utilities, reform champions need to be aware that strong institutions tend to become 'little republics' and can block further reform efforts later in the process. In any case, there seems to be no relationship between the entry points or the type of champions of reform and the success or failure of sector development. Reform champions must have certain autonomy and should not be released from their duties too early. Experience shows that guiding stakeholders in a new orientation and building resilience of a new system takes time (10–20 years) and requires convincing experts with sufficient knowledge. A particular problem is an ill-conceived institutional framework which generally can become a heavy burden for decades. The desire to avoid conflicts of interests and cater for specialized units sometimes compromises the principle of a lean structure. It has been shown that streamlining of a once created (bloated) institutional landscape will often fail for political reasons because interest groups have been established and powerful lobbies learned how to make use of the existing set-up and therefore resist necessary adjustments.

Potentials of the Sector Are Not Explored
The argument to hold water tariffs low is that people in the low-income countries cannot afford higher tariffs. This ignores the willingness to pay higher water bills by the rich and the rapidly growing middle income population in the countries. Unfortunately, in many low-income countries this self-financing potential of the sector lies fallow. Considering the huge and growing financing gap, it will only be a matter of time where governments of receiving countries will have to make the sector tap into this potential because external financing institutions (now providing the lion share for funding infrastructure development) will not much longer accept this ignorance. Interesting enough, the poorest of the four target countries is the only one where decision makers have already moved into this direction.

Tapping into the sector self-financing potential permits the countries to generate a surplus well above 100% O+M costs coverage which makes the sector more credible and able to attract additional donor contributions. Also, making use of the management potential the utility is in a position to cut costs of operation and thereby, can make room for paying back the loans received from donors. This has in the case of Burkina Faso ended the long standing off-loading of such payment from consumers to the taxpayers. For this, the support of the Ministry of Finance to enable ring-fencing surpluses for investments in the sector was crucial. Furthermore, the treasury was willing to extent the line of guarantees for the sector which adds to the sector credibility and opens the possibility to tap into new sources of funding at local level.

The professionalization of service provision and in the case of Burkina Faso also of asset development and national planning has attracted more qualified people to the sector. This has also helped water utilities to improve the effectiveness of fund mobilisation and to take on board additional functions such as onsite sanitation. Thereby, long term TA was helpful providing experience of internationally successful approaches which was blended with knowledge of local actors allowing for effective developments in different contexts. Such combination of experience and knowledge enhanced institutional learning. Nevertheless, funds channelled through public institutions are always subject to considerable corruption which should not deter donors from using national institutions because a well selected and integrated long-term TA can help to counteract misuse of funds by strengthening national forces committed to integrity.

Informality and Uninformed Representatives Prevent the Poor from Access
The inability of utilities to extent their services into the LIAs leaves room for the development of unacceptable informality in service provision and the persistent existence of the discriminatory two class water and sanitation systems in towns. According to many sector policies utility services have to replace informal providers but as long as the number of underserved people is still rising, they're lagging behind development. It is obvious that the sector in many low-income countries will have to struggle with the negative effects of informal service provision for some time to come. These countries need to be on guard not to allow a further increase in the number of informal providers.

The inclusion of the human rights to water and sanitation in the framework in the low-income countries was a bold move because it supported the desire to formalise service provision, increased the attention for the LIAs and helped to accept shared facilities for basic services. This provided many reforms with a strong poverty orientation which however, is often undermined by unprofessional sector steering and a misunderstood defense of the poor. There is a disparity of what the poor want and what the state provides often with the help of the pro-poor oriented international cooperation. The analysis in this book indicates that the rejection of water kiosks for water quality reasons by some international institutions is unfounded because all service levels meet the same risks and utilities can easily address water quality risks at the point of consumption with an adequate residuum chlorine level at their outlets. Equally questionable is the promotion of SIPs by some development partners which do not want to recognise the temporary character SIPs should have when filling the gap for formalised service provision.

Politicians deprive with insufficient tariffs utilities of the necessary funds to extent services to the poor. The accumulation of funds by the utility to develop costly infrastructure and attract with increased self-financing more investors is unfortunately seen as a result of exaggerated tariffs. The insistence on (social) household connections for all forces more and more people to remain with informal service provision, neighbourhood sales and the use of traditional but contaminated point sources. It is obvious that offering the limited number of the poor which have already access to utility outlets free water at kiosks or a subsidised house connection is less important for the development in the present stage than providing the many more underserved poor with access to a utility tap or sanitation chain that guarantees basic services. It has been indicated that in many LIAs universal access cannot be achieved with (social/subsidised) connection programs and when such initiatives start replacing water kiosks then in most of the cases many served dwellers are pushed back to informal service provision.

There is also need to understand that urban water and sanitation depend as much on infrastructure development than on solutions for social concerns. For the underserved poor who want to escape informality physical access even to shared facilities is a much higher concern than utility tariffs. It should not be forgotten that improvements in first mile infrastructure are, in the first place, to the benefit of the connected consumers and can only be passed on to the underserved poor with sufficient last mile infrastructure development. In the context of low-income countries, universal access cannot be achieved within an acceptable timeframe without low-cost technologies/shared facilities which however, should be seen as transitional solutions in a long term view of sustainable development.

Although there are very successful utilities in reaching the poor, it seems that generally they are the bottlenecks among the sector institutions. Many still don't want or don't know how to serve the poor and therefore pro-poor development requires a champion which exercises constant pressure on these monopolists. It is recognised that utilities face an uphill battle when moving into LIAs. Informal providers often organised in cartels are fighting the extension of formalised services. The poor in unplanned areas don't trust state institutions. Thus, the utilities cannot call upon the police but still need help to protect their infrastructure. All this indicates that the pro-poor enforcement in the sector is generally speaking still not effective enough and the support to build capacity to reach the poor needs to concentrate more on utilities than on other structures such as slum communities. Initial scepticism about low-cost technology can be overcome through study tours, peer learning, pilot projects and coherent up-scaling concepts. In addition, missing data deprive the utilities to adopt an effective strategy and report on progress in the different types of LIAs when trying to extend services to the poor.

Limited Professionalism and Weak Leadership Restricts Sector Credibility
Professionalization in the sector is often incomplete. While the operation of assets for piped water and sewer has now been transferred from civil service structures to

registered companies in almost all of the cases and regulation to institutions outside the ministries in many Sub-Saharan countries, the improvements in performance has been limited because of insufficient autonomy and the nomination of political appointees as BoDs. Another obstacle observed is the staff transferred by the ministry/municipal to the new institutions which often does not adapt a new corporate culture and carries on functioning like civil service. In addition, the small size of many utilities is not helping to achieve viability or attract qualified professionals. Although generally spoken there seems to be no optimal size for utilities, it is an advantage for the sector to define a minimum size and cluster utilities which fall under a certain threshold. Hence it can be said that professionalization in operation and in some countries in regulation is only partially achieved. Although this is unfortunate, it still has brought some improvements in the sector but limits further progress at some point in time.

Professionalization has in many cases not taken place in asset development and in onsite sanitation. It seems to be more difficult for ministries to outsource functions to autonomous professionals where a high amount of money is involved because of interests who do not necessarily serve sector development and the people. Burkina Faso proves that the delegation of asset development to professionals is possible and absolutely necessary to reach a high sector performance and to secure a continuous flow of funds for asset development. The experts interviewed suggest that the insufficient professionalization of asset development also limits the elaboration of necessary tools and procedures in the sector and is the main cause for the stagnation in access to urban water and sanitation despite ambitions reforms.

While the missing professionalization in asset development can be considered a sector internal problem, the missing professionalization of the onsite sanitation chain was due to the opposition to reforms by the health sector. The positive role professional water utilities can play in (onsite) sanitation development is often enough not understood or desired. There are indications that water utilities are willing to become engaged in on-site sanitation development as long as feasible concepts and financial and technical support is available.[2] Sector development is also limited because of insufficient leadership by the line ministry especially for stakeholder coordination and supervision of sector institution. The latter leaves room for unnecessary competition between sector institutions and the desire of some to enlarge their mandate in order to gain more importance or mobilise additional funds from donors, even when earmarked for activities outside their mandate.

Regulation Fosters Sector Resilience but Faces Limits

There is no doubt that the introduction of regulation brought many benefits to the sector. Especially in a situation where a high number of utilities operate in the market a regulator is very helpful and even simplified regulation can be effective where there are substantial failures in the market. Regulation has helped to make progress in transparency, increased knowledge and utility performance, improved customer care and orientation, etc. Especially useful for the regulator and the consumers is the

[2] Refer to UBSUP in Kenya.

creation of a substructure for the regulator operated by (supported) volunteers. It can act as a prolonged arm of the national regulator and provide feedback from the ground. This kind of consumer representation seems to be more effective than placing consumer representatives on national level nominated by ministers. The challenge with consumer representation groups is the selection of the members, their oversight, the provision of support and the containment of the risk that they are high jacked by politics.

Some regulators have become highly recognised professionals who are consulted with preference when water and sanitation issues are discussed. Where there is only one national utility and the need for a regulatory agency is not obvious it is very useful to introduce some of the standard regulatory instruments such as public hearings for tariff adjustments, accounting of utility to the public, etc. However, it has been noted that regulation has its limits and often regulators cannot count on the support of the line ministry especially when strong actions are needed such as the replacement of a BoDs in the case of a prolonged underperformance of a utility. In addition, regulators face limits when the number of utilities is high. Inspection cannot be carried out with the necessary depth and frequency which leads to insufficient data or its quality. Furthermore, the influence exercised by politicians or the owners of the utility (e.g. municipalities) can undermine the work of the regulator which is especially damaging when necessary tariff adjustments are prevented. It hampers utility efforts to accumulate funds for investments.

Nevertheless, regulators in the low-income countries still have to improve their own system by closing some gaps in the regulatory regime and obtain additional knowledge about the impact of their decisions on the sector. They need to find the balance between aiming for efficiency gains in operation and allowing the utilities sufficient surpluses for asset development and repayment of external funding. According to the expert interviews, the link of regulation to investments is not sufficiently developed which restrains the enforcement power of regulators. For this, they would need external support from donors, MoF, etc. Taking the specific context in Sub-Saharan countries better into consideration it would be helpful to extent classic (economic) regulation of water into areas like onsite sanitation and asset development.

Insufficient Data and Inadmissible Comparisons

Reforms improved significantly the quantity and to some extent the quality of data especially on utility performance, water supply and in certain cases also on water and sanitation related data in LIAs. There was less advancement in data on sanitation and in a number of countries almost none for asset development. This is certainly due to the missing professionalization/outsourcing of functions in these two areas. It has been observed that water ministries have difficulties to develop, maintain and make appropriate use of complex information systems. The regular reappearance of the demand to create one entire water sector information system at ministries neglects this fact and the feasibility concerns for information systems stretching across sub-sectors which are very diverse and each of them largely complex.

The outsourcing of data collection and analysis to professionals and linking of such functions to the obligation of reporting to the public has proved to be very helpful in two ways. First, the institutions understood that without an adequate information system and verification of data they cannot report regularly and build credibility in the eyes of the public. Second, the data are more scrutinised when institutions have to go public with their work. This puts pressure on the institutions to improve gradually the set and quality of data. Outsourcing such functions to autonomous institution backed up by legislation limits the pressure from ministries to produce success on paper. Updating of databases seems to be in general another challenge and the necessary mechanism needs to be part of the design of information systems.

Monitoring systems for the use of assets either do not exist or stop working soon after an investment project is ending. In the best case it is maintained until a post evaluation which usually takes place 2 or 3 years after project closure. This is insufficient for an effective analysis and obtaining lessons learned. For instance, many approaches are hailed as best practices such as delegated management with evaluation carried out just after the project has started operation. Observing such solutions over several more years would most likely reveal that results are very different in many cases. Continuous monitoring of results over longer periods would save the sector a lot of time and money. A precondition for this would be the anchorage of data collection and analysis at partner institutions by all involved development partners. There is still much room to improve data in the low-income countries despite substantial progress.

Another challenge is rising when data between countries are compared. This became very obvious when analysing access data. Without looking into different elements of such data and making adjustments the analysis produces misleading results. Unfortunately, there is as good as no literature on this issue but many studies derive conclusions from such incomparable data.

Sanitation Development Prone to Remain Neglected?

In general, the water sector has contributed little to the improvements in sanitation. This should be a very big concern because most likely, access to piped sewer systems in percentage of people is declining in all Sub-Saharan countries and there are very few initiatives to balance this decline with the development of a (decentralised) chain for onsite sanitation. It is crucial that development in these areas is progressing, which should start with the framework development for sanitation. Because there will be no progress towards universal access in medium terms without the proliferation of onsite sanitation infrastructure and the involvement of professionals in service provision and asset development, sector policies, strategies and legislation have to include these issues. The development of infrastructure and services for urban sanitation does not belong in the health sector. These functions should be covered by the line ministry of water and sanitation development and for implementation purposes added to the portfolio of the utilities. Today, few utilities are involved in the promotion of adequate household sanitation and even less in the chain for onsite sanitation. However, among these utilities some became outstanding examples demonstrating their potential of making substantial headway in sanitation development in the given contexts.

There are several reasons why it can be expected that sanitation development in the low-income countries will remain inadequate although it seems to be high on the agenda in the discourse and there is a wealth of technical solutions for improved household facilities. Firstly, sanitation projects concentrate mainly on household sanitation solutions for the poor by leaving aside the development of a formalised sanitation chain. Hence, many NGOs, academic and philanthropic organisations are testing ecosan solutions, cartridge systems or other approaches such as pee-poo bags, but unfortunately, each of them inventing their own sanitation chain generally established outside the formalised system. Secondly, because of the absence of cross-subsidisation options within a formalised system they have to make a business cases out of sanitation for the poor. As long as the idea of subsidies for the many poor is rejected, progress in onsite sanitation will be limited in the urban setting.

Thirdly, upscaling experience in urban sanitation for the poor is missing because pilot projects never reach sustainability when they grow beyond a few hundred households and collapse when depending on support from the municipalities. Even though civil society organisation regular criticise the absence of support from local administration, they seldom offer alternatives with the engagement of other state institutions. However, there are initiatives which provide ideas on how to up-scale low-cost sanitation within a formalised system and how it could work such as the UBSUP project in Kenya. Fourthly, onsite sanitation development is even more complex and fragmented[3] than last mile development for water. It needs additional time in implementation. Hence, the development of a national concept and the establishment of an institution on national level to support standard setting, provide oversight and guidance, ensure monitoring and capacity building at utilities, etc. are preconditions before implementation can generate numbers on scale. Donors shy away from such cumbersome projects where timeframes and results for indicators are difficult to predict and a lot of patience for results is required. Although the international cooperation is constantly emphasising on the need to double efforts in sanitation development donor engagement don't increase despite many utilities are willing to engage. Sanitation requires a high degree of flexibility, and intensive, long-term technical support, which few donors and politicians are willing to offer or support.

Weak Partnerships for Reforms and Infrastructure Development

The insufficient attention for the sector by receiving countries and donors alike is partly responsible for the huge and growing investment gap. Both need to do more for infrastructure development although donors provide already the lion share of funding in the Sub-Saharan countries. Apart of the need for more funds, the available financial resources could be used more effectively in respect to reaching more people and improving the framework for development. The latter is not an easy undertaking because it would require more engagement from the donors and another

[3] E.g. disbursements amounts are very small.

system of evaluating their contribution. Instead of measuring success by how much of the funds are disbursed without great difficulties, the impact on the sector development would have to be placed in the forefront. Such concerns would have to be taken in consideration already at the design and negotiation phases and would demand that the donor representatives in the partner countries leave their comfort zone and don't shy away from conflicts with the partners and within their own institutions if needed (expert interviews).

It is as good as unheard that donors offering funding insist that partner countries make full use of the sector potential with a sustainable system of self-financing or seldom engage in a continuous policy dialogue. The notable efforts undertaken by the development banks would be more effective if the recommendations of the Paris/Accra/Busan agenda would be more stringently applied. Unfortunately, these OECD initiatives have lost steam in the sector and would need a relaunch to pressure donors to use public funds more efficiently. Furthermore, support offered by donors often lack quality. This becomes obvious when looking at the negative effects of the decentralisation/devolution in the sector which is often strongly supported by donors. Also common platforms like regular donor-partner round tables[4] are missing where fresh money is linked to the results of the follow up of engagements signed by the different parties which, when absent, is limiting aid effectiveness considerably.

Reforms have helped to lift or ease some bottlenecks in the development of the sector such as low utility performance, missing pro-poor orientation, insufficient data/information, etc. and have thereby increased absorption capacity in a number of countries. This prepared the ground for accelerated infrastructure development and the use of fresh money to initiate and accompany further changes in the sector. Such a leap forward would need additional professional TA as partners for the FC and more engagement in national financing baskets. However, despite the many advantages basket financing is offering such as allocation of funds on a competitive base, donors seem to lose interest in this mode of funding. Equally, only few partner countries are ready to engage with substantial government funds in investment baskets where donors are involved. The latter could use their involvement in baskets to insist on more engagement by the partner countries and thereby increase transparency and accountability in the use of national funds. It would help to dry up the many 'black holes' government officials are using for diverting investment money to individuals and political parties instead of helping the underserved people. Next to these partnership issues there is generally an insufficient involvement of the Ministry of Finance in the sector which would be needed to overcome the limits in financial matters such as cost recovery (tariff adjustments), ring-fencing of sector income, introduction of basic financial management tools (e.g. financing modelling), repayment of loans (reduction of 'fiscal burden'), etc.

[4] Line ministry, utilities and donors.

Contributions Fall Short, Complexity and Specific Features Are Ignored

The analysis has shown that increased access is not necessarily the sum of improved utility performance and extra funding. Thus, there are other crucial factors which must enter into the equation and where more or less simultaneous development is required. Development in one crucial factor does not substitute development in other factors. This is important to point out because generally the complexity of the sector is not understood and therefore, the restructuring of utilities and increasing investments is in the forefront of attention when sector improvement is discussed. Decision makers of the international cooperation don't listen sufficiently to urban water and sanitation experts. It has also been shown that ignoring the specific features of the sector, for instance in the discussion of community involvement, can be very damaging to the sector development. Furthermore, transferring experience from the rural setting such as the use of single water sources and the community owned and managed solutions to the LIAs in towns extents the *'urban water and sanitation divide'* with its discrimination of the poor. There is also need to agree on global level what sustainable access to safe water and sanitation in towns should mean.

The Poorest Countries Can Be Best Performers

Expectations are always high when reform processes start, but are seldom fulfilled to the desired extend. The reasons are flaws in the design, inadequate priority setting and especially the inconsistent implementation of a new orientation. In the eyes of many, only missing money is holding back development. This has some justification because infrastructure is very important but would also suggest that wealthier countries invest more per capita in the sector than poorer countries. The result of the analysis in this work proves them wrong because Burkina Faso as poorest is the top performer among the four target countries and most likely finds itself among the best performers in Africa regarding urban water and sanitation development. Burkina Faso, having a much higher poverty level than the other three target countries, follows stringently its well-designed sector orientation in as good as all crucial factors for development and especially in cost recovery and poverty orientation. Such a discipline in the implementation of reforms secures an extended self-sufficiency of the sector and lead to an investment level per capita to be served which is much higher than the many better off countries reach. It is obvious that in such a positive environment created by the receiving country also donors support this orientation, link their investments better to good governance and reward progress with increased contributions due to a better leadership in the receiving country.

Chapter 9
Recommendations: Guidance to Master the Coming Wave of Challenges

Abstract The following messages are briefly outlined: Match the priority for the sector with its importance for the development of the individual and the society. Make reforms a permanent event guided by designated reform champions. Adopt an adequate sector orientation and ensure a buy-in at the highest possible political level. Find the institutional framework which suits the context and sustainable development requirements best. Agree on an effective development concept for universal access. Outsource as many functions as possible for W + S services from the state administration and guarantee autonomy for the management. This must also include the professionalization of asset development for W+S and the operation for the sanitation chain (piped and non-piped). Bring services for the onsite sanitation chain into the formalized system. Balance social and economic goals and promote good governance. Put a realistic financing concept in place and include services to reach the non-served people. Accept that the poor need subsidies to gain and maintain access and that they should be generated by the sector instead of making the sector dependent on handouts from politicians. Establish a baseline for low-income areas and ensure regular data updating and independent verification. Reform WATSAN structure on global level and relaunch the implementation of the Paris/Accra/Busan agenda. Take note of the areas of possible further research.

The following condensed recommendations should help experts and decision makers to evaluate, adjust and implement urban water and sanitation sector reforms in low-income and especially in the Sub-Saharan countries.

Match Sector Priority with Its Importance for Development

First and foremost, politicians should be made aware how important the sector is for the development of the countries and that social concerns should not limit the attention for sustainability of service provision and infrastructure development. This awareness must translate into a bundling of key functions for water and sanitation infrastructure development and services in the sector and to the provision of more money for asset development than in the past. Donor countries should remember how essential sustainable access to safe water and sanitation for everyone in towns

R. Werchota, *Empty Buckets and Overflowing Pits*, Springer Water,
https://doi.org/10.1007/978-3-030-31383-8_9

was for their development and pay more attention to the sector in the partner countries. Support for water and sanitation development should never be a component of cooperation programs in partner countries agreed for other sectors (health, governance, agriculture, etc.) because such components do in general not generate access for the underserved people on scale. Another preconditions for a successful development is the recognition that the sector in Sub-Saharan Africa is in a crises and a dramatic growth in demand for water and sanitation services in towns will take place during the next decades.

Make Reforms a Permanent Event Guided by Champions

The dynamic and volatile environment in the low-income countries requires a constant steering, streamlining and fine-tuning in the sector. Fine-tuning must also include the introduction of new tools the sector needs when moving to the next stage of reforms. Reaching universal access in the low-income countries will take much more time than anticipated in the past because sector development, especially last mile infrastructure development is a time consuming undertaking. Hence, the international cooperation should accept longer timeframes in their visions and engagements in the sector and national governments in the low-income countries need to recognise that reforms need constant efforts over a long period.

The ever-present risks of corruption and the (mis-) use of the sector by politics require strong determination to safeguard achievements of reforms and maintain a development process towards universal access. For this, there is need to rally the public and the highest decision makers in a country behind reform efforts. Making hard facts available to the public and thereby, concentrating on communication of proven achievements will ensure continuous support for further steps in the reforms. Curbing the risks of derailment in the development process will also need specialists on reforms who can guide the different sector institutions and advise the numerous national and external decision makers. Such national specialists are available in the low-income countries and their effectiveness can be reinforced with competent TA. Whom to select and where to place such a guardian of reforms is contextual. Important is that the person or group is established as a permanent poste or structure, full time available for the necessary functions, open for dialogue, sufficiently experienced to resists short lived fashions and has easy access to top decision makers.

Adopt an Adequate Sector Orientation and Secure a Strong Buy-in

Specialists on urban water and sanitation development should guide decision makers on national and global level. The sector should no longer be a playground for politics and incompetent stakeholders with individual interests. Hence, first a sector orientation based on the challenges the sector is facing should be lined out with sector principles. These principles should be adopted on the highest possible political level and guide the champions of reforms during implementation. In the framework documents and the implementation strategies of the sector institutions such principles need to be streamlined with detailed concepts at each sector institution. Furthermore, the sector needs a long-term vision for achieving universal access which must be based on local and academic knowledge as well as the lessons learned which gained global relevance. Urban water supply must be restricted to piped infrastructure where service provision is under regulation and utilities offer different service levels. Thus the use of single water points and small scale systems as well as neighbourhood sales should have no place in a modern supply system in towns and therefore be gradually outfaced. Business people, civil society, etc. must be given a defined role by the framework corresponding to their strength but limited by the containment of the risks to compromise the vision for development with their engagements.

Considering the huge amount of funds needed and the responsibility of donors in the use of public funds, the international cooperation would be well advised to consider aid effectiveness as the main criteria when allocating funds. Thereby, donors have an obligation to link their support to an appropriate sector orientation, framework development, good governance and adequate leadership in the partner countries. The support of donors must help partner countries in these areas because they have the mandate to support sustainable development. Evaluation of their contribution must include relevant indicators. Increasing access today is insufficient to justify an intervention when later assets prematurely degrade or access becomes stagnant because donor intervention did not assist in strengthening self-help of the sector in the receiving country.

Outsource Functions for W+S Services from State Administration
Experience suggests that transferring functions for or related to service provision from the state administration to professional sector institutions has a positive effect on sector performance. But outsourcing of function requires first, a reinforcement of the oversight (regulation) and second, securing autonomy for the institutions. With sufficient autonomy and protection of professional institutions, personnel from the state administration will find it more difficult to unduly interfere or act as a prolonged arm of politics. The regulator in cooperation with the civil society can functions as watch dog. Autonomy is especially needed in decision making, financing, management of staff and monitoring/reporting. When outsourcing of functions is combined with a transfer of personnel from the civil service structure then it is important that the top and middle management is competitively recruited from the open labour market in order to facilitate the development of a new corporate culture. Only the combination of outsourcing functions and securing autonomy can make sector institutions, especially utilities, perform like private sector enterprises. Therefore, outsourcing of functions and autonomy of sector institutions has to be addressed by sector policy and legislation. Whenever the limits of autonomy cannot be overcome, private sector participation might be a good alternative.

In addition, the regulator has to receive sufficient power to effectively intervene if the autonomy of utilities is endangered. For this, regulatory tools should be introduced such as corporate governance guidelines, special regulatory regime in the case of prolonged utility underperformance, etc. Regulation is important not only to safeguard the autonomy of utilities but also to professionalise tariff setting and to curb the inherited inefficiencies of a monopoly. Regulation must set the territory of the service areas for utilities according to population density and define standards on the number of consumers to be attributed to the different types of water outlets and sewer inlets. For this the regulator needs to do a regular reality check on the ground in order to avoid an overstatement of access by utilities.

The level of NRW seems to be the most appropriate indicator for the assessment of the management quality of utilities. Hence, regulation aiming at performance increases has to concentrate particularly on the introduction of technologies and procedures to verify and improve NRW. The success of new technologies of any kind depend on the level of the management quality. However, transferring functions from the ministry and municipality should go beyond the operation of assets and regulation and include asset development and support functions for the mobilisation of funds because sector performance depends heavily on infrastructure. In addition, the gap of the missing sanitation framework concentrating on the development of the chain for centralised and onsite sanitation must be closed in the sector with the involvement of professionals. Furthermore, for performance in the operation of assets, it is important to define the minimum size for utilities. Hence, a clustering of utilities might be needed in a situation where historical development (e.g. decentralisation) has led to many small sized management systems.

Shift Attention of Development to Underserved People

Sector progress in terms of access to regulated services depends on reaching the underserved people (mainly the poor) with utility services. In the development in most of the low-income countries attention must therefore shift more than in the past to LIAs where the poor can join the already connected population and escape the burden of informal service provision. Extending formalized services to a limited number of households in an area will lead to neighborhood sales which compromise minimum requirements of human rights. Hence, for several reasons explained, shared facilities placed in the proximity of households (allowing a 30-min cycle) as part of a formalised service ladder should be the first choice to be considered in order to speed up access for the poor. This shift of attention to basic services must be reflected in all proposed crucial factors. An adequate orientation must be factored into the framework documents and implementation concepts.

Reporting in the sector should be linked to a national baseline and demonstrate progress in the underserved LIAs. Although it is difficult for utilities in big towns to identify LIAs, update information and report on all of them, there is no other option but to monitor progress of utilities in these underserved areas. NGOs financed by donors and working closely with utilities and the regulator could help to overcome challenges the utilities are facing in reporting. Financing and asset development has to concentrate on last mile infrastructure as long as unplanned settlements exist and the number of underserved people is substantial. It seems that a ratio of one to ten of total investments should be earmarked for last mile investments in Sub-Saharan countries where urbanisation is rapidly growing and urban poverty is widespread. Furthermore, there should be recognition that a national up-scaling of basic services is best done with a national structure where donors, national government and NGOs are engaged.

Regulators should establish sub-structures to represent the poor and ensure that households in the LIAs depending on water kiosks can access the formalised complaint system of the utility. Any solution proposed as intermediate arrangements in LIAs should pass a realistic feasibility test before being introduced as an alternative to utility systems. If it is proven that an intermediate arrangement, such as community-owned systems in LIAs, is reinforcing the *'urban water and sanitation divide'* it must be discarded. Such intermediate solutions, if not avoidable at all, should be designed in a way that they can become part of a formal centralised system whenever utilities are able to extent their services into these LIAs or such systems can grow into a size where they can become viable utilities. In addition, there must be the recognition that generally informal service providers cannot be brought under regulation with acceptable efforts.

Find the Appropriate Institutional Framework

The design of the institutional framework must on one hand help to avoid conflict of interests and provide a certain level of specialisation and on the other hand secure that the framework remains as lean as possible. This balancing exercise with its outcome depends on the context. Considering the enormous difficulties of changing ill-conceived institutional structures once in place, it is of utmost importance that the framework is not tailor made to fashions of the day (e.g. private sector participation) or influenced by political interests or targeted towards benefits for individuals (e.g. awarding of contracts). The institutional framework should not be in the forefront when initial efforts are undertaken for solving problems. Its adjustment should only be considered after trying to solve problems by other means just as improving workflow management, etc. and once they have not produced the intended results. Furthermore, professional sector institutions receiving sufficient autonomy from the ministry need to be guided and controlled because becoming more competent includes the risk that the institutions will try to increase their power, enlarge their mandate and start unnecessarily to compete with each other. Hence, a strong sector oversight by the line ministry with the help of the reform champion is important but on the other hand should never compromise the autonomy of the institutions.

Agree on an Effective Development Concept for Universal Access

An effective development concept is based on a comprehensive view of the sector and knowledge of the existent challenges. Without apprehending the sector complexity decision makers will have difficulties to design such an effective concept. Effectiveness also depends on the best use of the potentials which sector players' are able to offer. Therefore, following the analysis in this book the establishment of centralised systems for water and combined systems for sanitation (onsite and offsite), their development and management by utilities and accessibility by all urban dwellers is the way forward in the low-income countries. The bottom line is that comprehensiveness of concepts and activities means to take at least the (six) proposed factors for development into account and aim for simultaneous development in all of them.

After obtaining the agreement on an effective development concept by the stakeholders it is crucial that the line ministry ensures the alignment of all stakeholders in the process of implementation. Hence, there must be platforms where stakeholders can be held accountable including the line ministry, other sector institutions and especially the donors and civil society organisations. For practical reasons, the platform for fund raising (financing infrastructure) should be limited to donors and selected national institutions (including treasury). The platform where civil society organisation participates should be used to make politicians aware of the urgency the sector is facing. NGOs can be very helpful to bring this message to the public and global level and rally support for the sector in the low-income countries. At the same time NGOs should not substitute utility services on the ground. Such platforms are also crucial to improve the often insufficient communication between ministries and the international cooperation in the low-income countries.

Balance Social and Economic Goals and Support Good Governance
Stakeholders, including donors and NGOs, should be made aware that social and economic goals can be achieved simultaneously in low-income countries. The assumption that poor countries cannot afford an average water tariff of 150% O + M costs is unjustified. Under general conditions, this would mean today that the average water tariff per cubic meter should be around one USD. In order to convince decision makers to move towards these objectives a peer exchange with successful utilities are helpful. It should help to convince national decision makers that utilities are able to cover O + M costs on such a level and thereby, secure a high percentage of self-financing and reimbursement of concessional loans contracted for infrastructure development. This will help to limit dependency on politics which comes with subsidies and hampers sector development. Only with financial sound utilities can the sector significantly contribute to development and progress in access for the poor.

However, decision makers in the sector should not be satisfied with a commercial success as long as many people, especially the poor, remain underserved. Regulators, Ministries of Finance, donors, etc. can help utilities to generate sufficient surpluses for infrastructure development, support their ring-fencing and use to offer benefits to the poor. As explained in the text before, cross-subsidisation between the high and middle income classes and the poor should be the way forward. An appropriate design of subsidies generated within the sector from rising block tariffs can help to avoid most of the risks and their negative effects often observed. It is undisputable that the poor need subsidies for water and sanitation in order to break through the poverty cycle despite the risks of ineffective distribution. In addition, international cooperation can help with TA to strengthen forces in the partner countries which are determined to practice good governance and thereby limit corruption in the sector.

More Attention to Professionalised Asset Development
Because donors in general provide the main source for financing new infrastructure the international cooperation has a great opportunity to help the sector to engage in reforms and build resilience. This is in line with their obligation to use public funds in order to foster self-help in receiving countries. Donors should be pro-active in securing the best use of concessional loans and grants, but equally of national funds for asset development and should never fill the gap created by the misuse of national funds. Bearing in mind the chronic investment gap, sector reforms in low income countries have to concentrate more on infrastructure development and fund mobilisation. It seems that under the present conditions in Sub-Saharan countries annual investments of around 15 USD per capita to be served would be needed in order to

achieve both objectives: increase the percentage of people served and reduce the number of the underserved people.

The present donor strategies in supporting partner countries have to change and this has to start with the support in developing an appropriate framework for fund mobilisation and investments. However, the FC cannot deliver such support without the help of a professional long term TA which should not be under the supervision of the FC but work with autonomy and in cooperation. Furthermore, regular round table events for sector financing combined with the support of TA placed at the national institutions are critical. Donors should also support the development of procedures to ensure transparency in awarding contracts, supervision of implementation and an after project monitoring. Participating in such round tables documents the determination to submit to accountability (donors and partner institutions).

Donor coordination and cooperation must become more effective and a premature ending of project must remain an option whenever the sustainability of the contribution or the repayment of loans is in doubt due to insufficient governance, etc. To avoid such a situation contributions should be financed in stages and funding should be provided under competition wherever possible. Aid effectiveness does not only mean that funds provide more people with access or increase the raw water availability for towns. It must also include the activation of the self-help potential in the receiving country with capacity building, better governance, increased self-financing through water bills and a battle against corruption. To achieve higher aid effectiveness in the urban water and sanitation sector, donors and national governments need to bundle their contributions in a sector financing basket. Investment projects closely overseen by donors must at least be part of a national investment plan, sector financing model and follow national standards. Taking the many risks of baskets into account, it is understandable that donors and receiving countries hesitate to work through baskets. Nevertheless, the failure of national baskets in the past is also a failure of donors to play their role in sustainable development.

Sector financing baskets have many advantages ranging from anchoring donor and NGO activities at national institutions, upscaling of last mile in water and sanitation as a national concept, long term monitoring, countrywide enforcement of national standards, capacity development of implementing receiving institutions, etc. right to the possibility to allocate funds under competition in a market where several utilities are operating. For the latter, selection criteria for utilities and projects must be clear from the start and a system of yellow and red flags as well as phased funding introduced. With these mechanism fresh money can be linked to past progress and good governance. Financing baskets must be placed an arm length from ministries. Once politicians have agreed on a development plan and on donor contributions then the field for asset development must be left to professionals. This separation of politics and investment funds is crucial but often very difficult to achieve as long as such an orientation is overruled by interests other than efficiency and public wellbeing. In addition, funding should be linked to regulation which helps to improve its enforcement. Furthermore, low-income countries have to find the appropriate mix of first and last mile infrastructure for their specific situation as long as not every household can have a connection to a formalised system. Experience indicates a ratio of investment spending towards 9 to 1.

Find a Realistic Financing Concept

Next to an inadequate asset development, fund mobilisation and a vision for sector financing are often the weakest elements in sector reforms. To this adds the often confusing debate on how to close the financing gap in the sector. Discussions seem to focus much more on 'innovative' solutions than on improving the often insufficiently explored 'classical' financing sources. This should change by concentrating in the first place on self-financing of the sector, a better use of existing national and donor funds and the improvement of fund mobilisation tools and mechanisms. Self-financing of the sector through water bills needs to reach a level which makes the sector attractive for investors. This includes the need to stop the drainage and wastage of funds from an already chronically underfunded sector by ring-fencing sector income for investments and better use of funds. The protection of accumulated funds is detrimental and can be achieved by establishing an account earmarked for investments and legal provisions for its management (e.g. regular auditing through independent experts). BoDs should be made personally liable for such a ring-fencing and a regular verification of an investment account by external auditors.

An increase in self-financing of the sector in the low-income countries should be rewarded with a rise in donor and national budget funding. The governments of receiving and donor countries should give urban water and sanitation more attention when allocating funds. With increased credibility the sector can attract more contributions from local commercial banks and thereby, adjust the mix of financing sources. However, considering the history of funding urban water and sanitation infrastructure in the industrialised world and the limited economic power of low-income countries it seems that commercial funding is likely to remain a niche product. Hence, low-income countries are well advised not to place too much expectation on 'innovative' financing proposals as long as they are not sufficiently tested and examples of up-scaling are available.

Strengthen Information Systems

As indicated, (reliable) information is a crucial factor for development. Hence, donors should support activities at professional sector institutions to establish or improve information systems and databases which should be made available to the public and for use by the state administration. To curb the risk of misinforming the public, sector institutions must have autonomy in reporting but at the same time have to submit to an external control in order to ensure data quality. The complexity of the information systems must match the capacity of the sector institutions and parallel systems anchored at different sector institutions must be avoided. The practice to produce favourable data and show progress on paper where proven decline is taking place (refer to Sect. 3.5 and Chap. 5) should not be supported with the commitment of fresh donor money for more investments especially if the exposure of such 'manipulation' is even punishable.[1]

Bring Onsite Sanitation Services into the Formalised System

Because the large majority of the urban population will not be able to access piped sewer systems for a long time to come, the low-income countries will have to concentrate on the development of onsite sanitation systems. In order to avoid conflicts the line ministries in the concerned sectors should demand a clear distribution of mandates across the different sectors by the higher political hierarchy. Thereafter, the water sector should ensure the elaboration of an enabling framework for its sanitation functions whereby onsite sanitation should be the centrepiece. The sanitation policy for the provision of services and its infrastructure development must not be hosted at the MoH. Next to the piped sewer systems, onsite sanitation development should become part of the utility's portfolio. There should also be recognition that households of the poor need support in the construction of improved sanitation facilities and subsidies to access the services of the onsite sanitation chain. Such subsidies should preferably be generated by the water and sanitation sector. Donors should match their expressed concern about the deplorable situation in the urban LIAs in respect to sanitation with more funding and support for up-scaling onsite sanitation. This will need more engagement in national institutions (financing baskets) and patience for results.

Reform WATSAN Structure on Global Level and Relaunch Paris/Accra/ Busan Agenda

Considering the persistent misleading messages on the status of the water and sanitation sector, the inadequate definitions of access and the practice that access to water and sanitation is repeatedly considered less important than other basic human needs, more room have to be given to urban water and sanitation service development expertise in the global discourse. The global monitoring and reporting system needs to be radically overhauled because the critical remarks of the General Assemble of the UN[2] has not yet been sufficiently taken on board by the drivers of the discourse. Such changes would require an autonomous WATSAN unit headed by water and sanitation development specialists (urban and rural) and the recognition that global monitoring for urban water and sanitation has fundamental weaknesses.[3] This would not only help in the global discourse on water and sanitation development but also ease the unnecessary competition between the MoHs and the ministries responsible for water and sanitation development in the low-income countries. It would also help to harmonise counting of access with the human rights

[1] https://www.reuters.com/article/us-tanzania-worldbank/tanzania-law-punishing-critics-of-statistics-deeply-concerning-world-bank-idUSKCN1MD17P and https://www.standardmedia.co.ke/article/2001298442/tanzania-law-punishing-critics-of-statistics-deeply-concerning-world-bank (last visited December 2018).

[2] The UN declaration A/C.3/70/L.55/Rev.1 from 18.11.2015, page 3/6

[3] E.g. the proxy of improved sources as defined presently should be abandoned for the urban setting and counting of access should only recognise sources from formalised services.

requirements and pursue more realistic policies in low-income countries. In order to push aid effectiveness to a higher level a relaunch of the OECD Paris/Accra/Busan initiative[4] would be very helpful. Such initiatives should be carried out in close cooperation with the UN institutions. Precise guidelines for donors and receiving countries, tailor made for the different sectors combined with a more effective monitoring and reporting system, would thereby be crucial.

Areas of Further Research

The most important limitations in the present work was that only four target countries in Sub-Saharan Africa were considered for the in-depth analysis and the quantitative date covered only 8 years. This restricted the application of statistical methods usually applied in socio-economic sciences. Hence, it would be helpful to enlarge the number of countries in future research on sector reforms.

In addition, data where missing or their quality was doubtful for some areas of interests which prevented a certain in-depth consideration in this book. Therefore, further research in urban water and sanitation development would be very useful in the following fields:

- Informal service provision in the low-income countries,[5] its magnitude and potential to be integrated in a sustainable development concept.
- Quality of utility water in the household at the point of consumption for all service levels and effective mitigation measures if water quality is compromised.[6]
- Importance of shared facilities as basic services in low-income countries for reaching universal access (acceptance, design, costs, etc.) and the effects of its delayed employment.
- Effects of formalised water and service provision and of modern sector reforms on solving gender inequalities and changing the traditional role of women in water and sanitation.
- Contribution of access to formalised services on poverty alleviation in the urban setting.
- Benefits and disadvantages when utilities delegate management in the LIAs.[7]
- Comparison of unique and block tariff systems related to effects on the poor and the performance of utilities.
- Influence of rapid urbanisation on formalised and informal water and sanitation services and its effect on the households (behaviour, spending, change from informal to formal services, etc.).

[4] OECD/DAC (2005/2008).

[5] An initiative by the regulator in Kenya to register and obtain basic information of informal service providers through the licensed utilities in several towns during the research work failed due to the resistance of the informal providers, including community owned and managed systems in the urban setting. This restricted the analysis to the formalised systems.

[6] Water quality will continue to be concern for all consumers as long as utilities have to carry out water rationing and cannot guarantee a continuous supply over 24 h.

[7] Cases of delegated management in Kisumu, Kenya and in Burkina Faso did not produce sufficient reliable data in order to include this experience in the analysis of this work.

- Links between failed PSP arrangements and the development in the six proposed factors for development which could reveal necessary steps to improve such arrangements.

Of particular interests would be further research in three other areas: sector financing and asset development in low-income countries especially for basic service levels in water, up-scaling of sanitation for the urban poor including the sanitation chain for onsite-sanitation and the influence of insufficient access to water and sanitation on the living condition of the poor in the rural and in the different types of the urban LIAs with its consequences for urbanisation as well as migration from low-income countries.

Considering the huge and persisting financing gap in the sector and constraints national governments are facing in providing sufficient public funding, sector financing and asset development in low-income countries should receive more attention in further research. This should also cover the issue of aid effectiveness including the motives in the allocation of funds by donor countries. Furthermore, there is extensive knowledge about technologies for safe household installations and treatment as well as the reuse of sludge. However, there is as good as no data available on the management of the chain for onsite sanitation. The first report of JMP, which included the notion of safely managed sanitation has been released mid-2017 only and indicates that for around half of the world population sanitation data is missing. This and the fact that the majority of people in the urban setting in the developing world will depend on onsite sanitation for the next decades to come would justify further research in this direction.

My hope is for this contribution of the knowledge for urban water and sanitation development in low-income countries to shake up decision makers, to irritate and thereby initiate support to overcome the sluggishness in the sector. It would benefit the billions who face empty water buckets and are forced to live with overflowing latrines. The purpose is to save millions of lives every year, especially those of helpless children.

Reference

OECD/DAC (2005/2008) Paris declaration on aid effectiveness and the Accra Agenda for Action. Available: http://www.oecd.org/dac/effectiveness/parisdeclarationandaccraagendaforaction.htm (last visited 12.2016)

Printed in the United States
by Baker & Taylor Publisher Services